So Ends This Day

The Portuguese in American Whaling

1765-1927

DONALD WARRIN

So Ends This Day

The Portuguese in American Whaling
1765-1927

LUSO-AMERICAN FOUNDATION

University of Massachusetts Dartmouth
Center for Portuguese Studies and Culture
North Dartmouth, Massachusetts
2010

PORTUGUESE IN THE AMERICAS SERIES 13

General Editor: Frank F. Sousa
Editorial Manager: Gina M. Reis
Manuscript Editor: Richard Larschan
Copyeditor: Michael Colvin
Graphic Designer: Spencer Ladd
Typesetter: Inês Sena

So Ends This Day: The Portuguese in American Whaling, 1765-1927

Printed by Signature Printing, East Providence, RI

The publication of this volume was made possible, in part, by generous support from the Luso-American Foundation and the New Bedford Whaling Museum.

Library of Congress Cataloging-in-Publication Data
Warrin, Donald.
So ends this day: The Portuguese in American whaling, 1765-1927 / Donald Warrin.
p. cm. -- (Portuguese in the Americas series ; 13)
Includes bibliographical references and index.
ISBN 1-933227-28-1 (alk. paper)
1. Whaling--New England--History. 2. Whaling--Social aspects--New England--History. 3. Portuguese Americans--New England--History. 4. Cape Verdean Americans--New England--History.
5. Azorean Americans--New England--History. 6. Whalers (Persons)--New England--History. 7. New England--Ethnic relations.
8. New England--Economic conditions. I. Title.
SH383.2.W376 2009
639.2'80974--dc22
2009014783

Table of Contents

Maps

Foreword

The economic and cultural significance of the American whaling industry has often been undervalued and underappreciated, both as a major feature of the emerging American economy in the late eighteenth and early nineteenth centuries and, later, as a unique manifestation of egalitarian employment opportunities for foreign immigrants and people of color. Deepwater whaling was already an important source of revenue in Colonial New England and New York, and had given birth to America's first industry: the manufacture of spermaceti candles. But it was in the immediate wake of the Revolution that the true value of whaling to the national interest began to show. Independence from Britain carried the penalty of closing British and British colonial ports to American trade, which put an end to the great advantages of Most Favored Nation status that the Colonists had previously enjoyed. In this new world order, so-called South Sea whaling—the deepwater hunt for sperm whales and right whales (which was an American invention anyway)—stood out as one of the few seaborne pursuits in which the former colonials already exceeded all others in scope, magnitude, expertise, and skill. Along with the tobacco and cotton trades, whaling was one of few up-and-running enterprises of which the products not only continued to be desirable in England but could be marketed outside the British sphere of influence. Whale oil was, in effect, a cash crop upon which the only limitations were the ability of the Americans to mobilize the resources to hunt it down and stave off competition.

The critical role of Portuguese Islanders in this American whaling milieu is also frequently underestimated. The Azores, and alternatively the Cape Verdes—Portuguese dependencies since the Renaissance—were customarily the first ports-of-call for outward-bound American whalers. Lying directly in the path of the Atlantic trade winds, they were thus directly on the voyage track, whether the ultimate destination be the South Atlantic, Pacific, or Indian Ocean whaling grounds. Not only were the Islands convenient in standpoint of navigation (almost to the point of being unavoidable), but, thanks to mild climate and fertile soil, they provided year-round markets for "recruits": fresh produce, water, flour, bread, cheese, livestock, and other provisions. By the middle nineteenth century—especially at Horta on Faial in the Azores, and partly as a result of frequent Yankee whaling in local waters—there also arose a commercial infrastructure by virtue of which whaling gear, ship stores, and other supplies could be acquired, repairs could be made, cargoes could be transshipped, and consular services obtained.

Perhaps most significantly of all, the Islands provided an ancillary labor pool—a reserve of willing hands to fill out the ranks of shorthanded crews. Some islanders wanted to escape limited local economic opportunities; others wanted to embrace the American Dream in the Land of Opportunity or to escape the much-dreaded conscription into the ranks of the Portuguese army or simply to opt for the adventure of a deepwater voyage. Many young men were signed on during these island landfalls, whereas many others stowed away. Often the captains were complicit in the draft evasion or illicit emigration; sometimes this was out of the goodness of their hearts (especially in the late nineteenth and early twentieth centuries, by which time some of the whaling captains were themselves of Azorean or Cape Verdean birth). Usually it must have been a combination of charitable compassion and practical necessity, as shorthanded crews could be filled fairly easily and without paying the stiff bounties charged for green hands recruited by agents back home. Whatever the specific circumstances, most of the islanders gathered in this way were landed and discharged two or three years later in New Bedford and other Stateside whaling ports, and if they wanted to return to the islands they had to ship out afresh.

Of course, many numbering in the thousands, made New England their home. Whaling was a natural first choice for a livelihood; families soon followed; and richly diverse communities of Portuguese Islanders—Azorean and Crioulo—were permanently planted in New England. Later in the century these communities expanded to California and Hawaii, in a Portuguese diaspora that was also occasioned by whaling. And as whaling declined, continuing waves of Azorean and Cape Verdean immigrants found their way into the fisheries, the textile mills and shoe factories in the East, and the canneries and agriculture in the West.

One of the truly dramatic results of the original westward migration of Portuguese islanders—coming as it did simultaneously with the whaling industry's influx of fugitive slaves, foreign immigrants, and miscellaneous hopeful young fellows from all over the nation and in the wake of the fundamentally egalitarian Quaker ethic upon which the Yankee whaling industry was originally founded—was the central position of Azoreans and Cape Verdeans in a uniquely polyglot, multicultural agglomeration of humanity in which many of the customary standards and habitual prejudices of the day were suspended, subverted, or ignored. Throughout most of the nineteenth century, New Bedford and a handful of whaling outports in the Northeast and California were the closest thing ever achieved to meritocracy in the labor force. Whaling was perhaps the only venue in the European and American world in which people of color—black, brown, yellow, and red—were supervising Caucasians in the workplace, and where all comers had at least a modicum of expectation of equal pay and genuine opportunities for social and economic advancement nearly at par with the sons of native-born white folks. Whatever draconian measures the powerful whaling merchants and notoriously unscrupulous crimps and land-sharks may have adopted to contain and repress the labor force, they tended to be applied comparatively evenly, without much thought of race, creed, color, or national origin. And it was in this unique atmosphere of authentic possibility that Azoreans and Cape Verdeans established footholds and maintained trajectories that propelled them successfully into the American mainstream.

The specifics of how these communities came to be, and how their fortunes unfolded in the most egalitarian of all industries up to that time, are the compelling subject of Donald Warrin's comprehensive study here. For more than a century and a half, whaling sustained New Bedford, spreading its tentacles all around what Melville called "the terraqueous globe," contributing substantially to the viability of the American economy. It is no small matter that mariners, shopkeepers, tradesmen, and artisans from the Azores and Cape Verde Islands played crucial roles throughout, many of them advancing to positions of authority and responsibility that were unprecedented among foreign-speaking newcomers to the American labor force; and no small matter that the Islands themselves became integral to this network of seaborne enterprise. It is gratifying that *So Ends This Day* not only undertakes a grand, historical overview of the Islands' involvement in the blubber-hunting business, but also characterizes many of the individual personalities, events, circumstances, and anecdotes that reveal the complexity of the whaling industry as a whole and the human character of Azorean and Cape Verdean involvement in it.

Stuart M. Frank
Senior Curator, New Bedford Whaling Museum

Preface

In the summer of 1862 a prospecting party of thirty-five men left the frontier settlement of Auburn, Oregon Territory, in a search of gold. Their leader was a man named George Grimes, and they were following Indian guides who had told them of a large valley in the mountains to the east. Among them were four Portuguese, one of whom was Manuel Filipe Cardoso, born on the island of Pico in the Azores. Known by that time as John Phillips, he would later become a folk hero in Wyoming.[1]

During the 1990s, as I was researching the history of Portuguese immigrants on the Far West frontier, I was struck by their participation in some of the more daring exploits of the time and in numbers that belied the modest populations of origin. The great majority were Azoreans, but the Cape Verdean presence was not lacking. It did not take me long to recognize that the whaling experience of these men—the hardships and dangers they survived, the exotic cultures they encountered—had conditioned them uniquely to be pioneers among the pioneers.

The expedition into what is now known as the Boise Basin in southern Idaho was tragic in that its leader, George Grimes, was killed in a fight with hostile Indians, but also notable, since it resulted in one of the richest gold discoveries of the era. Recalling that fateful day, Moses Splawn, a Missourian and chronicler of the expedition, noted somewhat condescendingly that, while he was asleep, the Portuguese "were making me a pair of pants out of seamless sacks."[2] Much later I would find many tales concerning this handicraft—of men sitting together in the forecastle of a whaleship, sewing and smoking. But at the time I wondered where they had acquired this skill. That is one of the questions that led me back to the whaling industry and to its fundamental character in shaping the personalities of these young men, as well as determining the settlement patterns that would endure through the years. It is a tale every bit as intriguing as that of the settlement of the American West.

This book could not have been written without the kind assistance of many people and institutions. I would like to begin by thanking my wife, Diane Beeson, without whose patience, support, and invaluable editorial skills I would never have completed this work. I am grateful to Dr. Stuart M. Frank, Senior Curator at the New Bedford Whaling Museum, for his insightful foreward and his instructive comments on the text. To my other two readers, Judith M. Downey, Head Librarian, Casa da Saudade Branch Library, New Bedford, and Michael P. Dyer, Maritime Historian at the New Bedford Whaling Museum, very special thanks for their gracious giving of time and expertise.

To my publisher, the Center for Portuguese Studies and Culture at the University of Massachusetts Dartmouth, and in particular to Dr. Frank F. Sousa, General Editor, I am deeply grateful for the continuing support, without which this book would not have been possible. I would like to express my appreciation, as well, to Editorial Manager Gina M. Reis for her expertise and also to Manuscript Editor Dr. Richard Larschan for his valuable editorial suggestions. Also a sincere note of appreciation to the New Bedford Whaling Museum, and its President, James P. Russell, for contributing the great majority of the photos used in the text. To the New Bedford Whaling Museum Research Library and its staff, especially Laura Pereira, Michael Lapides, and Kate Mello, many thanks for their wonderful assistance on this project; also to Judith Lund, whose compendium of whaling masters and voyages, and her personal suggestions greatly enhanced the comprehensiveness of this volume. I am also indebted to the book's cartographers, Garrett Bradford and Darin Jensen.

This book would have been impossible to write without the assistance of the librarians and archivists who were so helpful to me in my research at the many historical societies, libraries, and archives I frequented over the years. Beyond that there are other non-affiliated individuals to whom I owe sincere thanks, namely Elsie Casina Adams, Alex De Marban, Caroline Harris, Noah and Molly Itta, and Dorcas Stein (Alaska); Rafael Enos Aguirre (Chile); Francis Kartunnen (Nantucket); Mark Lovewell (Martha's Vineyard); David Cornell (Michigan); Ron Barboza, John Bockstoce, Daniel Rodrigues, and Jim Lopes (New Bedford area); Joan Druett, Peter Goomes, Tony Monteith, Rhys Richards, and Rosemary Taiaroa (New Zealand); George Bryant (Provincetown); Bill Pickelhaupt (San Francisco); and the De Castro family of Long Island. In addition, students Jeffrey Hyer, Michelle Mahler, and Oliver Surpless provided excellent assistance at key points in the process. Finally, a special note of appreciation is due the New Bedford Whaling National Historical Park for making some of its crew list information available for this book. And my apologies, of course, to anyone whose name has inadvertently been overlooked.

Donald Warrin
University of California, Berkeley

Note: Three works have been fundamental to the research undertaken for this book: Alexander Starbuck, *History of the American Whale Fishery* (1989 [1878]); Reginald B. Hegarty, *Returns of Whaling Vessels Sailing from American Ports, 1876-1928* (1959); and Judith Navas Lund, *Whaling Masters and Whaling Voyages Sailing from American Ports: A Compilation of Sources* (2001). Because of the frequency with which the whaling voyages and related data listed in these three volumes appear in the following text, and because these data are listed in either alphabetical or chronological order in these three works, these sources are not referenced in every case.

Notes

[1] Cf. Warrin and Gomes, *Land, As Far As the Eye Can See.*
[2] French, *History of Idaho*, 1:34.

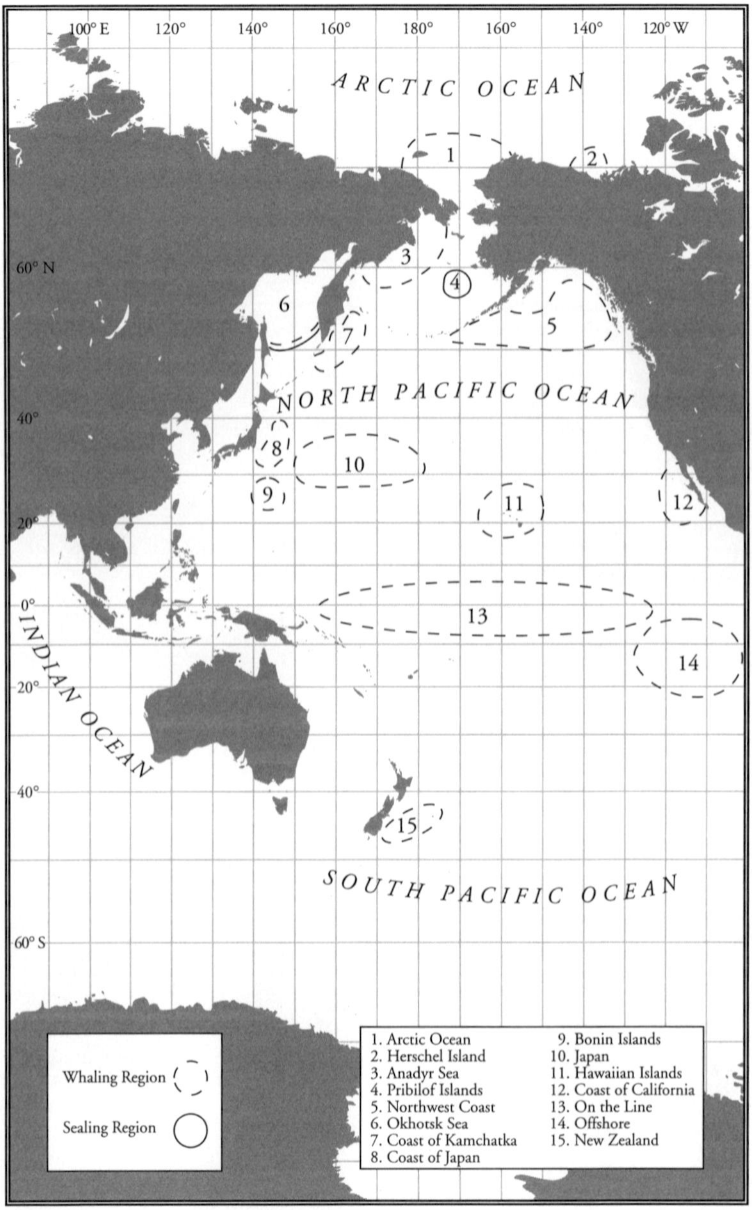

Whaling and Sealing Regions of the World

* These and other maps in this volume are not intended to be comprehensive but rather to indicate locations mentioned in the text.

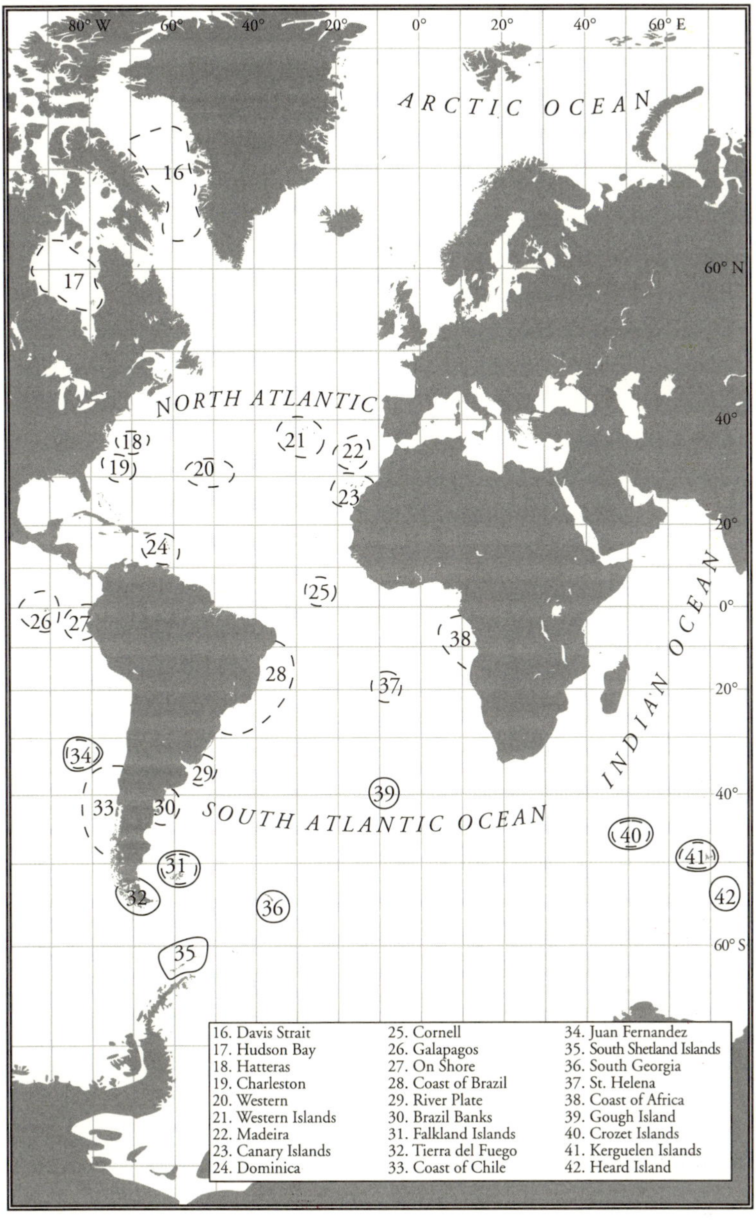

80° W
60°
40°
20°
0°
20°
40°
60° E
ARCTIC OCEAN
60° N
40°
20°
0°
20°
40°
60° S
NORTH ATLANTIC
SOUTH ATLANTIC OCEAN
INDIAN OCEAN
16
17
18
19
20
21
22
23
24
25
26
27
28
29
30
31
32
33
34
35
36
37
38
39
40
41
42
16. Davis Strait
17. Hudson Bay
18. Hatteras
19. Charleston
20. Western
21. Western Islands
22. Madeira
23. Canary Islands
24. Dominica
25. Cornell
26. Galapagos
27. On Shore
28. Coast of Brazil
29. River Plate
30. Brazil Banks
31. Falkland Islands
32. Tierra del Fuego
33. Coast of Chile
34. Juan Fernandez
35. South Shetland Islands
36. South Georgia
37. St. Helena
38. Coast of Africa
39. Gough Island
40. Crozet Islands
41. Kerguelen Islands
42. Heard Island

CHAPTER I

Background

> One notable thing in this age of iron and steel construction is that all the boats are made of wood, and a new whaling brig, now building, will be also of that material. The old whalers do not believe in iron ships. They claim that the wooden vessels ride the waves better, and, as a large part of the time is necessarily spent in cruising about in search of whales, the skippers prefer 'heart of oak' to the iron. (*Providence Sunday Journal*, 20 Mar 1910)

The passage above well evokes the increasingly anachronistic American whaling industry. One need only consider, during the apogee of the ocean liner, one of the last remnants of the American whaling fleet, the square-rigged bark *Wanderer* made its way out of Buzzards Bay in southeastern Massachusetts, on its ill-fated voyage in the summer of 1924, only to founder on the rocks at Sow and Pigs Reef, Cuttyhunk Island.

The following is a story of the American whaling industry, from its remote beginnings to its anticlimactic end, albeit told from a particular perspective, that of the Portuguese—at the time almost exclusively Azorean or Cape Verde islanders—who, over the course of the industry, assumed an increasingly important role. As T. Bentley Duncan has observed, "Small islands situated in the midst of seas and oceans, remote from the continental land masses, often possess a high importance in communications, navigation, and strategy—an importance out of all relation to their size and resources."[1]

Such was certainly the case with the Portuguese archipelagoes of the Azores, Cape Verde, and Madeira from the sixteenth through the nineteenth centuries. There is considerable evidence that the Azores and Madeira had been visited by Italian and Portuguese mariners in the

Bark *Wanderer* on the rocks in Buzzards Bay, August, 1924
Courtesy of the New Bedford Whaling Museum

fourteenth century.[2] But the traditional dates of discovery are: Madeira, ca. 1418-20; Azores, ca. 1427; and Cape Verde, ca. 1456. The settlement of the island of Madeira and neighboring Porto Santo began soon after discovery. That of the Azores took place during the 1440s, and Cape Verde, by the early 1460s. The settlement patterns of the Azores differed significantly from that of Cape Verde. To the former came not only Portuguese from the mainland but also a number of Flemings. On the other hand, under the Portuguese, the relatively infertile Cape Verde islands, lying some 300 miles west of the coast of Senegal, became deeply involved in the slave trade. Slaves from the African mainland were employed on the local plantations, and the archipelago became an entrepôt for the Portuguese slave trade with the Americas.

Through the intermixture of those of African origin with the relatively small number of individuals of European background, the population of Cape Verde became one of mixed race. Until the end of the conservative Salazar/Caetano regime in 1974, the Azores and Madeira were known as the "Adjacent Islands," showing their closer relationship with Lisbon than the so-called "Overseas Provinces," which included

Atlantic Ocean

Cape Verde—much neglected by the Portuguese authorities—along with Portugal's other colonies in Africa and Asia. Nevertheless, those arriving from the Cape Verde islands during the whaling years considered themselves Portuguese and, in fact, used that designation in an attempt to distance themselves from the extreme prejudice they saw directed against African Americans. In 1975, Cape Verde won its independence from Portugal, and the following year the Azores and Madeira became autonomous regions of the nation.

The Azores, spread across almost 400 miles in the North Atlantic, consist of three island groups. To the east are São Miguel and Santa Maria; in the center are Terceira, Graciosa, São Jorge, Faial, and Pico. At some distance to the west lie Flores and Corvo. Madeira, little touched by the nineteenth-century whaling experience, consists of two inhabited

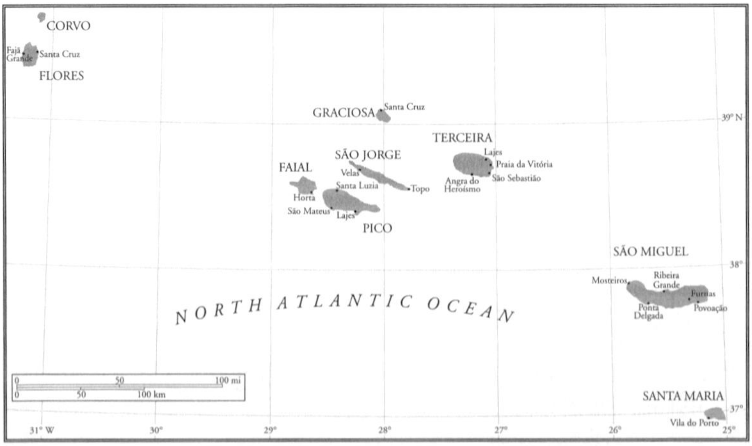

Azores Islands

islands—the main island of the same name and smaller Porto Santo, where Christopher Columbus spent a number of years and married the daughter of its governor.

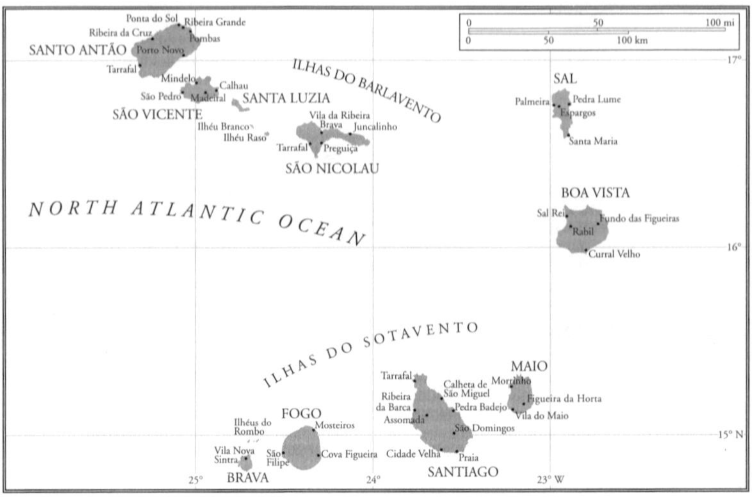

Cape Verde Islands

Cape Verde is composed of two groups, the Windward (Barlavento) to the north, consisting of the islands of Santo Antão, Boa Vista, São Nicolau, São Vicente, Sal, and Santa Luzia; and the Leeward (Sotavento) to the south, with Santiago, Fogo, Maio, and Brava. While nineteenth-century whaling vessels tended to call at the latter islands, especially Brava, records show a much greater distribution of residents throughout the archipelago among American whaling crews.

In the Age of Discovery sailing vessels could head south along the African coast, touching at Madeira and Cape Verde; but those returning to Portugal, following the prevailing winds, were forced to sail well westward toward the coast of South America and then turn eastward into the North Atlantic, thus passing by the Azores. These islands served a dual function at the time, to revictual ships after their long ocean voyage and also to offer protection from the pirates who sought their riches during the final leg of the voyage. Not only Portuguese ships paused at the islands, but also Spanish galleons returning from America. In fact, Christopher Columbus himself called at the island of Santa Maria on his return from his historic voyage of discovery. On his third voyage to America, in 1498, Columbus called at the Cape Verdes for provisions, as had Vasco da Gama the previous year—thus beginning a long tradition in which merchant and other vessels would head for these islands.

By the end of the fifteenth century the island of Madeira was the world's leading producer of sugarcane, while over the next few centuries the Azores became an important producer of pastel (which produced an important dye), cereals, wine grapes, and later oranges. During the seventeenth and eighteenth centuries, in particular, the Azores and Madeira became regular ports of call for American trading vessels. Ebenezer Bowditch Jr. left a record of a series of voyages to the Azores during the years 1753-1756 in the schooner *Eagle* of Salem, Massachusetts. In his log entries of August 1753, he notes that the vessel made Flores, Corvo, and Faial. And in September of the following year he called at São Jorge, Terceira, São Miguel, and Santa Maria. Departing the latter on October 2 for Madeira, he noted that they had "caught a large dolphin."[3] Presumably those on board would have been aware of the presence of whales in the vicinity. But it would take another decade before American whalers would explore the region.

The most important port in the Azores was that of Horta on the island of Faial with its deep protected anchorage of eighteen to twenty fathoms. There, beside the many whaling vessels, one might see Portuguese codfishers returning from Newfoundland, vessels outward bound for the West Indies, ships from the East coast of the United States on their way to Europe, and others from Brazil, etc. Secondary in importance to Horta in the seventeenth and eighteenth centuries, as a victualler to transatlantic ships, was the island of Flores.

The Cape Verde islands were only slightly less a thoroughfare. According to Duncan, "The pattern of North Atlantic winds made it likely that every European sailing ship, outward bound to destinations in the South Atlantic or in Eastern seas, would pass through the vicinity of the Cape Verdes."[4] There, in the good years, they would be provided with salted goat meat, sweet potatoes, yams, squashes, maize, beans and citrus fruits.[5] Salt as well—plentiful on the islands of Maio and Sal—for many years attracted ships beginning an extended voyage.

However, the Cape Verde islands are subject to severe periodic droughts. During the era in which American whalers frequented the islands, from the latter part of the eighteenth century until the early twentieth, famines were recorded every six to ten years. Given the frequency of contact with foreign vessels and the precariousness of the local economy, it is easy to see why young men from the archipelago would be eager to enter the service of visiting vessels.

Benjamin Hodge was born in Connecticut in 1753 and at the age of twenty-one left home to go to sea. In October of 1774 he sailed out of Providence, Rhode Island, on a whaling sloop under the command of Captain William White. The vessel carried a crew of thirteen—eleven whites, one black, and one Indian. Their first port of call was Cape Verde. As they passed Fogo, the island's volcano was spewing flames, fumes, and ash. As Hodge later recalled, "in the afternoon we See Fugo burning Mountain at night we passed to the leeward of the Island, the[y] are very Smoky and looked of a redish couler Smelt Smokey and Sulffery So we could hardly breath, the next morning we were all on well [unwell] took medeson and got well, the Sails and rigon all turned of a redish couler, Strike them with a Stick and they would Smoke like dust."[6]

At Boa Vista it was learned that the population was in the midst of a severe drought. Hodge noted that the people were "in a Starveing Situation perhaps a Thousand people all living in a thick town together about three or four miles from Shore." Ultimately some 22,000 individuals would perish during 1774-76. Yet there was some hope, for "they was expecting Some vessels there from Portigal to bring them provition."[7]

While at Boa Vista the captain added a local man whom Hodge recalled being named Mingo (Domingo). About six weeks after leaving Cape Verde the sloop arrived at St. Thomas in the Virgin Islands. There they saw two French ships with about 1400 slaves on board that had called at the island for water and provisions. The Cape Verdean soon left the vessel and when the captain attempted to force his return, the governor of the island refused to compel the man against his wishes. One day Mingo sent for Hodge and another crewman who went to visit him. Mingo informed them of his reasons for leaving the vessel, for someone had told him, perhaps in jest, that the captain was planning to sell him into slavery in the West Indies.

It would have been virtually impossible for any inhabitant of the Cape Verde Islands not to be aware of the slave trade, given the role of the archipelago at the time. But a native of the islands of mixed race would not have thought himself much subject to being enslaved, since he would not have considered himself to be black. And so it was that taunts by the white crew members aboard the sloop evidently gave him a new awareness of foreign racial attitudes. As he complained to Hodge and his companion, "Some of the hand[s] plaged [plagued] him too much and called him negro."[8]

In June of 1775 the sloop called at the island of Flores in the Azores. There they bought onions and wine, paying for them with fish they had caught. While there, Hodge and some of his crewmates "went to one of the nunnerys and saw the pale fased girls at work we tryed to talk with them but could not understand each other."[9] Soon they were on their way home and to the discovery that the Colonies were at war.

When the ship *Neptune* of New Haven called at Boa Vista in 1797 the keeper noted that "this is a poor place – plenty of Hogs, Goats & Pumpkins & Watermellons."[10] At the island of Santiago they were able to obtain pigs, poultry, eggs, and fruit: limes, bananas, and "very sweet" oranges. One of the pigs soon gave birth to a litter of piglets, which the author envisioned "growing for a pot pie finely."[11]

However, in 1810 a traveler, recently returned from the island of Maio, stated that the food was gone, and the mules had died and thus there was no way to harvest the salt that the vessels had wished to procure. During the drought years of 1831-33, in which some 30,000 perished, American vessels, including several whaleships out of New Bedford and

Fairhaven, Massachusetts, began to drop off supplies for the suffering and dying populace, which were sorely needed since Metropolitan Portugal was in the midst of a civil war and thus not prepared to send relief. When the whaling bark *Roscoe* sailed in August of 1833 it carried some one hundred barrels of bread, beef, etc., destined to help the people of Cape Verde. By 1834 conditions had improved, and the cycle began again.[12]

As the bark *Pantheon* of Fall River, Massachusetts, approached the island of Santiago in 1842 the keeper observed, "Afternoon drifted clost in to Land so clost that we Could smell the orranges growing a shore." On Christmas Day, two days later, he noted that they had purchased "25 Hogs 50 Fowls 10 Bushels of Banannas And 5,000 Oranges Besides Cocanutts sweet Potatoes."[13]

Hunting sperm whales from the bark *Dunbarton* of New Bedford, ca. 1854-55, sketch by Perry G. Wing. Courtesy of the New Bedford Whaling Museum.

The Hunters

> "There were probably no stronger ships ever built than the old whaleships that sailed from New Bedford. They were built upon honor, as the saying is, from the best selected live oak timber that grew in our forests, and it was a common thing for these ships to remain in service for a term of 50 or 75 years, and even longer."
> (L. A. Littlefield, "Fitting Out a Whaler")[14]

Hoisting the first blanket piece from a sperm whale, bark *Wanderer*, Antone T. Edwards, master. Photo by William H. Tripp, ca. 1922-23. Courtesy of the New Bedford Whaling Museum.

During virtually the entire history of American whaling the vessels employed in the industry were powered, essentially, by sail. While there were advances in design, and later auxiliary power, the basic vessel types did not change significantly. The smallest of these vessels (besides the whaleboats carried on board) was the single-masted sloop. Most whaling voyages in the early years of American whaling were made in these smaller vessels. Toward the end of the eighteenth century other larger types gained popularity: the schooner (usually two-masted—sometimes three—carrying two to three boats), the brig (two masts), and the ship (three or more masts).[15] By the early nineteenth century the bark, also three-masted, but of a configuration that required less hands to manage its sails, came into wide usage. Both bark and ship carried between four and five whaleboats.

In the earliest years of American whaling the blubber was stored in casks, to be tried out on shore at the end of the voyage. By the end of the eighteenth century whaling vessels were carrying tryworks on board, thus avoiding both the spoilage and reduction in volume that tended to occur with the old method. The tryworks consisted of a brick furnace in which two or three large iron trypots were placed and when heated

Whalebone (baleen), from the catch of the steam bark *William Baylies*, Capt. H. H. Bodfish, in the western Arctic, 1905; displayed in front of the offices of William Lewis & Son, New Bedford. Courtesy of the New Bedford Whaling Museum.

were used to render or "try out" the whale's blubber. The resulting oil was then stored below decks in large casks.

When a dead whale was brought alongside "cutting in" would begin. Usually the head was first severed and secured to the stern with a chain. A hook or toggle was then placed in the blubber, which would be lifted by block and tackle in "blanket pieces" as it was cut from the carcass while the latter slowly rotated. Each blanket piece would be lowered down the main hatch into the blubber room, where it was cut in "horse pieces" about a foot square. Sent back on deck, these were then sliced into fine "bible leaves," in a process called "mincing"; however, while still attached, however, to the whale skin. Then into the trypots they went to be rendered. When separated, the oil was placed in a copper tank until it was sufficiently cool to be stowed away in casks duly prepared by the cooper.

If the whale were a right or a bowhead, the valuable baleen—or "whalebone," as it was popularly known—consisting of fibrous hair-like plates that hang from the roof of the mouth, had to be cut from the upper jaw. If a sperm whale, after the head was hoisted on deck the first

Trying out on the bark *California*, William F. Joseph, master, voyage of 1903-04. Courtesy of the New Bedford Whaling Museum.

priority was to retrieve the spermaceti from the case in the upper part of the head. Commonly a hole was cut in the head and a bucket lowered within; sometimes a man might be lowered inside to bale out the spermaceti. The lower jaw was detached and set aside for its teeth, which became objects of scrimshaw art.

The processing of the whale carcass was one of the dirtiest moments of whaling. Oil permeated the decks and foul smoke filled the air. A slip from one of the cutting tools, a slide off the whale's back into shark-infested waters, even a splash from the trypot, could inflict serious injury or death.

Depending on the size of the vessel, it might carry from two to five whaleboats. These were light, swift, wooden craft measuring twenty-five to thirty feet in length, commonly made of native cedar and pointed at both ends so that they could quickly reverse course. They carried a small sail and the various implements necessary for the chase. Typically they were manned by four oarsmen, a boatsteerer (harpooner), and commanded by a mate. Upon closing in on the prey, the boatsteerer flung a harpoon into the whale. After this fastening, the boatsteerer changed places with the mate who, at the opportune moment, drove a lance into the whale's body to kill it. As will be seen, hunting these leviathans of the deep was anything but routine. Once dead, the whale would be towed—at times at great distance—to the ship to be tried out.

One notes, increasingly, the propensity for Azoreans and Cape Verdeans to be boatsteerers. This was a pivotal role and one that was reserved for veteran whalemen, especially the Portuguese. In one of his "old whaling yarns," G. Leroy Bradford explains humorously how they became such good harpooners. It begins thusly:

> When "Portagee Joe" from the Azores came,
> He sure give us Yanks the jitters;
> His whalin' iron was death on sparm,
> Cause he could hypnotize the critters.[16]

Courage was certainly a factor in taking on the role of boatsteerer, but experience weighed even more heavily. However, lack of education and limited ability in the English language were large factors in keeping these veteran whalemen from moving beyond this position, and so they often became very practiced at their art.

The Crew

The captain's quarters were located aft. Forward of that were those of the mates, of whom there were customarily from two to four, depending on the size and type of vessel. Somewhere amidships would be the cramped quarters for the several boatsteerers, the cooper, etc. At the bow of the vessel lay the forecastle (pronounced 'fo'c'sle'). It was here that seamen and green hands slept and passed many of their idle hours. Poorly illuminated by the light of one or two whale oil lamps, with the only ventilation from the booby hatch above and the odor of whale oil permeating all, it was a foul place to spend one's time.

The Hunted

There are essentially two types of whales, toothed and baleen. The latter, known as *Mysticeti*, have no teeth; rather they eat by straining krill and other small creatures through the baleen. The other order of whale (including dolphins and porpoises) is the *Odontoceti*, or toothed whale. Of the many varieties of whales, just five were commonly hunted commercially, and only one had teeth. The sperm whale, or cachalot, belongs to the order of *Odontoceti*. It inhabits most of the ocean waters, migrating toward the equator in autumn and the poles in spring. Males are larger than females, averaging fifty feet in length and weighing some forty tons. Females average thirty-six feet, while attaining a weight of some twenty-two tons.

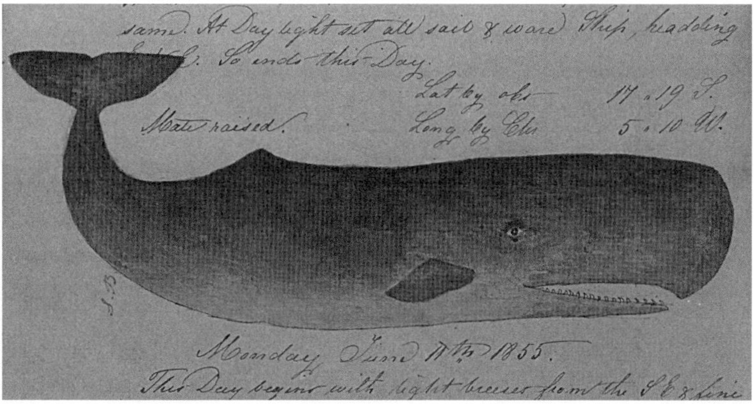

Sketch of a sperm whale by Perry G. Wing, bark *Dunbarton* of New Bedford, 1855. Courtesy of the New Bedford Whaling Museum.

Among the mysticetes the most hunted were the right whale and the bowhead. Both sexes of the right whale average fifty feet in length, with a weight of sixty tons. They inhabit the temperate zones of both the northern and southern hemispheres. Since they swam slowly, could be found not far from shore, were rich in oil, and had a tendency to float when dead, they were the "right" whales to catch. Unfortunately their stock now is quite depleted. The bowhead attains a length similar to that of the right whale, but weighs about twice as much. The range of this subspecies is limited to Arctic and sub-Arctic regions from the Arctic

Ocean and Sea of Okhotsk to Baffin and Hudson bays in eastern Canada. While the sperm whale is polygynous, the baleen whales are monogynous.

Humpback and gray whales, both of the order of *Mysticeti*, were also regularly hunted commercially. Whalemen would have liked to be able to hunt some of the larger and faster species, such as the blue whale (average eighty feet and one hundred tons), but their equipment at the time did not permit it. At times when whale sightings were scarce, whalemen did not hesitate to seek small cetaceans they called 'blackfish'—actually long- or short-finned pilot whales (*Globicephala melaena* or *macrorhynchus*)—which could be found most anywhere in tropical or temperate waters.

In the mid-nineteenth century whales were yielding several commercially important products: sperm oil and spermaceti from sperm whales, and whale-oil and whalebone (baleen) from the mysticetes. Ambergris was the occasional lucrative byproduct from the lower intestines of a sperm whale and was important to the production of perfume. Spermaceti, from the case (the upper part of the head of the sperm whale), yielded a substance unique in nature, a liquid wax that skilled refiners converted into superb candles, whereas sperm oil served as an excellent lubricant as well as a superior, and expensive, illuminant. On the other hand, the oil of the baleen whales, being much cheaper, was commonly used to light many an American home. Flexible whalebone served the functions that spring steel and plastic later assumed during the twentieth century: umbrella ribs, buggy whips, and most importantly, corset stays, among other uses. In the latter half of the nineteenth century, as petroleum products took the place of whale oil, the demand for whalebone continued strong into the first decade of the next century.

The Business of Whaling

> But whaling was not a passing boom, not a thing apart from all other interests, not local in nature and local in effect. Its influence as a social and economic factor was widespread. Whaling was a unit in a great whole—a part of the vast industrial interests of a growing country.
> (Walter S. Tower, *A History of the American Whale Fishery*)[17]

As whaling historian Walter Tower observed a century ago, "It seems safe to assert that no other industry so wide spread in its operation has been so closely restricted in the places from which it was carried on."[18] For, as evident in Appendix 1, there was great concentration in its ports of origin.

Clearly New Bedford and environs, as well as Nantucket, led significantly all other ports in the number of whaling voyages. The New London area ports were third, followed in order by those of Cape Cod, San Francisco, and Long Island. Other important ports in the Commonwealth were located in the Boston area and on Martha's Vineyard.

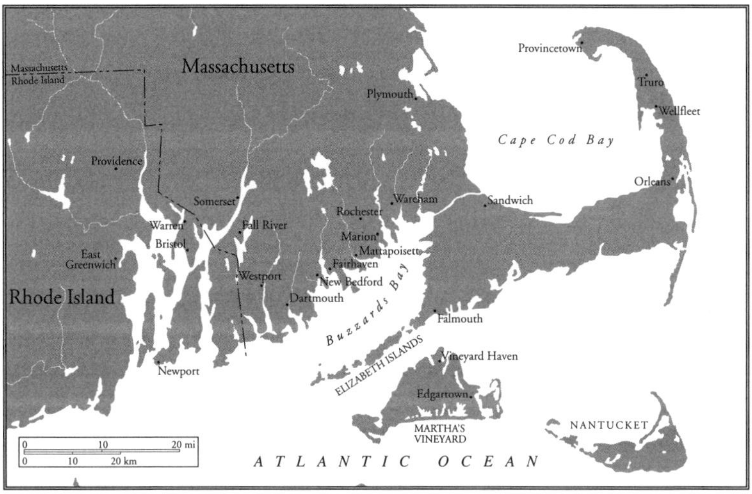

Whaling Ports: Massachusetts & Rhode Island

Rhode Island's most important ports were Warren, Newport, and Providence. Significant ports in the state of New York, besides Sag Harbor, included Greenport, New York City, and Cold Spring Harbor. Upriver, the town of Hudson sponsored a considerable number of voyages. Whaling out of Connecticut was centered around the New London area and, besides that port, included Stonington, Mystic, Norwich, and Groton. Finally, it should be noted that whaling took place on a minor scale in several eastern states, including New Hampshire, New Jersey, Delaware, Virginia, and North Carolina. None of these latter will be elaborated upon in this study.

All ports were not equal—that is, not only did they differ in relative importance but they tended to specialize. New Bedford and environs sent out vessels to the remotest corners of the earth without particular

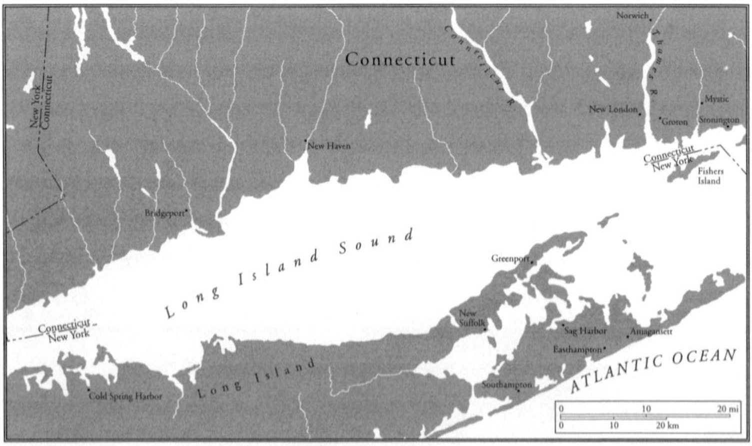

Whaling Ports: Connecticut and New York

regard to the type of whale. Sag Harbor hunted more of the right whale and also had a predilection for sealing. New London, along with Mystic and Stonington, developed a preference for elephant and fur seals—along with an occasional whale—that could be found in the barren and remote Crozet and Kerguelen islands of the South Indian Ocean, as well as sending many a vessel into Davis Strait and Hudson Bay in search of bowheads. And Provincetown, with a heavy Portuguese participation, opted for so-called "plum-pudding" voyages of shorter duration in the North and South Atlantic.

Being in its initial stages in North America essentially a Quaker industry, whaling originally took on an egalitarian, community aspect, in keeping with Quaker philosophy, as can be seen from its early years on Long Island and Nantucket. Nantucket whaling vessels were typically owned on shares, but curiously such was not the case with Sag Harbor. For a long time this latter port's development was somewhat hindered by its tradition of single ownership.

The role of the managing agent was crucial to the voyage. Most typically he was one of the principal owners of the vessel. Most of these agents were involved in some shoreside activity with a connection to the whaling industry, such as provision or dry goods merchants.[19] It was up to the agent to decide the type of whale to hunt and where. He had also to decide on an appropriate vessel and then to outfit it for the voyage. It was his ultimate responsibility to select those who would man the vessel, although the master—whom the agent would have chosen himself—would have a particular interest in the choice of the officers who would serve under him. The captain might have a hand in hiring his boatsteerers, cooper, cook, and steward, and even the foremast hands, if he was so inclined.

The managing agent and captain would work together planning the voyage, agreeing on the provisional extent and destination of the voyage, as well as on locations where correspondence between them could be received and dispatched. Traditionally the agent furnished to the master an initial charge in the form of a letter indicating the expectations for the voyage and often some paternalistic advice. In 1834, Charles W. Morgan, principal owner/agent of the ship *Condor* of New Bedford, began his advice to novice Captain George H. Dexter with these words: "As you have now taken the responsible Station of Master of a Ship & are a young man you will permit me to offer some advice. The greatest difficulty I have observed with young Masters, is either too great indulgence or too great Severity towards their crew."[20]

Instructions were often specific for the initial portion of the voyage; but it was largely up to the master of the vessel to determine the details of its course once upon the whaling grounds. The importance of both the Azores (or the "Western Islands" as they were known at the time) and Cape Verde are apparent in this missive from 1835 directed to Captain Tristram P. Swain of the ship *Isaac Howland* of New Bedford:

> You will proceed to sea first good time and of course will shape your course for the Western islands to lay in a Stock of Potatoes, etc. – we hope you may be so fortunate as to take some sperm oil about those Islands or the Cape de Verds in which case we should advise your shipping it home if an opportunity presents as it will give you more room on board and the Interest saved on the oil will pay the expenses of shipping it home – it can sometimes be left at Fayal with the counsel to be shipped when an opportunity presents – Sometimes there is an opportunity at the Cape de Verds to put it on board Some vessel bound to Boston – or some other port in New England....[21]

> A Captain of a vessel is besides a Lord, Judge, Doctor, Father and God knows what else (as the sailors generally say).
> (Manuel Silveira Andrade, Journal, bark *Pacific*, 1859-62, JAF)

During the voyage the captain was indeed lord and master over every aspect of his vessel. His was a heavy responsibility for which he was rewarded with pay that averaged three times that of a merchant captain.[22] He had to be a good seaman and navigator, and at least in the early years, was also expected to command a whaleboat during the hunt. Soon after sailing, usually after the morning-out muster, the captain would issue verbally his orders, which the seamen at times referred to as his "harangue." The logkeeper of the bark *Zone* of Fairhaven, Massachusetts, noted, one such moment soon after the vessel left port in September of 1855: "At 4 P.M. we choosed our watch and Boates crews, after which the Old Man gave us an harangue, after the following fashion: — 'Now Men. I have come Master of this Ship with the intention of trying to make a Voyage; not to fool my time in idleness, but to do my best to get some Oil. Now you shall have enough to eat and drink, and as long as you use me well, I shall endeavor to do the same by you; but when you do not, — look out.'"[23]

The agent lost direct control over the vessel once it left port. But communication through correspondence continued to be a possibility, despite crippling delays. Additionally, through correspondence with American consuls, local merchants, and even missionaries, the agent could assure that cash or credit would be available for the ship's captain at some distant port.

Whaling vessels customarily carried mail destined for others they might meet in the course of their travels. So agents could still relay messages, somewhat unreliably, to the captains of their vessels. These missals, along with letters from wives and family, were often deposited in ports where a particular whaleship was planning to call. Vessels often met at sea, with the officers and men visiting the other ship in what was called a "gam." Gamming was a way for whaling captains to share (if they so desired) news of their catch, etc., as well as letters that had been directed to them or others of the crew.[24] Often a wife would send copies of a letter aboard several vessels in the hope that one might reach her spouse somewhere at sea. Newspapers from home and fresh provisions might also be handed over during one of these meetings.

The Portuguese also tried to keep in contact with their family members through correspondence. Two letters from the island of Corvo eventually found their way to the ship *Hector* of New Bedford during its voyage of 1861-65. One was a long missal to Joaquim de Domingos, a crewman on board, from his father, dated October 20, 1862. The other, dated two days later, was destined for Manoel Ignacio Rafael, on board a bark identified as the *Mook*. Since the States were now at war, Manoel's father was particularly concerned that his son not become involved, begging him when he arrived in "Merica" not to go to war, nor to send money, for they were doing well.[25]

Half-Share Men

Where did the industry find the men to fill out its ship rosters for voyages that initially lasted only a few months, but by the beginning of the nineteenth century were stretching out to years? At first crews were made up of sons, nephews, and neighbors of whalemen and Native Americans, along with a significant number of African Americans. From the earliest times it seems that outfitters had a role in the industry, beginning with the so-called half-share men.

The idea of the half-share man has deep roots in colonial America. At its base is the feudal concept of inequality. One of the better-known examples is that of Nantucket in the mid-seventeenth century, in which the first settlers, soon realizing that they needed to draw craftsmen to the island, offered them half-shares in the town corporation. The plots of land of these half-share newcomers, as well, were only half the size of those of the full-share men. Neither were they afforded the same political representation. This practice soon led to resentment and a revolt of sorts on the island that was eventually settled amicably.[26]

The half-share has a long history in American whaling as well. For, early on, as colonists on Long Island began to depend on Native Americans for the bulk of their shore whaling activities, these men were paid just a half-share for their season's work. Such was the agreement in 1675 between James Loper and James Sceellener of Easthampton on the one hand and thirteen local Indians, each of the latter to receive one half-share of the profits—the other half destined, presumably, to Loper and his partner.[27]

Eventually, as the industry expanded in the nineteenth century, it became common practice for recruiting agents to provide the bulk of a whaling crew and to be paid a stipulated sum of money for this service at the time of sailing, the amount depending on factors such as the difficulty in obtaining these men and their station. But in colonial times on Nantucket, and even extending occasionally into the next century, the half-share was a popular method of contracting crew members from outside the local community for a whaling voyage. That meant, in effect, that one-half of a whaleman's final pay, based as it was on the profits of the voyage, was contracted to be paid to a middleman, who in turn would provide the sailor with the outfit he would need for the voyage.

As the industry grew rapidly in the years directly preceding the Revolutionary War, Nantucketers wishing to fill out whaling crews were forced to look beyond the limited supply of Wampanoags, free blacks, and local Yankee boys. Thus, not only were young men brought in from the neighboring mainland, but some foreigners, especially Portuguese, were contracted as well. It was here that the concept of the half-share was extended to these off-islanders, considered to be of unequal stature to the Nantucket Yankee.

Sundry examples of the half-share are seen in the journal/account books of William Rotch of Nantucket in the early 1770s.[28] Presumably, the individuals who were owed a half-share of the future income of the whalemen they placed on board were being reimbursed for not only making this arrangement, but also for providing the seaman with the proper outfit for the voyage. They were, in effect, the forerunners of the recruiting agents of the nineteenth century. One of them was early Portuguese entrepreneur Joseph Swazey, who earned eighteen pounds, four shillings, and sixpence for having placed a countryman, Ned Burgis (Borges), on Rotch's whaling schooner *Mermaid* in November of 1771.[29] Swazey also shipped a Benjamin Burgis (presumably a relative of Ned) on the same voyage, and Richard Edmunds (presumably a non-Portuguese) on the sloop *Friendship* that same month.[30] Other Portuguese engaged by the Rotch enterprises at the time include Joseph Thomas "Portugee" on the sloop *Loring* (April 1772); John Dyas on the ship *Mermaid* (October 1772); various "Portugees" (sloop *Sandwich* and brigantine *Temple*); and John Terra on the sloop *Delaware* (February 1774).[31] At times the master himself took on a half-share, as in the case of Captain Noah Pease of the sloop *Sandwich* in 1772, who received the half-share of "Portugee Will... who run away in the West Indies.... his Boy Antony Portugee was then put in Wills room remainder of the voyage."[32]

The use of the half-share disappeared with the resurgence of whaling after the American Revolution, but its occasional employment into the beginning of the next century is evidenced by the following agreement between Hezekiah Pinkham and Augustine Louesa (Agostinho Luiz; at times "de Sousa" is added), prior to the sailing of the schooner *Industry* of Nantucket on a whaling voyage in 1821:

> It is agreed by & between Hezekiah Pinkham & Augustine Louesa, that said Pinkham shall provide Said Louesa with outfits for a voyage to Brazil & else where in Schooner Industry, whereof Ch^{s}. Macy is present Master – in consideration whereof Said Pinkham Shall receive the one half of the voyage which Said Louesa may obtain – And said Pinkham is hereby authorized to receive of the Owners of Ship Samuel all debts & effects due to me said Louesa from them – and to keep the Same until my return
> Nantucket Novr 13. 1820
> [signed] Hezekiah Pinkham
> Agostinho Luiz Desza [de Sousa][33]

The Lay System

Unique among the maritime industries, remuneration for a whaling voyage was not based on a salary or wage. Instead, the lay system, as it was called, was based on a percentage of the final profits of a voyage. This system of varied shares was employed as early as the Viking era. The lay system addressed the risks of the voyage and was supposed to encourage productivity, although it did not always succeed. In the earliest days, when captain and crew were essentially of local origin, vessels small, and voyages relatively short, the difference between the shortest (captain's) and longest (green hand's) lay was relatively small. But as the industry evolved from a community activity to an industrial model, the distinction between these extremes grew exponentially, so that while a captain might earn a 1/12 lay (that is, 1/12 of the net profit), a green hand might be signed on at 1/200, or occasionally 1/300 or more. (In fact, when the ship *Governor Troup* of New Bedford signed on six green hands at the Cape Verde islands in 1844, one of them was given a lay of 1/500.)[34] And deducted from the crewman's share would be the advance he received at the start of the cruise, normally going into the pocket of the boardinghouse keeper and the crimp who sold him his outfit; cash advances during the voyage, including interest and insurance on the amount; items from the slop chest, and any medicine he might have needed. Also the man's share of the expenses for fitting and discharging the vessel would be deducted from his final settlement.

Nevertheless, in the 1830s it was still possible for a crewman to receive a decent return. When Azorean Frank Joseph returned from a three-year voyage on the ship *Golconda* of New Bedford in 1832, he received $518.20 for his 1/120 lay. In 1835, now evidently a boatsteerer at 1/80 lay, he received $797.66 for only eleven months service.[35]

As the industry entered its most prosperous period in the mid-1830s this disparity of income caused increasing difficulty in obtaining crew members. Perceiving the temptation of hiring foreign nationals to man American vessels, the Congress in 1817 had passed a law requiring that two-thirds of the crew on American ships be native-born Americans. In spite of such strictures a large number of foreigners regularly served on American whaling vessels, their nationality either left blank or falsified on crew documents, thus making any statistics on participation at best uncertain. For instance, during the voyage of the ship *Arnolda* between 1863 and 1867, several of the crew sailed under false names. On the release for payment of Abraham Place, it is noted "alias Antone Pena," and after that of Henry Carr, "alias Manuel Veara," the latter being their authentic Portuguese names.[36]

Besides Portuguese from the Azores and Cape Verde there were other Atlantic islanders and many from the Pacific islands, known colloquially as "Kanakas," as well as Europeans, South Americans, and others.

The Recruiters

'Tis advertised in Boston
New York and Albany
Five hundred young Americans
Are wanted for the sea

They take you down to Bedford
That famous whaling port

Green hands in New Bedford waiting to board, early 1900s.
Courtesy of the New Bedford Whaling Museum.

And give you to some landsharks
To board and fit you out
("Blow Ye Winds," old whaling chantey)[37]

Initially, in the late eighteenth and early nineteenth centuries, it was not too difficult to find men for a whaling voyage. At most an ad in the local newspaper would suffice to entice men to sign on. But as the number of voyages increased substantially after the 1830s, and the subsequent westward movement began to attract the adventurous, more drastic measures had to be relied upon. Such became the role of the recruiting agent, popularly known as a "crimp" or "shark," whose job it was to beguile the unwary into committing themselves for the equivalent of voluntary servitude for a period of several years aboard a whaler. The men who filled the role of recruiting agent (or "boarding master") were normally local businessmen such as boardinghouse owners or the proprietors of outfitting stores catering to the maritime trade.

Boston was a popular spot to find unwary men to ship as green hands. Agents in New London often had recourse to Manhattan's Washington

Market, where they would generally find unemployed seamen who, perhaps, were less informed about the rigors of an extended whaling voyage. Once at the port of embarkation the man would be placed in a boardinghouse where he could be watched over until sailing time, in case he might get cold feet over his commitment. Then it was off to the outfitter, where he was furnished with clothing often quite inadequate for the voyage and for which he might be charged several times its retail value. But the exploitation did not stop there. As William B. Whitecar Jr. explained in 1860, the sailor would be given a set of dungarees, when what he really needed for a voyage to the high latitudes was a substantial outfit of woolen fabric:

> One of these suits will last him about a week; but as he gets into high southern latitudes he finds that he requires woolen clothing, and goes to the slop-chest, imagining that he can get what he wants at a reasonable price. If he inquires how much such an article is valued at, the captain will tell him that he does not know; but, nevertheless, he must have the clothes, and therefore takes them, and thus his account goes on increasing during the voyage. Just before the ship returns home, his bill is handed to him by the captain, and what is his dismay to discover that he is indebted to the owners of the slop-chest, one hundred dollars, or more, independent of the outfitter's bill.[38]

At times such a sailor in debt was even forced to make a subsequent voyage in order to reimburse the whaleship owners. More commonly he was paid off with a small sum.

The unwary mariner was preyed upon by the recruiters not only on shore but also on board ship. As soon as an incoming whaler was spotted coming into the harbor the crimp headed for a boat that would take him aboard the inbound ship. There, sporting a smile, a good selection of cigars, and a flask or two of whisky, he competed with his many rivals for the sailors' attention with offers of new clothing and a great boardinghouse. In San Francisco crimps could be expected soon after—if not before—the ship entered the Golden Gate. In New Bedford in 1859, the newly formed Outfitters Association, in an attempt to control the more egregious actions of these individuals, promulgated a series of regulations:

> [These men] being desirous of so conducting said business as to avoid the necessity of night watching for the arrival of ships at this port without losing the chances of a fair and honorable competition in the same, have united themselves together.... No ship or vessel arriving at this port, or that of Fairhaven, shall be boarded by any member thereof, or by any person in his behalf, at any time between sunset and sunrise, in any part of the bay river or harbor, until after the arrival of such ship or vessel in the bay river or harbor shall have been announced by signal or otherwise, and the party boarding the same shall not start from the shore, for the purpose of boarding such ship or vessel at any point farther south than the north side of Hathaway & Luce's wharf at the foot of Walnut Street.[39]

Only a few days before, the Association had agreed as well to a prohibition against the carrying of alcoholic beverages "in boats used by the association." It was also decreed that any member who used the term "shark" would be fined the sum of twenty-five cents. By 1873, however, the Association ceased to exist.[40]

Life Aboard Ship

Soon after sailing, the crew was called aft and was divided into two watches—starboard and larboard. Crew member Charles Kennon gave a lively description of life aboard the ship *Laurens* of New London in October 1855, about a month out of port, as the vessel headed southwest toward the Cape of Good Hope:

> A whale ship is a busy little city floating over the blue waters.... our watches consist of fifteen men each. One of these watches has to keep constantly on deck, night and day. If the sails want shifting all hands of the deck watch leave their employment and trim sail; one man is kept constantly at the fore and another at the main masthead looking out for whales. Another man is kept at the wheel. About the ship in different parts are the green hands plaiting ropes, making sennet or mats for the oar muffles, etc. In one corner is the blacksmith with his anvil and forge, and a face as black as any factory shore man. Then there are the two coopers, with all the parephenalia of their craft making tubs, buck-

> ets, and barrels; the boatsteerers are grinding their harpoons and lances of which they have many. Whilst some are cutting out sheaths for them, others making lance poles, some making the different little fixings for the boats to which they belong. The carpenter stands knee deep in shavings, making anything and everything that belongs to woodwork, from a wedge to a mast. The cook sits picking chickens, his mate is busy with potatoes, whilst the steward passes to and fro generally with a saucer in his hand. The Captain walks, Lord of all he surveys, taking more notice of the chickens and pigs than the men; the pigs lie scattered about in all directions, and the chickens peck about among them, all of one family. The men have got the pigs' names from something peculiar, "little pete," "black leg," etc., and after one another.[41]

Indeed, various animals were generally kept on deck until the moment to add them to the day's menu. While the bark *George and Mary* of Westport was at the Cape Verde Islands on its outbound voyage in October of 1852, seaman Joseph Silva noted in his journal that one of the boats "came on bord bearing 18 pigs a lot of fouls Pumpkin Bananas and oringes, and 2 Portugues."[42]

Particularly on a warm and calm evening in the tropics one might find all nature of entertainment taking place. The steward of the ship *Eliza Adams* of New Bedford described just such a scene in 1852 as he:

> went forward to night to hear some music, found the fidler playing the fourth of July. Evans keeping time with the bones, the blacksmith playing Juber on the banjo, Goss was playing Bonapart crosing the alps on the fife, and Kimble was whistling Yankee Dodle do. The Portugues was singing a song of their own, and some of the rest was singing old Dan Tucker is come to town. Came as far aft as the stearage, found the fiddle there and accordian in full blast, one singing when I can read my title clear, another O Miss Lucy Neal – then went into the cabin and found the old man [Captain] ratlin away at the Symnoscope, the rest trying very hard to go to sleep....[43]

An essential element of maritime travel is food. Because of their extended voyages whaling vessels often fell far short in providing the

Dinnertime on the John R. Manta, Henry Mandly Sr., master, 1906.
Courtesy of the New Bedford Whaling Museum.

nutrients necessary for maintaining the basic health of crew members. While calling at ports such as those in the Azores and Cape Verde Islands might provide, for a short time, fresh fruit and vegetables, it was not long before a return to the basic staples of beans, salt junk (salted beef or pork), hardtack (a hard biscuit), and the occasional duff (basically a stiff flour pudding to which fruits such as dried plums might be added). According to one writer, "The bill of fare was generally as follows: Breakfast, a meal of food called souse [lobscouse], consisting of hard tack and salt junk cooked together, and coffee; dinner, pea or bean soup, salt junk duff now and then and coffee; supper, meat hash made from the left over, with hard-tack, sometimes with soft biscuit and tea; which as a whole was not a very sumptuous fare."[44]

Upon the exhilarating cry from above of "There blows," boats were lowered and the chase began. The most hazardous moment occurred immediately after the boatsteerer had flung his harpoon and penetrated the whale. Immediately the call would be made to back off in order to avoid a blow from the flukes or tail of the whale. But sometimes this was not accomplished quickly enough, with dire results for the whaleboat occupants, who at times could be smashed right through the bottom of the boat and never seen again. Such an occurrence is recounted in a narrative of the voyage of the ship *Planter* of Nantucket in the late 1840s as it cruised among the Gilbert Islands:

> Quickly the boates moved over the water towards their prey. The whale was now motionless. The boat of the Captain being ahead and close on to the whale, he ordered his boatsteerer to stand up, Just then the whale raised its huge square head high out of the water, and the Captain shouted, 'Now is your time! Give it to him!' The next moment two irons went into the huge creature just behind the hump. 'Stern all!' cried Capt Hussey; 'Stern all, I tell you!' but the men were not quick enough, and a moment later the whales tail came crashing through the bow of the boat, Killing one man and badly injuring the boatsteerer. The whale then struck the boat aft and smashed it to pieces. By this time we were along side in the bow boat. We took in the boatsteerer and the body of our unfortunate Comrade, and the rest of the crew of the demolished boat climed into our boat. Then we pulled back to the ship, hoisted our boat up, and attended to the needs of the injured boatsteerer. The body of the man who was Killed was laid out on a matteress on the quarter deck....
>
> I shall never forget the sight of my poor shipmate as we pulled up to what remained of the Captains boat. After supper that night the Sailors gathered around his cold and lifeless body, Subdued and broken in spirit. Poor fellow! We watched over his remains through the night, and in the morning we washed, shaved, and dressed in clean clothes, and then carefully sewed him up in strong canvas, and secured a bag of stones at his feet. An air of profound Sorrow reigned throughout the ship.
>
> It was indeed an hour of sadness when the ship was hove to, to perform the rites of burial. At 8 A.M. all hands were called aft and stationed in the Starboard gangway; the American ensign was run up at half-mast, and the body was placed on a board in the gangway. After this was done Capt Hussey commenced the reading of the Service of the Episcopal Church. There was a moments pause as he came to the Sentence, "We now commit his body to the deep." It was read, a deep splash was heard, and the body of our poor shipmate sank beneath the blue wave, there to rest until the sea shall give up its dead. He was a Portuguese, and belonged to the Western Islands.[45]

Death, being only an instant away from those at sea on a whaler, was usually a moment for the crew to conjoin, as in the above. Or when a Cape Verdean boatsteerer died on the ship *Norman* of Nantucket in 1849 while the vessel was sailing in the South Indian Ocean, upon which Captain John J. Gardner penned these words in his journal:

> At 10 PM A Man by the Name of Altoneo Pena Sylva Died after an illness of 5 days with inflammation of the Bowels aged about 25 years, A Native of the Island of Bravo one of the Cape Verds he Shipped at Bravo in the Room of Peter Charles the 16 of October 1848 and Was to have the Same lay of Peter Charles Middle & Latter part light Winds from ESE to EWE Stood to the SE at half past Nine AM backed the head yards at 10 A prayr was red Over the Corps and wee consigned it to a watera grave there to awate the Reserection When the Sea Shall Give up its dead, all was Sadness and Sorrow, many tears was Shed over the Corpse of our depar[t] ed Brother and all Seamed to feal the Solemnity of the Occation....[46]

That this feeling of community was not always shared is evidenced by the following comments from a green hand on the ship *South Boston* of Fairhaven in 1851. Joseph Jocking (Joaquim, aka Joe King), a seaman from the Azores, had just died of dropsy, and he was about to be given a deep-sea burial:

> ... at 9 A M brought the corpse aft on deck an sowed it up in canvass and put a bag of Bricks at its feet and Laid it on a board called all hands and took it in the waist when the mate began to read a prayer. The crew were all gathered all around with their hats off. When Mr. Heath (the mate) had partly read the prayer the signal was given and the corpese was Launched forever from view in the raging ocean and when the prayer was finished the crew returned to their places not a word being spoken by any of them but I suppose they would have felt worse had he not been a portugee.[47]

It was not only during the whale chase that men risked life and limb. Climbing in the rigging, maneuvering the decks slippery with oil, avoiding the boiling trypots—all required constant vigilance. Such an unhappy moment was noted by the log keeper of the bark *Prudent* of Stonington as it was cruising in the South Atlantic in 1848:

> At 11 Am a Most Horrable accedente Accred [occurred] in Aowlle [while] reaffing the foar Top sail [reefing the fore topsail] a man fell from the yard whilst notting the Aoist [knotting the hoist], was herd to strike the rail or Bows. The Cry of a man over Bard was instantly Made All Convend All Hands was Called Backen [backed] the main Top Sail Lowered a Boat an Made all Possible serch returned with out finding him. The wether was thick and rainy ate the time we Think that he was Killed instantly ded in falling as Thare was no sound or ado made by him we examend the rail and Bows but cud not se whare he Struck. His name was Mar Sails Josa [Marcelo José]. A Portigues shipped at the Western Islands Flowres [Flores]. he was a bowt 21 years of age and belonge to Floures. He has Left no Wife father Mother Brother nor sister to Morn his Loss.[48]

Others had better luck. On the ship *Potomac* in 1845, "a Portuge fell overboard and he Could not swim he was going down and one of the Boat steerers Jumped over and saved him he Came prity near being lost."[49] Even more fortunate was Manuel Antone in March of 1851: "The captain of the Euphrates, whaler, which arrived at this port last evening, relates the following: On the 10th inst. Manuel Antone, a seaman of his ship, fell from the fore-top gallant yard, struck on the main top sail yard, and alighted on the slings of the main yard, without suffering the smallest injury. He was not even bruised, and immediately returned to his duty."[50]

Sometimes a modest suggestion could lead to tragedy. While the bark *Noble* of New Suffolk, New York, was cruising near the Fiji Islands, "one boat got a drift from along side & no boats [were free] to lower for her by the advice of Mr Luyge a Portuguese Emanuel jumped over after her. Got an old boat from over head quick as possible & pulled as we thought in a right direction for her, but very dark & rugged & saw nor heard nothing. So here are we in horrid suspense as to the fates of the man."[51] Early the next afternoon, in rugged weather, the boat was finally located, "but horrid to say the poor Soul was not in her. Probably he was devoured by the Sharks before he reached here as we saw some in our wake soon after he went overboard.

> His tomb is in the ocean bed
> Far from his Island Shore

Hungry Sharks may have on him fed
But his home he'l see no more."[52]

In May of 1845 the *New Bedford Weekly Mercury* announced the death of seaman John de Gomez (Gomes), "a native of the Western Islands." He had in fact died the year previous in August of 1844. His ship, the *Huntress* of New Bedford, was in the South Atlantic, and the crew was stowing casks into the after hold. Gomes had been standing on the deckhouse when he slipped and fell into the hold.[53]

Unfortunately, as the keeper noted, Gomez "cut his hed open six ore eight inches brekeing the Scull." The following day another entry indicated that he had "brok his scul so that it does not mech by half and inch."[54] Gomez survived almost two weeks before succumbing to his injury. No mention was made in the log of any medical treatment he might have received in the interim. Nevertheless, the log keeper noted that "the Capteon could not halve Done more for him iff hee Had bin hiz brother."[55]

The Captain as Physician

The master of an American whaling vessel was expected to treat the sick and injured, no matter how serious the situation, unless there happened to be a foreign ship or a port nearby. How well he performed this responsibility depended largely on his experience and attitude. When Joseph Thomas was badly injured during a voyage of the ship *Clematis* of New London, in 1854, Captain William Benjamin noted later, "Joseph Thomas got very badly gamed with the [whale's] head fetching away hove his hip out of Joint I set it as wel as I could weather I got it back in its place or knot I don't know for certainty sum hurt and som sick and so it goes."[56]

British and French whaleships were required by law to carry a physician onboard, but not the Americans. Upon sailing, each American vessel was furnished with a medicine chest and a book of instructions detailing various ailments that might commonly be encountered at sea and their suggested remedies. Each of these remedies was given a number. One such list of instructions was composed by Dr. E. D. Skinner of Greenport, New York. For example: "No. 8 Laudanum—An excellent medicine for relieving pain and procuring rest; it often prevents the fever liable to be induced from wounds and broken limbs, 30 or 40 drops taken in water....”; and “No. 17 Extract of lead.—Excellent in case of inflammations, swelling, Etc. Keep cloths wet with the preparations made with one tea spoonful of the extract mixed with one gill of rain or other soft water, constantly on the part affected.”[57]

Besides heavy metals such as lead and mercury, and drugs such as opium, many of the remedies were natural ones still in use today, including essence of peppermint, chamomile, mustard seed, and flaxseed powder.[58] Thus a seaman's recovery depended in large part on a captain's diagnosis and his interpretation of the medicine chest instructions. Of course there was always the threat of venereal disease and its debilitating effects, to be countered with consecrated nineteenth-century remedies. In 1851, while the ship *Montreal* was cruising in the Pacific, one of the crew noted, “no less than ten of the crew turned up with the lady's fever, most of them brim full of mercury, so that at present we could not man four boats....”[59]

The master of the vessel most often simply dispensed the indicated medicine—the expense of which would ultimately be deducted from the man's payout at the end of the voyage. But he might also be called upon to set a broken bone or to perform some other surgical procedure. This was particularly true in the more remote regions where there was no access to a port or a European vessel. At times the call for treatment might be sudden and serious, as occurred on the bark *Josephine* of New Bedford as it made its way through the Indian Ocean toward Capetown in September of 1904. As Captain Horace P. Smith noted in his log that evening, “at 7 30 A.M. the Cooper C.J. Olsen and a boatsteerer Antone Santos had some words Santos struck Cooper, and then Cooper carved Sants face with a drawing knife severing the ear and cheeks nearly off his head I had quite a job to get it stiched in place a gain.”[60]

The captain's ill preparation in the matter of medicine is reflected in the much recited legend of the inventive master who, in a moment of necessity, is called upon to improvise. The apocryphal tale of the Portuguese boatsteerer Antone is related in “Dead Reckoning”:

> Captain Josh Winters looked at Antone, boatsteerer on Mr. Smith, the second mate's boat. Writhing in pain and partially delirious, the boat steerer lay in his bunk with several awed shipmates peering over the shoulders of the captain and second mate. “Tony” was apparently critically ill. Captain

Winters left the bunk of the sick man, climbed up the narrow companionway and with a measured tread, which always indicated deep concentration on his part, walked aft to his quarters and opened the medicine chest.

Pulling out a dog eared pamphlet titled "Prompt Diagnosis for Master Mariners," the Cap'n sat down beneath the swaying sperm oil lamp and thumbed over the pages. There it was — "Acute pains in lower abdomen. Give patient six drops of No. 14 in small amount of water. Repeat every hour or until pains subside."

He returned to the medicine cabinet and from the neat array of numbered bottles and boxes, selected bottle No. 14. He shook it, then held it up to the dim light. Dry! Some apothecary clerk in refilling the medicine chest when the *Midnight Sun* was being fitted out, had failed to check his work. Now here he was with a mighty sick man on his hands and none of the No. 14 prescribed by the book.

The skipper of the ship pushed back his hair, the better to think. It was because he "had a square head on his shoulders," the agents of the ship had told him, they had decided on him as master of the ship. And here he was — 1,000 miles from port, on his first voyage as captain, faced by the prospect of losing a valuable boatsteerer.

Captain Winters' deep study was interrupted by his second mate, who knocked at the cabin door, and then, without waiting for an invitation, stepped in. "He's much worse sir, he's in terrible pain." Cap'n Winters didn't hesitate further. They had given him command of the ship because be knew how to use his head. He wasn't going to lose it in the first emergency!

Stepping up to the medicine chest he drew out bottles numbered 9 and 5, carefully measured out three drops of each into a small quantity of water and handed the concoction to the mate. "Nine and five's fourteen, that ought to do as good as No. 14," muttered the skipper to himself as the mate hurried for'ard with the medicine.

"And so dear Lord, we commit his earthly remains to your care." Captain Winters closed his bible with a snap, the signal for the two men standing by the rail to tilt the inboard end of the wooden slide they were balancing on the ship's rail. A form, shrouded in sail cloth, slid from the board, there was a splash and those near the side of the ship saw a few bubbles rise from the spot. "Nine and five, I'm sure that made fourteen," mused Cap'n Winters.[61]

Another Antone, aboard the bark *Pacific* in 1867, was more wary of the master's remedies. As the keeper noted in the vessel's log, "Antone has been off duty about a week. Capt Allen has been doctrien him Some time for the bad disorder this morning gave him A Pill Saw him take it in to His mouth after wards found it tied up in a handkerchief So Ends."[62]

Sexually transmitted diseases were a common problem aboard whaleships, with their exotic ports of call. Again it customarily fell to the master of the vessel to initiate treatment. Joseph P. Benton (Bento) was a Cape Verdean from the island of Maio.[63] As captain of the schooner *Clara L. Sparks* of New Bedford during its voyage of 1891-95, he was forced to deal—this time more decisively—with the case of a crewman who refused to follow a regimen of treatment. The keeper observed:

> Robert John Pinis is getting worse he is very doarty and careless with himself he wont clean unless he is told too the mate has to go forward three times a day to see him wash and dress it his Pinis. today the Capt call him aft and told him he must cut the end of the foreskin of [off] he said no he didnot wunt it cut the Capt told some of the men to hold him he rysist he said he was feeling his own pain that he would rather dye then to have it cut – these men were call to witness:
>
> Charles William X his cross
> John Moyse X his cross
> Joseph Jones X his cross
>
> these men has touch the bible all was done in present of the Afficers and Boatsteers."[64]

Fortunately, if a vessel was not far from land it could call at a port. Because whaleships often tended to spend time around the so-called Western Grounds, near the Azores, they could bring in an ailing crewman for treatment. The log of the ship *Franklin* of Sag Harbor noted in 1847, at the island of Faial, that "at 6 PM the Capt. went on shore at 10 came of again taking the Carpenter who had a sore leg and another man – Lake who had inflammation on the lungs, ashore to be examined by the doctor."[65] Two days later both men were discharged from the vessel after the doctor "said they were not fit to go."[66] A week later boatsteerer Henry Babcock was taken ashore at Terceira, and after being examined by a doctor was declared unfit for the voyage.[67]

Filling out a Crew in the Islands

One of the ways to fill out a crew was to sail shorthanded and pick up men on the voyage out. Since it was common for American whaling vessels to call at the Azores for supplies on the outward voyage, it early became the practice to sign on men, especially from the western group of the islands: Flores, Corvo, Faial, Pico, and São Jorge; as well as occasionally from those islands to the east. Young men were picked up from Cape Verde as well, Brava and São Nicolau becoming the most common points, although other islands were often visited. Other islanders from the South Pacific and later the West Indies helped to fill out a whaler's crew. Limitations of language and education, and some Yankee prejudice, would most generally keep them from a mate's berth in the early years. But, nevertheless, they might sign on again and again in the hope of accumulating a nest egg to sustain them once again in their native land. Sailing on an American whaler, especially for the Portuguese, was also a way to follow their countrymen into new settlements in New England and California.

William B. Whitecar served from 1855 to 1859 aboard the bark *Pacific* of New Bedford and had ample time to observe Azorean whalemen. He later wrote about his experience sailing the seven seas:

> Great numbers of young men are carried off from these islands annually, by American whaleships, the government demanding of each young man, born in the islands, a certain amount of military duty in Europe. To emancipate themselves from this irksome service they join whalers.... When these people first come aboard the ship they are indifferently dressed, and invariably barefooted; when those we shipped were supplied with an outfit of sea clothes, they were greatly astonished and delighted. They are a very economical people, and by dint of washing for others, patching, at which in a short time they become adept, and other little jobs, they soon become possessed of a large amount of clothing, which they hoard up and gloat over as a miser would his gold. They are shipped for little or nothing as regards remuneration, scarcely anything being said about a lay on either side; but the captain, if generous, will always make them a liberal allowance on the ship's arriving at New Bedford. They are generally strong and able-bodied, and make good working-hands to pull

> and haul, but, except in rare instances, do not rise in position above steering a boat; although there are several ships at present sailing out of New Bedford whose masters are Portuguese by birth, yet in each instance, I am informed by good authority, they were taken from the islands at a very early age, and sent to school in America between voyages.[68]

Actually, a good part of the education of these future masters most likely was acquired on board. For these latter youngsters, having been signed on sometimes as young as nine, and serving as cabin boys, would have had a unique opportunity to commingle with the more-educated officers and to have received some tutoring in the English language, perhaps even from the captain or his wife, that would not generally have been available to the foremast hands.

It should be pointed out that chores such as washing and sewing formed a normal part of shipboard life for the forecastle hand. As a crewman aboard the ship *Illinois* of Sag Harbor noted in 1846, as the hands were busy quilting, "The jacks both old & small, brought their needles & their palms & sitting around on chests & buckets & barges, surrounded the quilting by the number of ten—Twas an odd sight to see these 'heavy dogs' use the needle & quilt—some smoking others chewing, while in rotation they spun their yarns."[69]

Sewing and patching were simply one of many leisure-time activities aboard a whaling vessel. As Scotsman Robert Ferguson, boatsteerer on the bark *Kathleen* of New Bedford during the early 1880s, noted in his journal, "There has not been much to do lately. Some of the men have been doing scrimshaw work and making canes. Others have been making gimcracks of ivory, paper knives, cribbage boards, ditty boxes to hold buttons, needles, thimbles, scissors, spools of thread and things for sewing. Some have been patching their clothes. I am making a cap with a good-sized peak or sunshade for the eyes. It seems strange to see the men making clothes, light duck breeches, jumpers and even canvas shoes, and making good jobs of them."[70]

In *The American Whaleman*, Elmo Hohman makes particular reference to the Portuguese islanders' parsimony and industriousness:

> Certainly many of them were veritable marvels of thrift and of parsimony. Their slop-chest accounts, showing the supplies drawn during the

> course of a voyage, were often negligible in amount, and far below the average of the other hands.
>
> Some of the Portuguese hands eked out their earnings by washing clothes for their less thrifty shipmates. Since the whalemen were often bathed in perspiration and exposed to innumerable forms of dirt and grease, and since their clothing was periodically saturated with whale oils and salt water, the task of washing such garments was hardly agreeable. It was rendered all the more laborious, too, because the scrubbing had to be done in salt water; for fresh water was available for cleansing purposes only when rain water could be caught upon deck. It was not surprising, therefore, that at times certain members of a crew were willing to pay trifling sums to escape such tedious laundry work.[71]

Sometimes these chores were done for other than direct remuneration. Boatsteerer Robert Ferguson noted in March of 1881 that, "Whenever I could find the time, I have been giving Frank Gomez (Gomes) and John Brown their lessons. Frank pulls an oar for me and is a very good fellow in a boat. He is very anxious to learn to read and write. To offset the lessons, he does little odd jobs for me, like washing my clothes and helping me out at any time."[72] Frank Gomez was from the Azores and John Brown hailed from Brava. Both were quite anxious to learn to read, as Ferguson later observed: "I held school for an hour today with Frank Gomez and John Brown. Frank is getting along fine. John can read some small words but the larger ones puzzle him, so he comes to me at dog watch and asks the meaning of them. I have to explain, but he has a good memory."[73]

Asked by one of the other boatsteerers what he planned to do when he returned to New Bedford, John Brown told him that

> if he had five hundred dollars coming to him, he would go home to Brava. He has a piece of land there, five acres, that he bought with money from his last voyage. There is a little two-room house on it built of stone, which he has rented until he gets back. He said that there was plenty of wood on the island and that he could live there a long time by selling chickens at ten or fifteen cents apiece, and having a goat for milk, and raising some pigs. He said he was going to buy an old boat, fix it up so he could fish, as there are plenty of fine fish to be caught. He would not need much clothing as there is no cold weather, and there is plenty of fruit of all kinds.[74]

Harbor of Horta, Faial, Azores, ca. 1880.
Courtesy of the New Bedford Whaling Museum.

When the ship *Hannibal* of Sag Harbor anchored at Horta on the island of Faial in 1827, Captain Henry J. Green took aboard not only 142 bushels of potatoes, 150 cabbages, and 48 pumpkins, but also "one Portugee with me by the name of Domingo Silvia."[75] Procuring seamen was not always a one-way transaction, as can be seen from an entry in the log of the ship *Delta* of Greenport in 1834: "At Fayall. Monday, Aug. 4th came to Anchor and furled the Sails. At dark got out vegeateables on board put out Sperm oil in the Lanch to bee Sent home we got one man and one run away."[76]

Elisha Dexter, in his 1848 memoir of service aboard the whaling brig *William and Joseph*, was particularly impressed with the Azores and its people: "The scenery of these islands is also very beautiful," he wrote, "As you sail by them, at the distance of two or three miles, in the season when the orange groves are in full bloom, and the fields green and verdant, nothing can be more picturesque and delightful than their appearance. The contrast exhibited by the lofty mountains and the deep valleys, is grand and striking."[77] Dexter's opinion of the inhabitants was very positive, but with the common American whaleman's caveat regarding their religion:

> As to the inhabitants of these islands, I speak but the common opinion of all who have ever visited them, when I say that they are a very kind

and generous people. Any person may enter their houses, without any invitation whatever, and receive an apparent, and, I doubt not, an actual welcome. Sometimes, with a graceful motion of the hand, in the skilful use of which they resemble the Asiatic races, they will say, "enter." When you have entered, they will, sometimes present you with a "Jesu Christo," which is an image of our Saviour, made of dough. In presenting this, they will kiss it, and you are expected to receive it with great veneration. It is indeed a pity that so gentle and hospitable a people should not possess a purer and a better faith. I speak now of the lower classes, not having access to the higher. They are very partial to the Americans. The word "Americano," sounds well in their ears.[78]

Dexter succinctly describes the typical function of these Western Islands at the time: "On the 2nd of September [1840] we ran into Fayal, one of the Western Islands, or Azores, for the required complement of men, and for the purpose of obtaining the usual supply of vegetables, which are there very cheap."[79]

A highlight of a visit to the port of Horta, for those fortunate enough, was a visit to the estate of the American Consul, Charles W. Dabney, with its gardens, walks, and observatory. In the summer of 1835 John Swift Jr. of the ship *Janus I*, of New Bedford, remarked with enthusiasm on such a visit: "At 8 AM went ashore at Fayal rode out to the Counsels County Seat and dined in great splendour in a fine shady grove."[80]

The history of the Dabney family on the island of Faial encompasses virtually all of the nineteenth century. John Bass Dabney arrived in Faial as American Consul in 1806. He soon established an important business in the Pico wine trade, in ship supply and repair, and the transshipping of whale products, among other ventures. Charles Dabney became Consul upon his father's death in 1826. Elisha Dexter noted that the whaling fleet of 1839 had deposited "nearly 6,000 barrels of sperm oil" with Charles Dabney that year.[81] Charles was succeeded by his son Samuel W. Dabney in 1872. However, when it was decreed that American consuls could no longer operate independent businesses, Samuel returned to the U.S. in 1892.

Whaling left a deep impact on the Portuguese Atlantic islands. The constant passing of foreign vessels, the departure and eventual return of so many of the young, created a degree of cosmopolitanism not to be seen, for example, in an American farm boy from the Midwest. For a long time it was the most significant means of eventually starting a new life in America.[82]

For Azorean poet Pedro da Silveira distances, in his youth, had little relevance:

> When I was young [on the island of Flores] and saw the ships that
> passed and afterward disappeared in the distance
> I only knew the name of two places where they might be going.
> If they were going south their destination was Faial
> and if it was to the northwest the name was America;
> but the distances were exactly equal.[83]

Listening to the tales of his well-traveled elders, an island youth could almost palpably experience life in the most exotic of settings and look forward to his own adventures away from his circumscribed present:

> You were ninety years old, Uncle Joe from Pico, and you were like a child
> seated in the corner of the kitchen recounting the events of your life:
> eleven times at sea,
> eight times shipwrecked,
> wandering from Honolulu, Sydney, Punta Arenas,
> lover of the queen of Samoa…
> (Your hand sketches in space,
> accompanying your soft speech,
> the marvelous adventure you lived).[84]

Whaling, and its influence on family life, was no less of a reality in the Cape Verde Islands. As Cape Verdean writer Teobaldo Virgínio mused:

> Time that passes
> ship that departs
> story of the whale
> of the adventure of the seas…
>
> Roots transplanted
> streams that are reborn
> in the field of *Saudade*!

Ship that returns
(from how many misfortunes?)
to beloved Rincão![85]

In the novel *Chiquinho* by Cape Verdean novelist Baltasar Lopes, one can glimpse the influence of the whaling experience on both the young and old. When the harpooner Chico Zepa of the bark *Wanderer* drops by to greet his family, "At each moment he inserted American words in his conversation… Nhô João Joana asked about America: 'Do they still have a *light ship* at the entrance to *Betefete* [New Bedford]?' 'Still. But now the government has placed a signal buoy near the *Stream*, *you know*….'"[86]

And for those long absent, family life adjusted. As Chiquinho, the narrator of the same novel, reminisced, "In spite of living at 102 South Second Street [New Bedford], Dad was always with us." This phenomenon led to what were known in the islands as "American widows" (*viúvas americanas*), the result of long absences of family men from their native land, during which wives would don the black clothes of the true, grieving widow.[87]

Attitudes toward Portuguese Atlantic Islanders

What was it like to be Portuguese aboard an American whaleship? Most likely in the early days, when there might have been only one or two on a voyage, it would have been an opportunity to learn quickly the rudiments of English and to adapt to the predominant culture of the majority of officers and crew. On the other hand, by the end of the nineteenth century and early in the twentieth, at a time when the majority of those aboard an American whaler might hail from either the Azores or Cape Verde, there would have been comfort in the ability to maintain one's own language and culture. In between, however, during the heyday of American whaling, when it was not uncommon to find a handful of Portuguese on each voyage, it would have been particularly difficult, for while there were sufficient compatriots to encourage continued use of the native language, as well as the maintenance of elements of the culture, there were strong pressures against their manifestation, particularly that of language.

On the bark *Pacific* of New Bedford in March of 1868 the keeper noted, "when we came down to dinner to day as Mr davis was coming from aloft he heard a man by the name of —— talking Portuguese which was against the rules of the Ship on decks Mr davis asked him what he was talking that way for he replied that was his lingo and he said I be damed if I wont talk it for all any body So after Dinner call him aft and put him in irons at dark put him and the other one in the run." Two days later he added, "after Supper let the two men out of irons they promised to do their duty and not talk Portugese on decks."[88]

The poverty of the Portuguese islanders, their lack of education, and even their religion added to their being discriminated against. As one American crewman opined about the Azoreans, "the natives are mostly poor made so by the Queen of Portugal & Catholic Priests...."[89] While this may appear to be an anti-Papal diatribe, there is truth to the young man's observation, since Lisbon for centuries practiced a policy of benign neglect when it came to its Atlantic possessions. And the philosophy of the Catholic Church in many regions under its hegemony was to encourage education only for an elite or for those entering the priesthood. The result during the nineteenth century was a population generally illiterate and ill prepared to function in positions of authority where the command of written language was a necessity.[90] This would have a deep impact on their upward mobility within the whaling industry.

Of the many observations on Portuguese crewmen during the nineteenth century probably none is more balanced and fair than that of veteran whaleman Charles Nordhoff:

> A great many Western Island Portuguese find employment in American whalemen; almost every vessel sailing from New Bedford carrying more or less of them. They are a quiet, peaceful, inoffensive people, sober and industrious, penurious, almost to a fault, and I believe excellent whalemen. They are held in great esteem by ship owners and captains, but are often despised by their shipmates in the forecastle, who seeing them of such different habits to their own, choose to decry them as sneaks, and tale-bearers.
>
> I found them quite the reverse; and with one or two exceptions, those with us were the only individuals of the crew with whom I could associate with any degree of pleasure. Brought up in the most abject poverty, it is natural that they should be saving their hard earned money for trifles or in dissipation, as is the fashion with sailors in general.

> They have moreover an object in life, which is never lost sight of in all their wanderings and toils. It is their hope some day to be able to settle down on their native islands, among their friends and kindred, and with the savings of years of hard labor, to spend their latter years in peaceful retirement. Very many, I have been informed, have lived to realize this day-dream and making the few hundreds of dollars, which is the sum of their savings, have returned to live at ease in the home of their youth. It is not possible that men who cherish such recollections, and enter on life with such hopes and determinations, should fall to the depths of depravity and vice in which whalemen generally lose themselves.[91]

Given the prejudice of the time against people of color, it is no wonder that Cape Verdeans in particular felt isolated on board.[92] As a crewman aboard the English whaleship *Warrens* of London, England, noted, while the vessel was cruising off the coast of Borneo in 1838:

> Fine moonlight night. breeze continuing-The crew amusing themselves with various entertaining occupations "hunt the slipper. hanging by cords &c - while the 4 Portuguese amused themselves / no one seemed to take any notice of them forward with a more refined enjoyment viz playing on the guitar. this being more in union with my feelings I amused myself for 1/2 an hour or so talking with them between while playing a tune & giving the Spanish Fandango—they feel pleased by being taken notice of—and because they are darkies as the sailors call them – they do not associate with them.- such unkind feelings I never can cherish – and whatsoer the cast it pleases me to join in their conversation - & to take off the uncharitable feeling which many persons cherish towards others. I feel for them - fancy how I should feel under similar circumstances.[93]

Bigotry is apparent in log entries such as that of the bark *Zone* of Fairhaven while at the island of Fogo in 1855: "In an hour came off with 10 hogs, 4 goats, a dozen chickens and a bunch of bananas —capt. came back again at 10 p.m. bringing three new Portoguese, to be added to the rest of the live stock!"[94] The next day the keeper added, "We have got plenty of music on deck now for there is nothing else to be heard but a pig squealing, or a goat blaating; and if that is not enough to satisfy, go forward and you can hear a half dozen of 'Dagos' jabbering Portuguese. Maybe that will cause you to express a wish to get to the masthead or somewhere else, where you cannot hear."[95]

One notes as well the anonymity into which Atlantic and Pacific islanders often fell. This is seen, for instance, in the notation of a tragedy that occurred one afternoon in December of 1854 while the ship *India* of New Bedford was lying off and on the island of Rarotonga in the South Pacific. The ship had sent two boats ashore after provisions. The first boat returned safely; the second was seen to pull out through the reef and set her sail, and that was the last seen of her. The ship, lying about four miles to leeward, set a lantern that night; and the next day a search was made, to no avail. As the *Whalemen's Shipping List* noted, "The boat's crew consisted of seven men, Charles F Little of New Bedford, 5 Portuguese, and a native of Rorotonga."[96]

John Thompson, a fugitive American slave who served aboard the bark *Milwood* of New Bedford in the early 1840s, recalled a situation of prejudice against his Portuguese crew mates while in the Indian Ocean:

> We remained in Madagascar three weeks to repair the ship, which was damaged at sea. While lying in port four of the crew escaped, and were concealed on shore by the natives; who afterwards came and betrayed them to the captain for a price. The mate, with a boat's crew of Portuguese, was sent for them, with whom they not only refused to return, but severely cut and bruised them. Afterwards the captain, with the captains of five other vessels, then lying in port, went for them, conducted by the natives, who knew their place of concealment, in a native hut.
>
> When he discovered them, the captain calmly told them he wished them to return with him to their duty on board the vessel, to which they readily gave their assent, saying they would have gone before had he sent Americans for them, but that they would not willingly submit to be fettered by Portuguese.[97]

The Whaling Life

Come all you young Americans
And listen to my ditty
It's all about a whaling bark

That left New Bedford city
The bark *Gay Head* it is her name
She's known both far and near
Her rules and regulations
They are most awful queer.
("The Bark *Gay Head*," old whaling chantey)[98]

Life in the American whaling industry was indeed unique; and, even prior to the Revolutionary War, it was not restricted to Americans. While merchant vessels headed directly to their next port, whaleships cruised about, at times without a fixed destination, for months at a time, with long stretches of boredom to be broken by the cry of "There blows!," followed by the most dangerous moments in the life of the whalemen. Dwarfed in their twenty-five to thirty-foot boats by the largest living creature in the world, they risked life and limb to harpoon the leviathan, to lance it, and to pull it to the mother ship for processing, which in itself presented a myriad of hazards.

Unique also was the lay system, whereby the common seaman and his shipmates shared the profits at the end of the voyage as their sole remuneration. Here, what began as a system to reward community members who shared common interests, evolved by the nineteenth century into a means of exploiting the lowest members of the hierarchy. Hiring on eager foreigners on the outward voyage made this task much simpler. And thus began the long tradition of American whalers calling at the Azores and Cape Verde to provision and fill out their crews.

For their nature of being foreigners of a different tongue, with a modicum of education, and not infrequently being people of color, Portuguese Atlantic islanders faced varying degrees of discrimination. Nevertheless, in spite of the bigotry of the day, which was actually increasing during the nineteenth century, abetted by the pseudo-science that Herman Melville attempted to criticize in his oft-maligned essay on "The Gees," people of color from Cape Verde, as well as those from St. Helena and the West Indies, were increasingly able to move up to the rank of officer aboard American whalers.[99]

Notes

[1] Duncan, *Atlantic Islands*, 1.

[2] Cf. Fernández-Armesto, "Medieval Atlantic Exploration" and Verlinden, "Portuguese Discovery Era."

[3] Schooner *Eagle*, Salem, Mass., 2 August 1753-21 May 1756, Log 1753E, PEM.

[4] Duncan, *Atlantic Islands*, 158.

[5] Ibid., 171.

[6] Note: In order to remain faithful to the originals and to convey a sense of the era, this and succeeding quotations have not been corrected for grammar or spelling. Benjamin Hodge, "Reminiscences of early life, including a whaling voyage, and a voyage to the West Indies in 1775," Benjamin Hodge Papers, MSS HG 63. This item is reproduced by permission of the HLCA. Hodge states that the vessel was owned by John Brown but never gives its name. He is not to be confused with the Benjamin Hodges cited in the next chapter.

[7] Ibid; Lobban and Lopes, *Historical Dictionary*, 83.

[8] Hodge, "Reminiscences."

[9] Ibid.

[10] Ebenezer Townsend, Jr., Journal, ship *Neptune*, Greene, New Haven, Conn., 1 Feb 1797, Log 869, KWM.

[11] Ibid.

[12] *New Bedford Daily Mercury*, 10 Oct 1832, 12 Jul and 9 Aug 1833, 28 Feb 1834; Lobban and Lopes, *Historical Dictionary*, 83; Albuquerque and Santos, *História Geral de Cabo Verde*, 1:13.

[13] Bark *Pantheon*, Borden, Fall River, Mass., 23 and 25 Dec 1842, Log 769, MSM.

[14] Littlefield, "Fitting Out a Whaler," 6.

[15] Occasionally there are references to a snow (pron. 'snoo'), a variation of the brig.

[16] "Sparm" = sperm whales. Bradford, *Barnacles and Bilge Water*, 12.

[17] Tower, *History*, iii.

[18] Ibid., 55.

[19] Cf. Davis, Gallman, and Gleiter, *In Pursuit of Leviathan*, 402.

[20] Charles W. Morgan, Letterbook, 1833-37, 26 Aug 1834, Morgan Business Records, BLH.

[21] Isaac Howland Jr. and Co., Letter Book, 9 April 1835-19 Oct 1838, MSS 79.22, ser. 3, box 6, f.2, Carlton D. Morse Whaling Coll., 1782-1964.

[22] Davis, Gallman, and Gleiter, *In Pursuit of Leviathan*,16.

[23] Bark *Zone*, Fish, Fairhaven, Mass., 4 Sep 1855, Log 278, KWM.

[24] A most bizarre communication was received on board the ship *Stonington* of New London in 1830 while it was cruising north of the Falklands. The mate noted, "at 11 Shot an Albatross With A vial tied to his neck from the Electra he reports her with 3 whales in the latitude of 40 Long 46.27 [lat.]." (Ship *Stonington*, Pierson, New London, Conn., Log, 21 Oct 1830, MSS 540, Whaling Logs Collection, YUL.)

[25] Ship *Hector*, Chase, New Bedford, Mass., 1861-65, Miscellaneous Whaling Ships, Nicholson Whaling Coll., s-g 13, box 7, Ser. 7-B, S-s 2, f. 1, PPL. Since there was no American whaling vessel with a name approximating that of the *Mook*, it is very unlikely that Manuel ever received this letter.

[26] Cf. Byers, *Nation of Nantucket*, 63-73.

[27] Hedges, *Records of the Town of East-Hampton*, 1:378-79.

[28] Cf. Vickers, "Nantucket Whalemen," 277-96.

[29] Note: While these individuals are clearly identified as Portuguese there is no further indication of their origin. This lack of precise identification is found at times throughout the years of the whaling industry. As for Joseph Swazey, his name appears alternately in various historical documents using a "z" or an "s." In the William Rotch Journal it was "Swasey." For the sake of uniformity the Swasey/Swazey family name will appear always in this text with a "z."

[30] William Rotch Journal, Rotch Family Papers, MSS 2, s-g 1, Ser. B, S-s 1, v. 1:431, ODHS.

[31] Rotch Journal, 2:492, 583, 607, 789.

[32] Sloop *Sandwich*, Rotch Journal, 2:504.

[33] "Memorandum for Hezekiah Pinkham to profit outfitting to Augustine Luisa, crew member," Nantucket, Mass., 13 Nov 1820, Edouard A. Stackpole, Coll., MSS 335, f. 994, NHA. Listed in the 1820 census as Augustine Lewis, he was living at the time at the residence of Mr. Pinkham (Census of the Town of Nantucket, Mass., 1820, NAL).

[34] Ship *Governor Troup*, Jenney, New Bedford, Mass., Whalemen's Shipping Paper, mic 182, NBL.

[35] Ship *Golconda*, Baylies, New Bedford, Mass., Howland Family Papers, MSS 7, ODHS.

[36] Ship *Arnolda*, Hawes, New Bedford, J. B. Wood & Co. Records, MSS 75, Ser. A, S-s 1, v. 1, ODHS.

[37] Huntington, *Songs the Whalemen Sang*, 42.

[38] Whitecar, *Four Years aboard the Whaler*, 409.

[39] Records of the Outfitters Association of New Bedford, 7 Mar 1859, Business Records Coll., MSS 56, Ser. O, S-s 5, ODHS.

[40] Records of the Outfitters Association, 2 Mar 1859; *New Bedford Morning Mercury*, 29 Jul 1916.

[41] Charles Kennon, Journal, ship *Laurens*, Smith, New London, Conn., 24 Oct 1855, NLHS.

[42] Joseph Silva, Journal, bark *George and Mary*, Manchester, Westport, Mass., 5 Oct 1852, Log 615, ODHS.

[43] "Six Months Outward Bound: John Jones, Steward, Ship *Eliza Adams* of New Bedford, 1852," Frank, *Meditations from Steerage*, 15. As Frank explains, "Symnoscope" is presumed to be a corruption of the trade name of a parlor organ of a type not infrequently found in the aftercabin of a Yankee whaler." (Ibid.).

[44] Lemuel D. Eldred, "The Voyage and Eventful Return of the Bark Jacob A. Howland of New Bedford in 1880," Pam B III-116, ODHS.

[45] William [no surname], "Narrative of the Voyage of the ship *Planter*, Nantucket, Mass., 1847-51," Log 830, KWM.

[46] Captain John J. Gardner, Journal, ship *Norman*, Nantucket, Mass., 14 Feb 1849, Log 234, Ships' Logs Coll., MSS 220, NHA.

[47] Samuel B. Morgan, Journal, ship *South Boston*, Williams, Fairhaven, Mass., 4 Dec 1851, Log 761, ODHS.

[48] Bark *Prudent*, Nash, Stonington, Conn., 4 Sep 1848, Log 820, MSM. The author of another account claimed that the deceased "Left a Mother and 2 sisters to morn the deth." (Jared S. Crandall, Journal, bark *Prudent*, Nash, Stonington, Conn., 4 Sep 1848, Log 541, KWM).

[49] Thomas Atkinson, "A Journal of a Sperm Whaling Voyage to the Pacific Ocean in the Ship Potomac," MSS HM 68457, HLCA.

[50] *New Bedford Weekly Mercury*, 21 Mar 1851; ship *Euphrates*, Crosby, New Bedford, Mass.,1849-51, Whalemen's Shipping Paper, mic 180, NBL.

[51] Brown, *In the Wake of Whales*, n. p. (10 Jul 1842).

[52] Ibid., 11 Jul 1842.

[53] *New Bedford Weekly Mercury*, 16 May 1845.

[54] Ship *Huntress*, Shearman, New Bedford, Mass., 17-18 Aug 1844, Log 164, ODHS.

[55] Ship *Huntress*, Log, 30 Aug 1844.

[56] Ship *Clematis*, Benjamin, New London, Conn., 12 Jan 1854, Log Wh C615.1919L, Nicholson Whaling Coll., PPL.

[57] E. E. D. Skinner, "Directions for the Use of the Medicine Contained in the Medicine Chest of the Ship Roanoke," FML.

[58] Ibid.

[59] W. F. Gloucestershire, Journal, ship *Montreal*, Fish, New Bedford, Mass., 3 Nov 1851, Log 506, ODHS.

[60] Bark *Josephine*, Smith, New Bedford, Mass., 30 Sep 1904, Log 633A, ODHS.

[61] Bradford, *Barnacles and Bilge Water*, 46-47.

[62] Bark *Pacific*, Allen, New Bedford, Mass., 19 Dec 1867, Log 677A, ODHS.

[63] Joseph commonly referred to himself as "Bento"; but officially he is listed as "Benton," as is Anthony, presumably his older brother, and a whaling master himself.

[64] Schooner *Clara L. Sparks*, Benton, New Bedford, Mass., 23 Nov 1892, Log 73, ODHS.

[65] Ship *Franklin*, Cooper, Sag Harbor, N.Y., 12 Aug 1847, Log 1074, ODHS.

[66] Ibid., 14 Aug 1847.

[67] Ibid., 21 Aug 1847.

[68] Whitecar, *Four Years aboard the Whaler*, 40-41.

[69] Elias Trotter, Journal, bark *Illinois*, Jagger, Sag Harbor, N.Y., 3 Dec 1846. Log 1005 A and B, ODHS.

[70] Ferguson, *Harpooner*, 52.

[71] Hohman, *American Whaleman*, 54.

[72] Ferguson, *Harpooner*, 52.

[73] Ibid., 71.

[74] Ferguson, *Harpooner*, 109.

[75] Ship *Hannibal*, Green, Sag Harbor, N.Y., 9 Aug 1827, Log 204, MSM.

[76] Ship *Delta*, Sayre, Greenport, N.Y., 1832-35, Log, SHM.

[77] Dexter, *Narrative*, 12.

[78] Ibid., 13.

[79] Ibid., 6.

[80] John Swift Jr., Journal, ship *Janus I*, Taber, New Bedford, Mass., 13 Jul 1835, Log 225, MSM.

[81] Dexter, *Narrative*, 9.

[82] One advantage of sailing on an American whaling vessel was that the time aboard counted as residency in the United States, so that a couple of extended cruises could put the immigrant who so desired it well on the path to citizenship. (Cf. *New Bedford Weekly Mercury*, 10 Jan 1840.)

[83] Silveira, "Retrospectiva," *Corografias*, 57. Translated by the author.

[84] Silveira, "Memória," (dedicated "To the memory of José Laureano da Silveira [1828-1901] and António José de Mendonça [1836-1911], pioneers of the Gold Rush, my grandfathers"), *A Ilha e o Mundo*, 41-42. Translated by the author.

[85] Rincão = Ponta Rincão on the island of Santiago, Cape Verde. Teobaldo Virgínio, "Terra Longe," in *Cimboa* (Spring 1999), 28. Translated by the author.

[86] Lopes, *Chiquinho*, 67-68. Translated by the author.

[87] Ibid., 19. Cf. Carreira, *People of the Cape Verde Islands*, 50.

[88] Bark *Pacific*, Log, 25 and 27 Mar 1868.

[89] Jeremiah S. Wells to his friends at home, ship *Henry Kneeland*, Whalon, New Bedford, Mass., 5 Sep 1854, Log 725, KWM.

[90] Cf. For example in the period 1899-1910 the illiteracy rate among Portuguese immigrants to the U.S. (mostly from the Azores) was 68%, the highest among European nations. (Department of Commerce, Bureau of the Census, *Fourteenth Census of the Population, 1920*, 2, Population, 805.)

[91] Nordhoff, *Whaling and Fishing*, 60-61.

[92] The misconception of Azoreans as people of color, common in the whaling era, has not totally disappeared, viz. "Portuguese from the Azores, Kanakas from the Sandwich Islands, and blacks from the mainland increased the community's population of 'free people of color' from 274 in 1820 to 571 in 1840." Byers, *Nation of Nantucket*, 298; also, "Azoreans spoke Portuguese and were light enough in skin color to mix in white society." Busch, *Whaling Will Never Do for Me*, 42.

[93] Robert Smith Owen, Journal, ship *Warrens*, Grey, London, 5 Mar 1838, Log 98, KWM.

[94] Bark *Zone*, Fish, Fairhaven, Mass., 17 Oct 1855, Log 278, KWM.

[95] Bark *Zone*, Log, 18 Oct 1855. While the word "dago" tends today to be a disparaging term for Italians, it historically referred to Portuguese and Spaniards as well (Dago < Diego).

[96] *Whalemen's Shipping List and Merchant's Transcript (New Bedford)*, 8 May 1855; referred to hereafter as *WSL*.

[97] Thompson, *Life of John Thompson*, 124.

[98] Huntington, *Songs the Whalemen Sang*, 34-35.

[99] Melville, "'Gees,'" 507-09.

CHAPTER 2

Beginnings of the American Whaling Industry

> The Cape Codder was the first to attempt this fishery sporadically. The Long-Islander deserves the full credit for first making it an organized business. But to the Nantucketer will always be the glory of having founded a great national industry.
> (Clifford Ashley, *The Yankee Whaler*)[1]

Whaling by the colonists of the Northeast began not long after their settlement in America. It is mentioned in the Charter for the Massachusetts Bay Company in 1628.[2] The trying out of cast-up whales was noted by Governor John Winthrop of the Massachusetts Bay Colony in 1635, with the suggestion that the practice had been carried on already for several years.[3] But for the earliest record of organized whaling by colonial settlers we must look to Long Island. Extending eastward some 120 miles from Manhattan, it is the largest island in the continental United States. While there was some early settlement adjacent to Manhattan in the early seventeenth century, it fell to the Puritans of New England to begin the development of the then remote eastern tip. And so it was that a group of colonists from Lynn, in the Massachusetts Colony, founded the town of Southampton in 1640, soon to be followed by the creation of nearby Easthampton.

To what extent Native Americans of the Northeast might have sought whales on their own before the arrival of the Europeans is unclear. But very early they appear to have been engaged in this pursuit at the instigation of the Basques. In 1602, when English explorer Bartholomew Gosnold and his crew touched at the coast of southern Maine in his small bark *Concord*, after a run across the Atlantic, including calling at the island of Santa Maria in the Azores, they encountered Indians, one of whom was dressed in European attire, manning a Basque shallop, with an "iron grapple, and a kettle of Copper."[4]

Three years later, when another Englishman, George Waymouth, visited the same region he commented in some detail on a native whale chase:

> One especial thing is their maner of killing the Whale, which they call Powdawe ["he blows"]; and will describe his forme; how he bloweth vp the water ; and that he is 12 fathoms long ; and that they go in company of their King with a multitude of their boats, and strike him with a bone made in fashion of a harping iron fastened to a rope, which they make great and strong of the barke of trees, which they veare out after him: then all their boats come about him, and as he riseth aboue water, with their arrowes they shoot him to death; when they haue killed him & dragged him to shore, they call all their chiefe lords together, & sing a song of joy: and those chiefe lords, whom they call Sagamos, divide the spoile, and giue to euery man a share, which pieces so distributed they hang vp about their houses for prouision....[5]

The existence of native whaling on Long Island, at least in the early years of the colony, is suggested by an order of the General Court of Southampton in 1643 forbidding a certain Robert Bond from making "for any Indian or Indians any harping Irons [harpoons] or fishing Irons which are knowne to be dangerous weapons to offende the English."[6] Whatever might have been their original experience, it soon fell to local Indians to form the bulk of the crews of the English shore whalers.

Contrary to the preferences of the colonists, which were for the oil, the Indians of Long Island considered the tail and flukes a great culinary delicacy. Similarly, at first the colonists only occupied themselves with those cetaceans that had been washed up onshore, as they had in Cape Cod. In 1644 the town of Southampton was divided into four wards, with eleven persons responsible in each for the cutting up of whales cast-off on the local beaches.[7] The succeeding decades show a continuing development of the industry, initially with shore whaling. In January of 1650, John Ogden Sr. was given the sole right "to kill whales upon the South sea [of Long Island] at or within any part of the bounds of the saiede towne for the space of seaven yeares...."[8] Presumably, Ogden hired local Indians for much of his crews.

The type of vessel in which these Native Americans were to be occupied is not mentioned, nor is the length of their obligation. But already at least some planning had been taking place for whaling at a distance from Long Island. In 1666, a license was granted by Peter Carteret, the assistant governor of the North Carolina colony, to John Cooper and Humphrey Hues of Southampton and Nicolas Steevens of Boston, to whale along the North Carolina coast. The following year an agreement was signed between these three, and various minor partners, to outfit the sloop *Speedwell* of Southampton for a voyage of six to eight months, with a crew of "thirteen men and a boy" to travel "to Roanoak or those parts upon the designe of killing or getting whales or great fish for ye procurement of oyle."[9] Unfortunately there is no confirmation that this voyage actually took place. But it is known that shore whaling was also being practiced intermittently along the North Carolina coast about this time. And in 1688 there occurs the first record of the intention to obtain sperm oil. In that year the captain of the brig *Happy Return* of New York requested permission from the governor to hunt sperm whales "about the Bohames Islands, And Cap florida."[10]

Nantucket and Martha's Vineyard

> With the landless gull, that at sunset folds her wings and is rocked to sleep between billows; so at nightfall, the Nantucketer, out of sight of land, furls his sails, and lays him to his rest, while under his very pillow rush herds of walruses and whales.
> (Herman Melville, *Moby Dick*, Chap. 14)

Whaling would eventually become the major industry on the island of Nantucket, but its inhabitants were slower to realize the importance of whales than those of the neighboring island of Martha's Vineyard. Settlement by the English colonists commenced on the latter by 1641. At the time both islands had considerable populations of native Wampanoags. On Martha's Vineyard the harvest of beached whales is first mentioned in 1652, and shore whaling developed not long after. Nantucket lies some twenty-five miles south of Cape Cod, Massachusetts, significantly more distant from the mainland than Martha's Vineyard. It is also smaller and much less fertile than Martha's Vineyar, and so it was with great enthusiasm that Nantucketers eventually adopted whaling as their primary industry.

Around 1671 a whale strayed into the harbor at Nantucket and, with difficulty, was killed. The residents, realizing their need for orientation to this new industry, made an agreement with the previously mentioned James Loper to spend two years on the island teaching them the art of whaling.[11] Most historians doubt that he made the trip. Instead, credit is given to a whaleman from Cape Cod by the name of Ichabod Paddock, who arrived in Nantucket in 1690 to train the residents in the art of whale catching.

At the end of the seventeenth century the Long Island whale fishery was still the most prominent, but it was soon to lose its hegemony to Nantucket. The eighteenth century saw a great expansion in the industry, as well as the ports that sustained it. By mid-century pelagic whaling was being practiced in Massachusetts from both Nantucket and Martha's Vineyard. Additionally, Newport, Rhode Island, as well as Sag Harbor on Long Island, and New London, Connecticut, were all active whaling ports.

During the 1730s most colonial vessels headed north and entered Davis Strait, while the presence of French and Spanish privateers limited their ability to move farther east into the Atlantic. Toward the end of this period, around 1760, the British themselves began to restrict New England's access to the coasts of Labrador and Newfoundland. Nantucket historian Obed Macy gives the following dates for the extension of the Nantucket whale fishery: Davis Strait, 1746; Gulf of St. Lawrence, 1761; coast of Guinea, 1763; Azores, 1765; and the "Coast of Brazils," 1774.[12]

Already in the 1750s Nantucket vessels had begun whaling around the West Indies and the Gulf of Mexico. Following the equatorial counter-current from these grounds, by 1763 they were cruising off the coast of Guinea and by the end of the decade had discovered whaling grounds off the coast of Angola and present-day Namibia.[13] It was not until 1765, however, that American whaling vessels began to cruise the so-called Western Grounds, located southwest of the Azores.

A notice from Providence in the fall of 1766 reported that one whaling vessel from the port of Warren, Rhode Island, had traveled as far as the Azores and returned with 310 barrels of oil.[14] And in August of 1767 two sloops arrived at Cape Cod after having whaled around the islands, each with 400 barrels of oil. They reported some twenty or thirty New England whalers having been active there that season.[15]

By this time whale products—oil, baleen and sperm-oil candles—formed a major component of colonial exports, both to Great Britain and the West Indies. In fact, during the years 1768 to 1772 oil produced from whales "accounted for 53 percent of all sterling earned by direct exports to Great Britain from the northern colonies."[16]

In 1768, the governor of the Azores, Dom Antão de Almeida, informed the minister of the Kingdom of Portugal that the English (the New England colonies were still English) had been cruising "all summer in sight of these same coasts with two hundred small vessels of one mast," obtaining not only whale oil but "especially spermaceti and some ambergris."[17] (He is most likely referring not to the whaling vessels themselves, at this time mainly sloops, but rather to their boats.)[18] Furthermore, "based on notices and news that he said were gathered on the island of Terceira, from some captains from New England… the year before, the English ships, numbering seventy, had obtained in the waters of the [Azorean] archipelago about 10,000 barrels of oil, subsequently sold in America."[19]

Early on in their visits to the Azores, American whaling vessels were using the island ports to obtain supplies. The ship *Dighton* sailed from Nantucket on April 4, 1769, bound for the Western Grounds. One of the crew, Benjamin Hodges, kept a journal of the voyage. On July 3 he noted, "We made Carvo & Flores at 4 after noon." And again on July 21, they "went a Shore upon Flores & got sume wine & unyarns." On the 28th they espied the island of Pico, and on the following day "made fiall & St. George & Graciosha." On August 1, they "went a Shore apon St. George but we cold git nothing but a fue grapes & Inyarns." On the third they "went a shore a pon fial & got Sume Wine Inyarns and Cabbige." The *Dighton* returned to its home port that same year, on October 16, with 203 barrels of sperm oil.[20]

An incident with the brig *Polley* at the island of Terceira in 1774 was a portent of more serious troubles to come involving whaling vessels trading in the Azores. Evidently the captain of the vessel offered to trade some cheese (possibly obtained on another of the islands) for local wine but ran into trouble with the authorities:

> [After the vessel] went into the harbor and went a Shore With the Starbord boat and the Governor and two more com off With the Captain

> and mate and the Capt and mate went a Shore again and was treted well and in the After noon Eight Portege Gentlemen Came of with the Capt. and Bought a Cheses and ware to Pay for it in wine and the Capt and mate and Willim Green with a boats Crue went a Shore for to Git the pay and was treted Very Well with wine but when they would Com on board the Portegees took Hold of the Capt and Said he must Go to Gale [jail] but Green Run Down to the boat that Lay off Clear of the Land and Said hand a Lans [lance], which we did and Run up for to Clear the Capt and mate but they fought their way through and Com to the Boat but they took Green and Put him in Gale and they followed the Capt & mate to the bote but Could not kitch them and hove Stons at the bote the Capt had his Coat tore of his back. So we Com on board very mad and left Green a Shore in Gale[21]

The following day an officer came on board the vessel and said if the captain payed two and a half dollars they would release the prisoner and the seventeen gallons of wine that had been bargained for. After a delay of three more days, Green was finally released, and the brig was permitted to sail.[22]

The First Portuguese Whalemen

> No small number of these whaling seamen belong to the Azores, where the outward bound Nantucket whalers frequently touch to augment their crews from the hardy peasants of those rocky shores.... How it is, there is no telling, but Islanders seem to make the best whalemen.
> (Herman Melville, *Moby Dick*, Chap. 27)

Not only were Portuguese sailing out of Nantucket on the whaling vessels of William Rotch in the early 1770s, but also they were participating even earlier on the vessels of Aaron Lopez out of Newport, Rhode Island. Lopez was a Portuguese Jew who had fled the Portuguese mainland in 1752 amid the last remnants of the Inquisition, eventually to become the wealthiest and, in the words of his biographer, "the most distinguished merchant-shipper" of Newport, Rhode Island.[23]

Vessels from Newport and Nantucket frequently visited neighboring Martha's Vineyard to complete their outfitting and to ship the remainder of their crew. Thus it is that we find such an early settlement of Portuguese at the latter island. Of these, the Dias and Swazey families are the best known. The three Swazey brothers, Joseph (1743-1807), Anthony (1752-1835), and Manuel, apparently all arrived on the Vineyard in the 1760s. Joseph Swazey is the earliest documented Portuguese to participate in the American whaling industry. He is listed as receiving "¾ of his Share" of the profits of an unnamed whaling vessel out of Newport, Rhode Island, under the command of William Roberts and belonging to Aaron Lopez. The vessel had returned from a successful voyage in the fall of 1765.[24] The following year Joseph, who later served as a whaling agent, married Susannah Pease. Over the course of the next twenty-five years the couple had thirteen children.[25] Anthony married Jerusha Dunham in 1774 and was living in Edgartown in 1805. The couple appear to have had seventeen children.[26] All three brothers apparently served on the American side in the Revolutionary War. Manuel Swazey was less fortunate than his two brothers, being captured on the privateer sloop *Charming Sally* in January 1777, and carried, along with his crewmates, to Mill Prison in Plymouth, England. Fortunately he was able to return, and in 1795 married Elizabeth Basset in Nantucket. According to the family the original name was de Sousa, and the men were born in Lisbon.[27]

Joseph Swazey became a master mariner and owned his own vessel. Family history relates that Joseph's father was a Lisbon wine merchant, and there is mention of the father's vineyard. Joseph is supposed to have arrived in America on a vessel carrying his father's wine.

The early presence of Portuguese on Martha's Vineyard is signaled as well in the account book of Aaron Lopez's ship *Jacob*, bound on a whaling voyage. On 4 May 1775, Antonio and Francisco were paid for two days of loading the vessel; and a note on June 4 of that year requests, "Mr. Lopaz Sir please to Settle with Frank pourtegas," for nine days work aboard the *Jacob* (presumably on shore).[28]

Other early Portuguese on the Vineyard include Joseph Frederick, who was born in Portugal in 1748. He is supposed to have left a privateer commanded by John Paul Jones when it docked in Holmes Hole (now Vineyard Haven) in 1763. The following year he was married to Jerusha Pease. Frederick later served as boatswain's mate aboard the frigate *Alliance* from April 1779 to August 1780, first under

Captain Peter Landais and then Captain John Paul Jones.[29] Emanuel Silvara Noans (Nunes, known commonly as Emmanuel Silvara) married Sarah Fish in 1776 and died six years later. In a later document Silvara was identified as "a man of collour"; perhaps thus he was an early Cape Verdean.[30]

In February, 1775, the British Parliament passed the New England Trade and Fisheries Act, popularly known as the Restraining Act, since it restricted the New England colonies to trade with Great Britain, Ireland, and the West Indies. Furthermore, it prohibited fishing (and whaling) off Newfoundland or any other place in North America. By the next year the Revolutionary War would limit American whaling even further, as the vessels now became prey to British privateers. However, Portuguese participation in American maritime activities seems to have increased, for it is about this time that they begin to appear on American privateers and men-of-war. The most notable example of the latter is the *Bonhomme Richard*, under the command of John Paul Jones. Some twenty-six Portuguese sailors served on this celebrated vessel in 1779. They included all sorts of ranks, from apprentice to gunner's mate, paymaster, and quartermaster. With the exception of one individual, Jean Silveyra [João Silveira], a "boy" from the island of Faial, their provenance is not indicated.[31]

Joseph Dias, an Azorean, was one of the first Portuguese to settle on the island of Martha's Vineyard. He, too, melded with the island establishment, marrying Sarah Manter in January of 1780. Dias was subsequently involved in hostilities with the British. While it has been suggested that he served as a Revolutionary soldier, it seems more likely that his action was on shipboard, quite possibly on a privateer. Sometime that same year he was captured but released in time to return home in December 1780. At that time it is recorded that he converted to the Baptist faith. Soon he was off again, this time to be captured and confined on the infamous prison ship *Jersey*, in New York harbor. It was customary for the British to offer prisoners taken at sea the option of serving with the British Navy; if this option had been offered to Dias it had been refused, and he died that same year still on the *Jersey*.[32] In May of the following year a son was born to Sarah Manter Dias; he was given the name of his father, without "junior" appended, since he was now "senior."

Following in his father's footsteps, Joseph joined the crew of the privateer brigantine *Yankee* of Bristol, Rhode Island, during the War of 1812. The vessel carried a crew of one hundred or so men, with seven prize masters, one of whom was Joseph Dias. During the year of 1813, Dias earned a total of $103.74 as prize master of the captured vessels *Katie* and *Marquis*. However, for the rest of the war years his prize money increased dramatically, amounting to $4249. The *Yankee* was, in fact, the most successful privateer in American history. During the war, it captured forty vessels from the British, worth five million dollars.[33]

It is interesting that Joseph Dias shared prize money with several others aboard the *Yankee* who possibly had Portuguese origins, including Francisco Furgonda, Joseph Anthony, Lewis Joseph, Joseph Gonsalve, Samuel Swasey and Perry Rose.[34] Although apparently he never whaled, Joseph Dias continued in the maritime service as a harbor pilot in his hometown of Tisbury, Martha's Vineyard.[35]

Aaron Lopez, the "Merchant Prince of Newport"[36]

It was not long after his arrival that Lopez was dealing with sperm oil and spermaceti. By the beginning of the next decade it had become evident that manufacturing capacity had outstripped supply. Therefore, in November of 1761 Lopez joined several other firms in establishing the United Company of Spermaceti Candles, which was one of the first trusts in the Colonies. The declared purpose of the trust was to discourage competition from any outsiders. It also stipulated that if the price of head matter were too high the trust would outfit a number of whalers to add to the supply.[37]

It was about this time that Lopez entered the whaling business himself. By 1762 he had begun building ships. And, around that time, in company with Francis Rotch of Nantucket, he began sending out whaling vessels, evidently initially around the West Indies. By the late 1760s, Lopez is reported to have owned twenty-four vessels outright and several more in partnership with others.[38] At least by 1775 the Lopez-Rotch vessels were sailing near the Azores, but to what extent Lopez himself was instrumental in bringing Azoreans on board his whaleships is unknown.[39] It is more likely that this was a spontaneous result of the many American whaling vessels calling there beginning in the mid-1760s. Lopez's vessels were regularly touching as well at the Cape Verde Islands by the late 1760s, but this was in their role as slave vessels rather than whalers.[40]

Besides the *Leviathan* (see below), other whaling vessels sent out by Lopez, presumably from Newport, included the sloops *Endeavor* (various voyages between 1765 and 1770), *Acteon* (1766), *Mary* (1767, 1768, 1769), *Dolphin* (1769), and *Industry* (1769, 1771); the schooner *Eleanor* (1767); and brig *Royal Charlotte* (1772).[41]

In April 1773 Aaron Lopez's brig *Leviathan* set out from Newport on a whaling cruise. In his letter of instruction, Lopez told Captain Thomas Lothrop that, after completing his outfitting at Martha's Vineyard, he should proceed directly to the Canary Islands, where he was to begin whaling. In the event of a lack of prey in the area, Lothrop was instructed to proceed to "any Other Part of the Ocean."[42] Evidently, Lothrop did this, as later that year found him off the coast of Brazil.

Whales had been noted off the Brazilian coast by Portuguese navigators as early as 1532. By the beginning of the seventeenth century whaling was being conducted in Brazilian waters by Basque whalers encouraged by the Portuguese authorities. In 1612 whaling was declared a royal monopoly and tightly controlled. Shore whaling concessions were established in Bahia and Rio de Janeiro and later to the south in Santa Catarina and a few other locations. These shore whalers hunted the right or the humpback whale, both baleen whales. However, it was the oil that the Portuguese needed for their sugar mills—the whalebone was generally discarded. It is estimated that by the 1770s the whaling station at Santa Catarina was processing over 500 whales per year.[43]

But Portuguese authorities were evidently intrigued by reports of sperm whales beyond the reach of the whaleboats used in the industry, whose crews were hesitant to travel out of sight of land. And so the news that a whaleship from the British colonies of North America was being detained in the harbor of Rio de Janeiro must have excited those who desired expanding the Brazilian fishery.[44]

Finding himself with a leaky vessel somewhere off the coast of Brazil, Captain Lothrop had decided to make for Rio. Unfortunately he missed the entrance to Guanabara Bay and sent a boat ashore just to the north. The authorities were soon alerted, and the next day, September 22, 1773, the *Leviathan* was brought into Rio under guard, for at the time it was unlawful for unauthorized foreign vessels to enter into Brazilian waters.

The men were put into prison, whereas the captain and mate were confined on board. Evidently there was suspicion that they were engaged in smuggling, rather than whaling. And so the *Leviathan* was condemned.[45]

Intrigued by the stories of the sperm whales that the *Leviathan* had caught off the coast of southern Brazil, Portuguese officials arranged to have several vessels outfitted in the manner of the *Leviathan*, and its former crew members sent along to instruct the locals in the art of sperm whaling. By September of the following year twenty-two barrels of spermaceti had been sent to Lisbon where it would undergo analysis for its quality.[46]

Captain Lothrop submitted a report of the incident in December of 1773, but it is not known when it may have been available to Lopez. However, he received a letter in late spring of 1774 from Samuel Hart Jr. of Barbados, informing him that Captain William Mooers of Francis Rotch's brig *Bedford* had chanced upon a Portuguese snow off the coast of southern Brazil that was carrying Thomas Lothrop on a whaling voyage. Evidently there were several vessels operating, each with whaleboats in tow, and presumably with the American crew distributed among them. Lothrop and the mate subsequently succumbed to smallpox, and the remainder of the crew was still apparently under confinement as late as 1776. As a consequence, when Lopez and Francis Rotch sent a whaling fleet into the South Atlantic in the fall of 1775, they explicitly warned the captains not to venture within four leagues of the Brazilian coast.[47]

Difficulties with England for the American whaling fleet had begun gradually. In 1766 a duty was placed on colonial whalebone and oil imported into Britain. Additionally, the British had established a bounty on whale oil products, and now the colonial trade was excepted. Yet another serious impediment occurred when, in 1775, the Continental Congress passed acts banning whaling. In order to work around the ban, Francis Rotch and Aaron Lopez planned to establish a port of rendezvous in the English-controlled Falklands, in order to maintain their oil trade with London. First they obtained an exception to sail from the General Court of Massachusetts Bay; and, in late summer of 1775 they sent off a whaling fleet of some sixteen vessels, gambling that their connections with England would shield their enterprise. Unfortunately five of the vessels were seized by the British near the Azores and taken to England. By this time Francis Rotch was in London, and pleaded for the return of the seized whalers and protection for the others, some of which evidently had already arrived in the Falklands. The five captured whaleships eventually were released; and at least two made sail for the Falklands under the English flag, although with American crews. There they whaled successfully for some time, anticipating by a few years the establishment of the British Southern Whale Fishery. On the other hand, there is little extant documentation detailing the wartime activities of Rotch and Lopez's Falkland fleet, but apparently it had little success.[48]

The War severely limited Lopez's business activities, as in 1777 he was forced to move his family inland to Leicester, Massachusetts. Five years later, in 1782, he met a tragic death at the age of fifty-seven when the horse pulling the sulky he was driving went into the water at Scott's Pond, near Providence. Lopez, ironically, could not swim and soon drowned.[49]

The Birth of New Bedford

In 1652 a group of settlers purchased a site in southeastern Massachusetts from the Wampanoag Indians. They called it Dartmouth, and its area encompassed modern-day Acushnet, Dartmouth, Fairhaven, and New Bedford. Dartmouth was incorporated as a town in 1664. After having been destroyed by the Native Americans during King Philip's War in 1675, the town once again began to develop. By the end of the century the majority of the population consisted of Quakers. And by the 1750s, Joseph Russell, the founder of New Bedford, was sending whaling vessels out into the Atlantic on voyages of six to eight weeks.[50]

In 1765, Quaker whaling merchant Joseph Rotch of Nantucket purchased ten acres of land that would eventually become part of the city of New Bedford. Rotch envisioned a grand future for whaling along the Acushnet River leading into Buzzards Bay. Soon the Dartmouth area—and specifically New Bedford and neighboring Fairhaven, across the river—would be rivaling Nantucket. In fact, between 1771 and 1775, before the outbreak of the war that would devastate the industry, Nantucket averaged sixty-five vessels outfitted per year, while Dartmouth was sending out sixty.[51] By now New Bedford had become the second largest town in population in the colony.

This virtually all came to an end with the advent of the Revolutionary War in 1775. Three years later a British raiding party torched the city. From then until the end of the war in 1783, only Nantucket braved the seas and the exposure to the English, who ultimately seized 134 Nantucket vessels; so that there was little success for the industry during those years.[52] In 1774 the American whaling fleet had consisted of some 360 vessels. By the end of the war, the industry was almost non-existent. And it was not until the end of the decade that whaling began to attain its pre-war momentum—the beginning of a great expansion for the once-dominant industry.

But now the traditional market for sperm oil was closed to Americans as the British imposed an alien duty of eighteen pounds per ton, exceeding the cost of production.[53] In order to alleviate the situation, Massachusetts instituted a bounty on whale oil, to take effect in January of 1786. The result, unfortunately, was an overstimulation of production, given the fact that demand remained low—for the public, after years of scarcity, had generally adopted tallow candles.[54] Given this state of affairs, Joseph Rotch moved his business to Dunkirk, France, where the French established a duty-free market for American whale products. In the next few years others from Nantucket established whale fisheries in Nova Scotia, in Wales, and in the town of Hudson, well up the river from New York City.

An Invitation to the Portuguese Colonies

The Portuguese had expected that their Brazilian whale fishery would continue without interruption, but by the 1790s the increased international activity nearby in the South Atlantic began to reduce sharply the catch of the Brazilian shore stations. It was in this context that Portuguese authorities in Lisbon received a report from a young visitor to the United States. In 1798 young Hipólito José da Costa, a native of Brazil who had recently graduated from the University of Coimbra, was sent by the Portuguese government on a mission to North America to "seek products and methods of agricultural production in the United States and Mexico."[55] During the time he spent in the young American republic, Costa visited twelve of its fourteen states, studying and later reporting on animal husbandry and fishing in New England; the growing of wheat, corn, and other cereals along the central seaboard; tobacco in Maryland and Virginia; rice, cotton, etc. in Georgia and the Carolinas;

and hemp and minerals to the west.[56] But it was with the whale fishery that Costa saw perhaps the greatest promise, and his report ends with a bold suggestion, conditioned with youthful optimism:

> Finally, it remains for me to discuss the fisheries. Ever since the Dutch ceased their deep-sea fishing because of the risk their ships ran of being seized by the English, the Americans have been supplying Europe with fish oil, spermaceti, and whalebone, in addition to the great quantity of salted fish that they export to Portugal, Spain, and the ports of the Mediterranean.
>
> With regard to whaling, if we decide to establish the same regulations and laws that the Americans have, there is no doubt that this extensive branch of commerce will fall exclusively into our hands, for we have these advantages over them:
>
> 1. Our sailors receive lower wages. Whereas among our people a sufficient number of sailors can be found (willing to sail) at 8 or 10 *mil-reis* per month, it is necessary in the ports of the United States to pay sailors 16, and as much as 24, *mil-reis*, and even then it is difficult to obtain them.
>
> 2. The Americans have to sail from America to the coast of Brazil, which is where they do most of their fishing, and then have to sail back. Since we fish along our own coasts, we save on the delay, the expense, the risk, and the tying up of money.
>
> 3. Our ships that fish along the coasts of Brazil have our own ports there, where they can easily take shelter for repairs or for necessary provisions, a convenience that the Americans lack, for they must remain under sail from the time they put out until the time they finish their fishing.
>
> The Americans, however, are experienced to such an extent in this trade that my plan would be to invite a number of families of fishermen in America, especially from among those who live in Nantucket, have them settle in two different places in Brazil, add Portuguese [i.e. Brazilian] sailors to them, and have merchants of the country enter into partnership with them. If after this is done we provide fishermen with the same liberalities and exemptions that the Americans do, there can be not the slightest doubt that in two years, and no more, the whaling trade will be entirely in the hands of Portugal.[57]

Unfortunately for the Portuguese this suggestion did not bear fruit, for although several Nantucketers were invited to Brazil to assess the possibilities, they did not stay long. The fishery was now moving well beyond the Brazil Banks into the Pacific, and thus the idea had lost much of its value. But as has been observed, quite possibly, "had the whalefishery continued to center on the Brazil Banks the experiment would have developed as predicted by Hipólito."[58]

A more concrete step toward establishing an American (viz. Nantucket) presence in Portuguese whaling was taken, however, when in 1805 four Nantucketers, Timothy Folger, Francis Paddack, William Hull, and John Hillman, traveled to Portuguese Mozambique on the southeast coast of Africa with the intention of establishing a whale fishery there. However, the enterprise was short-lived, for in 1810 news was received that the four had become ill there and died.[59]

Into the Pacific

As the industry revived and competition from the English increased, it became apparent that new whaling grounds must be discovered. In 1787 the British whaleship *Emilia* rounded Cape Horn into the Pacific Ocean. And four years later the ship *Beaver,* several other Nantucket vessels and the *Rebecca* of New Bedford ventured into the Pacific.

Throughout the new century whaling was not a monolithic industry. Trade was to become an increasingly important element. When whaling vessels commenced operating in the South Atlantic and soon after in the eastern Pacific, they began to exploit the commercial value of other marine mammals, notably the fur and elephant seal, the sea otter, and the walrus. Seals were found to produce a good quality oil and the fine sea otter pelts were especially valuable in the China trade.[60]

The ship *Barclay* sailed out of New Bedford in October 1797. In March of the following year the vessel arrived at Más Afuera Island (now known as Alejandro Selkirk), one of the Juan Fernández Islands, and remained there for the southern winter. On the fourteenth of April it was noted that the catch for the week was 1800 sealskins. When the *Barclay* returned to New Bedford in June of 1799 it was carrying 21,000 sealskins. It has been estimated that there were at one time some 500,000-700,000 seals on Más Afuera before their rapid decimation.[61]

While the American whaling fleet continued to grow, there were other difficulties. The Embargo Act of 1807, prohibiting all maritime exports, had an especially negative effect on the industry, with its dependence on the oil trade with England—and vessels sat idle in their ports. With the outbreak of war with England in 1812 came a virtual cessation of American whaling activities, with the exception of a few vessels that ventured out in desperation from Nantucket. Of the thirty-one whaling vessels that sailed from there in 1811, eleven were eventually captured.[62] A general resumption of whaling activities, and further expansion of the industry, would await the end of hostilities three years later.

Crew lists for the final decades of the eighteenth century and beginning of the nineteenth are quite incomplete; and so Portuguese participation during this period is difficult to establish. A John Swazey, "Portuguese," is listed as having died aboard the brig *Polly* in October of 1796. Serving on the ship *Maria* out of New Bedford, during its 1808-10 voyage into the Pacific, was Emanuel Joseph, a seventeen-year-old from the island of Pico in the Azores. Sailing out of the same port that year was the ship *Sally*, with Joseph Rose, of the same age, a native of Flores, who measured a scant three feet eleven inches. Rose was already a resident of Nantucket and married there in 1813.[63]

From Nantucket we can only conjecture on the nationality of Francis Joseph (ship *Manilla*, 1793-94). Later, Antonio Joze sailed on the ship *Thomas* (1815-17) and Joseph Rose on the *John Adams* in the same year (1815-17). Two years later Jonathan Swazey, son of Anthony, sailed on the ship *Independence* (1817-19) along with Manuel Jose. As has already been noted, in 1821 Agostinho Luís de Sousa spent the year on the Brazil Banks with the schooner *Industry*.[64] As for the provenance of these men—when such details are given—the western islands of the Azores (namely Faial, Flores, Pico, and São Jorge) predominate at this early date.

While Cape Verdeans cannot be numbered among the very first arrivals, some can be seen to have reached Nantucket by the beginning of the nineteenth century. One of the earliest was Michael Douglass, "a Cape di Verde Portuguese Negro" born in the 1760s who settled on Nantucket somewhat after 1800 and in 1811 married Mary Boston.[65] When the ship *John and James* sailed from Nantucket in 1811 it carried a seaman from Cape Verde named Simon Duggins.[66] José da Silva was born on the island of Brava in 1794, and three decades later, in 1824,

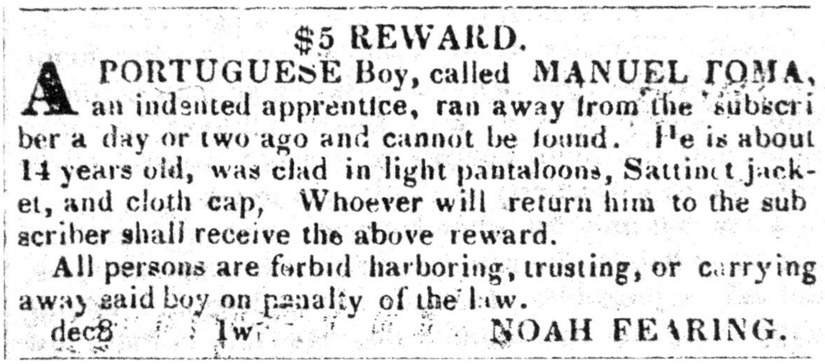

$5 REWARD.

A PORTUGUESE Boy, called MANUEL TOMA, an indented apprentice, ran away from the subscriber a day or two ago and cannot be found. He is about 14 years old, was clad in light pantaloons, Sattinet jacket, and cloth cap, Whoever will return him to the subscriber shall receive the above reward.

All persons are forbid harboring, trusting, or carrying away said boy on penalty of the law.

dec8 1w NOAH FEARING.

New Bedford Morning Mercury, 1836.

became the first Cape Verdean to be naturalized on Nantucket. Eventually, as in the case of the majority of the da Silvas of New England, his name was changed to Sylvia.[67]

The whale fishery, having sent virtually no vessels out in 1813 and 1814 due to the British blockade, rebounded quickly at the end of the war. The following year Nantucket sent out fifty vessels; New Bedford and environs, thirteen; Hudson, two; and Sag Harbor, three.[68] In 1816 vessels were sailing from the Massachusetts ports of Nantucket, New Bedford, Rochester, Holmes Hole (Vineyard Haven), Wareham, Fairhaven, Edgartown, and Boston. And by 1820 the port of New Bedford was growing rapidly in importance, sending out thirty-six vessels to forty-four for Nantucket.

One must note, parenthetically, the existence of a hierarchy among whaling vessels and ports. First, especially among the Nantucketers, the sperm whale held supreme over the right whale. As Ishmael points out in *Moby Dick*, regarding the latter, "all hands commonly disdained the capture of those inferior creatures...."[69] These same whalemen, during the heyday of the nineteenth century, looked down as well on those competitors from New London, where sealing voyages were common, and Sag Harbor, with its generally less adventuresome voyages and preference for right whales and seals.

Indentured Servitude: The Nature of a Whaleman's Contract

Indentured servitude was still actively practiced in the early nineteenth century. Among the indentured servants in New England at the time was Joseph Silver, a native of Portugal residing in Edgartown, Martha's Vineyard, who, "having no Parent within the United States nor legal Guardian," was indentured there in 1802 at the age of fourteen for the period of seven years. For a young man with no means, the indenture was not without its benefits, since the contract stipulated that the illiterate Silver was to be taught to read and write "& Cipher as far as the Rule of three" (i.e. to learn the rudimentaries of algebra).[70]

In New Bedford, a Cape Verdean by the name of Joseph Antone was indentured in 1813, at the age of fourteen, as a servant to Jeremiah Mayhew, "to do farming work, gardening & housework, and to go to sea," until he reached the age of twenty-one. Mayhew was most likely a Quaker, and the contract was far from one-sided. The master obligated himself to teach the young man to read and write and, at the end of the term, was to provide Antone with "two suits of wearing apparel, one suitable for the Lords days and the other for working days, and twenty dollars in money."[71] Whether Antone ever went to sea is not known. The education he received, however, he later put to good use, as a founder of the African Christian Church of New Bedford.[72]

The term "indentured servant" may sound harsh, but that was essentially what a crew member of a whaleship agreed to upon signing ship's articles, for he was then obligated to remain with the vessel until the end of the voyage, which represented at that time around a two-year commitment. Until this requirement was abolished by the Seamen's Act of 1915, desertion would subject the mariner to imprisonment or some other form of punishment. On the other hand, as the time drew close to return to the home port, it was not uncommon—and neither discouraged by the owners—for the captain to employ various tactics to encourage crewmen to abandon the vessel and thus leave more profits to the owners and officers. This abuse, relatively uncommon at the early stages of the industry, became quite prevalent as the nineteenth century progressed.

By the 1820s there is regular participation not only by Azoreans but also Cape Verdeans, plus an occasional Madeiran or individual from Lisbon. For instance, in 1825 there were fourteen Azoreans, six Cape Verdeans, and one individual each from Lisbon and Madeira aboard outbound New Bedford whalers. These Cape Verdeans were natives of several islands of the archipelago, namely Boa Vista (1), Brava (1), Maio (1), São Nicolau (2), and unstated (1). The Azoreans hailed from Flores (2), Pico

(1), São Jorge (3), São Miguel (1), Terceira (1), and unstated (6). One notes here already that the idea of the latter coming from only the western group of the Azores, or the former mainly from Brava, is an over-simplification. During the decade of the 1820s some 266 Portuguese islanders served aboard whalers sailing out of New Bedford. They hailed from all the islands of the Azores save Santa Maria. From Cape Verde sailed men from the islands just mentioned plus Santiago and São Vicente. By far the largest segment was from São Nicolau, followed by Brava.[73]

The number of Portuguese crewmen continued to grow in the succeeding years, along with the whaling fleet. In 1830 seventy-six Portuguese sailed on New Bedford whalers. More than half (forty-five) were Azoreans, followed by eleven Cape Verdeans, one Madeiran, and nineteen others listed simply as "Portuguese." In all, between 1822 and 1834 over 700 Portuguese sailed out of New Bedford area ports on whaleships; and that figure does not include the many who were recruited at the islands.[74] The percentage of Portuguese among the crews of American whaling vessels would continue to increase throughout the century.

New London, Connecticut, was founded in 1646. Blessed with an excellent natural harbor, it sent out its first whaling vessel, the sloop *Society*, in 1718. But a century would pass before the port became very active in whaling. Overall, New London was to become one of the major whaling ports, making a total of 1116 voyages from its beginning until 1909.[75] Generally less wide-ranging than vessels from the other major whaling ports, those from New London and vicinity nevertheless made it into the South Atlantic by 1829; the Pacific, the following year; and the Indian Ocean, by 1833—their preferred grounds being close to Antarctica, where sealing was also possible.

Portuguese were slower to sail out of the New London whaling ports. In 1818 there was just one; but by 1821 there were ten. The ship *Commodore Perry*, on its voyage of 1824-25 to the South Seas, counted three Portuguese among the crew. In July 1829, the ship *Wabash* sailed from New London with five men from the Azores (two from Faial and three from Flores), and shipped three more once it reached the islands. One of these men was the second mate, Nicholas C. Anthony, who would eventually become one of the earliest Portuguese whaling masters. That year nineteen Portuguese shipped out of the New London ports. By 1832 the number had grown to forty-one, and three years later to fifty-eight. It is not until 1833 that the first Cape Verdean is identified. Manuel Deararde [*sic*], from Brava, sailed that year on the ship *Electra* out of New London. After that Cape Verdeans appeared regularly along with Azoreans.[76]

On May 20, 1834, the ship *Aeronaut*, under the command of Alvin Mallory, sailed from the nearby port of Groton carrying seven Portuguese. Joseph Folger, of Pico, was mate and Manuel Bartholomew, from Faial, was third mate. Additionally there were three Azoreans, one man from Cape Verde, and two Portuguese whose origins were not unidentified. By mid-century Portuguese were a major element on whaling vessels out of the Connecticut ports, with the Cape Verdean segment increasing. In 1850, for example, the ship *George and Mary* of New London had in its crew seven Cape Verdeans and two Azoreans.[77]

The village of Sag Harbor was established early in the eighteenth century in order to export the agricultural products of the region, principally to neighboring New England. Its protected harbor between the north and south forks of Long Island made it ideal for the nascent local whaling industry. In 1760 three sloops left the port in search of whales; but development of the industry was slow at first, in part because of the system of ownership at the time, by which vessels were owned individually, in family partnerships or by firms. Some say that credit is due to James Fennimore Cooper, principal owner of the whaleship *Union*, who encouraged the owning of whaling vessels on shares, a system long in usage in Nantucket.[78]

It is unsure when Portuguese began to serve on Sag Harbor whaling vessels, but by 1827 the ship *Hannibal* counted among its crew three Azoreans, one of whom, Domingo Silvia, entered the vessel at Faial while it was provisioning for its voyage at the port of Horta.[79] By that time a few Portuguese had settled in the region, for in 1829, due to the lack of a Catholic church, mass was being held in a private home for fourteen Irish families and one Portuguese family.[80]

Adaptation to a New Culture

From the very first years one notes the trend toward the Anglicization of Portuguese surnames. Generally the adaptation was no more than a simplification of a foreign-sounding name or the use of the English equivalent, such as "Joseph" for "José." Soon these became formulaic, so that "Pereira" became "Perry" and "Silva" and "Silveira" were modified to

"Sylvia." Another phenomenon was the occasional translation from Portuguese to English, such as when "Neves" became "Snow," or "Madeira" was transformed into "Wood."

Since many Portuguese were picked up at a very early age, even before their teens, and started out as cabin boys with a close relationship to the vessel's captain, they occasionally adopted the master's surname. Early examples of this are Joseph Folger and Henry Starbuck, both natives of the Azores residing in Nantucket and serving on board the whaleship *Timoleon* of New Bedford on its voyage of 1824.[81] Folger, a native of the island of Pico, had signed on originally as third mate, but became the first Portuguese to take command of an American whaling vessel when Captain Charles Starbuck died the following year, whereupon Folger brought the vessel in from the South Atlantic.[82] Other surnames that resonate in New Bedford whaling are represented by Frank Howland, from the island of São Jorge, who sailed on the ship *Hydaspe* of New Bedford (1826-27) and Charles Coggeshall, from Corvo, on the bark *Lydia*, also of New Bedford (1880-84).[83]

There were, as well, the imaginative names given to Portuguese crewmen, some of their origins being more decipherable than others. Joseph Azores served on the *Ann Alexander* in 1824-25, while John Friday was a crew member on the ship *Pacific* in 1824-27.[84] Some of the names of crew members aboard the ship *George and Martha* of New Bedford are more fanciful. Perhaps the name of the vessel suggested a modicum of patriotism as one Azorean from the island of São Jorge became Thomas Jefferson in 1823. On the succeeding voyage the following year Jefferson had the honor of sailing with James Madison, another native of the same island. In 1825, Jefferson shipped again with a compatriot from his island named John Quincy Adams.[85]

There are still others with consecrated names: Andrew Jackson, a native of the island of Faial, and well-known in whaling circles, served aboard the ship *Maine* of Fairhaven in 1834-35.[86] It should not be assumed, moreover, that these surnames were necessarily a jest of the moment, and that the individual soon would revert to a more appropriate appellation. Thomas Jefferson, for instance, served on at least seven whaling voyages out of New Bedford/Fairhaven between 1823 and 1831. Henry Clay, an Azorean whose original name was Acquilla Rodrigues, began serving on American whalers during the 1850s. He

Columbian Courier (New Bedford), 1804.

later commanded several whaling vessels out of New Bedford, besides becoming an important New Bedford whaling agent and ship owner.[87]

Views of the Azores

As has been noted, American whaling vessels on their outbound voyage commonly put into the Azores and Cape Verde archipelagoes in order to complete their provisioning and fill out their crew. To give an idea of the volume of this activity, from June 1831 to June 1832, seventy-two American whalers anchored in the harbor of Horta, on the island of Faial; and another eighty-two, avoiding port fees, cruised in the vicinity, obtaining provisions and presumably crewmen. Ten years later one hundred sixty-five dropped anchor there.[88]

An early observer aboard the brig *Aurora* of New York City in 1820, off on a sealing voyage to the South Atlantic, conveys a rather common contemporary impression of the Azores and its inhabitants—the former looked upon as positive; the latter, in a negative light:

> About 17 o'clock came to anchor in Fyal road—I found the island (contrary to my expectation) one of the most beautiful places to appearance I ever saw— It rises to a considerable hight in acclivity and the land on the side of the hills is laid out in very small lots divided by hedges of small reeds, which at a distance give it the appearance [of] a garden—

Street scene, Flamengos, Faial Island, Azores, ca. 1896. Courtesy of the New Bedford Whaling Museum.

The road at this season is a good one, but it is exposed to a N.N.E. & southerly wind— There is a small town of about 2000 inhabitants—it is a place of considerable trade in wine, most of which are brought here in small boats from Pico—

It is an excellent place for shipping to stop for Vegitables—Potatoes & onions and some other culinary production are excellent in quality & reasonable, in price—The water is not of the first quality, but palatable and is procured with facility—

In clear weather there is a grand view from the roads of Fayal, St. George, and Pico— The celebrated peak on the latter lies about thirty miles distant & is elevated about 9000 feet above the surface of the sea— The whole scene combined would afford a fine subject for the pencil—

We were under the necessity of landing our dory in order to repair our sheathing, which we accomplished as well as circumstances would admit—Mr Dabany our consul shew'd us much civility but I had little time to avail myself of his profered hospitality—

> After all that can be said in praise of the place, we still keep in mind that it is in the possession of *Portugues* — which is sufficient to disgust most people who are acquainted with their character—against this or any of their dominions.[89]

One would assume that the vitriol here is due to an acute anti-Catholicism, very prevalent at the time in New England.

Over the course of the nineteenth and early twentieth centuries the Azores often represented deliverance for those naïve green hands who had signed on expecting adventure and some degree of comfortable living aboard. The ship *Swift*, under Captain Frederick Arthur, had sailed from Nantucket on the fourth of June, 1823, and arrived in just under three weeks at the island of Flores to obtain provisions. For two barrels of whale oil the captain received "hogs fouls potatoes eggs." But unfortunately, five men decided that they had had enough of the whaling business, at least in that ship, and disappeared.[90]

While one sees constant references to American whaling vessels touching at Flores and Faial, as well as the neighboring islands of Pico and São Jorge, it was not uncommon for them to make port at various of the other islands farther to the east. For example, little over a week later, on the same voyage, the *Swift* spent several days at São Miguel provisioning.[91] Curiously though, over the course of the century it does not appear that whaling tempted a great many men from Terceira, São Miguel or Santa Maria.

Ever since the days of the gold- and silver-laden Spanish galleons, ships traveling in the vicinity of the Azores had been subject to pirates who attempted from time to time to intercept vessels crossing the Atlantic. Whaleships were no exception, although the pickings might be somewhat slim. In July of 1828 the ship *Peru* of Nantucket was boarded by pirates off the island of Pico and was robbed of "a large quantity of provisions, blocks, spyglass, and other articles."[92]

As has been noted, the international aspect of whaling implied realities that did not touch domestic industries. Even normal maritime commerce, with its more regular routes and schedules, did not suffer to the same extent the vagaries of international strife. Once a whaling captain left port he was pretty much the arbiter of the destiny of the vessel. Elihu Russell, the novice master of the ship *Galatea*, seems to have had more than one lapse in judgment early in the voyage. The *Galatea* left New Bedford on June 1, 1828, destined for the Brazil Banks, with the intention of provisioning in the Azores. There were a total of twenty-two men on board, all Americans except for one African, one Scot, and one Azorean, twenty-two-year-old Manuel Antone.[93] Their early fortune in whaling was mixed. Two weeks after sailing they lowered three boats after a pod of sperm whales, subsequently striking and killing one. However, one of the boats was swamped; and in the process of saving the men the whale was lost. A week later they succeeded in taking nine blackfish and the following week made the island of Flores. On June 27, after being out for nearly a month, they passed by Faial, Pico, and São Jorge on their way to making port at Graciosa, where they traded "some oil for Onions and Pigs and finding the potatoes too small." Then, as the captain later explained, having been several times at Terceira, he decided to call there for supplies.[94]

When the Portuguese King Dom João I died in 1826 it set off a dramatic struggle between his two sons, Dom Miguel, representing the conservative faction and Dom Pedro, the liberal, that would lead to a civil war not resolved until 1834. The Azores early became a strategic locus of the conflict, with its islands dividing their allegiance between the two factions, particularly São Miguel for the *Miguelistas* and Terceira backing the forces of Dom Pedro. Thus it was that the *Galatea* found itself in the midst of this nascent conflict in mid-1828.

On the morning of July 4, 1828, the *Galatea* was cruising some ten miles off Terceira. The timing was unfortunate, for less than two weeks before, a liberal uprising had placed a new government on the island; and Dom Miguel's forces, operating out of São Miguel, were eager to defeat it. And so a brig from the conservative armada sent a shot across the whaler's bow, subsequently dispatching a boat with an officer to inquire of the ship's destination and business. Captain Russell was then informed that the island was under blockade by the forces of Dom Miguel and that he should turn about and bear away. By this time the ship was only about five miles from land. Soon a frigate appeared and took possession of the *Galatea*, directing it to São Miguel. Also brought in at the same time was the brig *Gleaner* of Portland, Maine, which had been accompanying the *Galatea* but was evidently not whaling.[95]

In his protest filed with Thomas Hickling, the U.S. Vice-Consul at São Miguel, Elihu Russell claimed to be ignorant of the blockade; but it seems hardly credible, having made port at Flores and Graciosa—the latter so close to Terceira—that someone would not have advised the *Galatea*'s master of the situation. On the twenty-sixth of July, with his vessel detained at anchor at São Miguel, Captain Russell, after having dined aboard the *Gleaner* with its captain and those of two impounded English vessels, headed for the *Galatea*, which he apparently had not been aboard since soon after its arrival. According to Russell he had his papers in order to sail, but the Portuguese evidently were not of the same mind. A number of the original crew members were aboard the *Galatea*. As the boat carrying the captain, several of the crew, and the master of one of the English vessels—who claimed he was along simply to wish goodbye to the American captain—approached the ship, the mate, who was on board, cried out for them not to approach. As they pulled away back toward the *Gleaner*, a Portuguese soldier on board the *Galatea* began to fire into the boat, seriously wounding the ship's second mate and the English captain.[96]

The next day twelve of the crew of the *Galatea* were put in prison under a charge of mutiny and sedition. According to the Portuguese, Captain Russell had been attempting to retake his ship and sail away without permission. The prisoners eventually were sent in irons to Lisbon and incarcerated there until their release on December 21 of that year. Elihu Russell was more fortunate, as he was able to sail from São Miguel to New York that fall. In spite of this ill-fated first voyage, he spent the next four decades as master of a series of whaling vessels.[97]

The *Galatea* arrived in Lisbon on the fifteenth of September flying American colors but without a single American aboard. Rechristened the *Felicidade*, she was taken at Lisbon by the French in 1831 and presumably burned. In March of 1837, almost a decade after the incident, the U.S. State Department received payment from the Portuguese government for all claims plus interest to settle demands of the owners and crew of the *Galatea*, and several other impounded vessels.[98]

However, not long after there was great movement of the whaling fleet in the area. On August 4, 1834, the logkeeper of the *Commodore Perry* of New London noted, "Their was 21 Ships in to fayal that Day and five brigs."[99]

Hawaii

While island chains farther west in the Pacific were discovered by Europeans as early as the sixteenth century, it was not until 1778 that English explorer James Cook touched upon the Hawaiian Islands, which he named after one of the sponsors of his voyage, the Earl of Sandwich.[100] The Portuguese presence in the islands, modest at first, began even before the arrival of the first missionaries or American whaling vessels in 1819. Ebenezer Townsend, Jr. kept a journal aboard the ship *Neptune* of New Haven. When it arrived in Hawaii in August 1798, he noted that two Portuguese were helping to build a vessel for the king of the island of Kauai.[101]

Antonio Ferre (Ferreira), known locally as Aikona, was a Portuguese who married the daughter of the Spaniard Francisco de Paula Marín. Under the direction of his father-in-law he built a stone house in Honolulu for King Kamehameha I in 1809-1810.[102] A most eccentric character was John Eliot de Castro, supposed to be Portuguese or half Portuguese, who arrived in the islands around 1812 aboard the *Hazard* of Boston.[103] There he served as physician to Kamehameha I, and later as supercargo of the Russian American Company ship *Ilmen*. Subsequently he was captured by the Spanish in California for illegal trading. Later released, de Castro again sailed to Hawaii where he served as physician and secretary to the King until the latter's death in 1819. Then he was off to Manila and a career as commander of a vessel in the Spanish colonial navy.[104]

In 1818, the so-called "Offshore Grounds" were discovered by Capt. George Washington Gardner in the ship *Globe* of Nantucket. Lying midway between Peru and the Marquesas, these grounds gave new impetus to Pacific whaling. The following year the ships *Equator* of Nantucket and *Balaena* of New Bedford sailed north past the Galapagos Islands and along the coast of California, then turned west to Oahu, where they became the first American whalers to call at the Sandwich (Hawaiian) Islands.[105] The discovery in 1820 of the Japan grounds, located in the North Pacific Basin east of Japan and the Bonin Islands, further encouraged Pacific whaling.

In 1822, a mere three years after the initial visit by the *Equator* and the *Balaena*, some sixty American whalers visited Hawaii.[106] In 1830, ninety-five whalers called at Honolulu and sixty-two at Lahaina on the island of Maui.[107] That year the British consul in Honolulu described the customary routes followed by the whalers: "The vessels engaged in whaling... generally arrive here in the months of March or April, and Sail toward the coast of Japan in May, where they cruise until the beginning of September (often in sight of the coast), when they leave and return to these islands, where they arrive early in October and remain until the latter end of November.... Those not full proceed toward the Equator and cruise between the parallels of 5° N and 10° S until February, when they proceed towards these Islands to refit previous to their proceeding to the coast of Japan."[108]

Whaleships, with their demand on the agricultural products of the islands, especially potatoes (both sweet and Irish) and fresh beef, were becoming the major factor driving the economy. As Ralph S. Kuykendall has noted, "During the long reign of Kamehameha III [1824-54], the whaleship came to be the symbol of Hawaii's economy."[109] Later in the century the production of sugarcane would become a major element with significant participation from Portuguese Atlantic islanders. According to Hawaiian historian Edgar C. Knowlton, they were early involved in this industry as well, "By the year 1828 there had appeared in the Hawaiian islands a Portuguese by the name of Antonio or Antone Silva or Sylva (died Febuary 9, 1854 in Honolulu); local tradition has it that he was the first to grow sugar cane on the island of Maui.... He lived in Waikapu, where he built a sugar mill in 1828."[110]

Organized whaling in colonial America had its beginnings on Long Island, but by the early eighteenth century the emphasis had shifted to Nantucket. And, by the end of the century, led by its enterprising Quakers, American whaling was beginning to spread across the globe. The industry suffered severe hardships during the wars with England, but emerged with renewed vigor soon afterward. While Portuguese participation dates back at least to 1765, documentation for the years between the wars is scanty. Nevertheless, it is clear that a number of Azoreans and Cape Verdeans were not only serving on American whalers at the time but were also settling in the vessels' home ports, especially New Bedford, Martha's Vineyard, and Nantucket. By the 1830s, however, there is a marked increase in the number of Portuguese whalemen, with some of them already moving up to the higher ranks aboard ship.

Notes

[1] Ashley, *Yankee Whaler*, 29.

[2] Braginton-Smith and Oliver, *Cape Cod Shore Whaling*, 5.

[3] Ibid., 5-6.

[4] Quinn and Quinn, *English New England Voyages*, 145.

[5] Ibid., 303-304.

[6] Hedges, Pelletreau, and Foster, *Records of the Town of Southampton*, 1:30.

[7] Ibid., 1:31-32.

[8] Ibid., 1:70.

[9] Ibid., 2:50-53; Simpson and Simpson, *Whaling on the North Carolina Coast*, 6.

[10] Starbuck, *History*, 15.

[11] See p. 37.

[12] Macy, *History of Nantucket*, 54.

[13] Ibid.

[14] *New-London Gazette*, 10 Oct 1766.

[15] *New-Hampshire Gazette and Historical Chronicle*, 28 Aug 1767.

[16] Kugler, *Whale Oil Trade, 1750-1775*, 3.

[17] Gomes, *Ilha das Flores*, 655. Translated by the author.

[18] Cf. Afonso, "Baleação pelos Açores," 1281.

[19] Ibid.

[20] Benjamin Hodges, Journal, ship *Dighton*, Paddock, Nantucket, Mass., Log 778, ODHS, 4-5 and 7.

[21] Brig *Polley*, Barnard, Nantucket, Mass., 30 May 1774, Log Wh P773 1774j, Nicholson Whaling Coll., PPL.

[22] Ibid.

[23] Chyet, *Lopez of Newport*, 9.

[24] "Captain William Roberts, Bill of Accounting Headmatter," Aaron Lopez Coll., MSS P-11, box 12, f. 4, AJHS.

[25] *Vital Records of Edgartown*, 78 and 183; Charles E. Banks, "Portuguese Immigrants to Martha's Vineyard before 1800," MSS, Portuguese on Martha's Vineyard, Record Unit 224, MVHS.

[26] *Vital Records of Edgartown*, 182; Banks "Portuguese Immigrants"; *Vital Records of Nantucket*, 2:577.

[27] Herbert, *Relic of the Revolution*, 247; *Vital Records of Nantucket*, 4:455; Banks, "Portuguese Immigrants."

[28] Ship *Jacob*, Wiswall, Newport, R.I., Aaron Lopez Papers, MSS P-11, box 3, f. 18, AJHS.

[29] Banks, "Portuguese Immigrants."

[30] *Vital Records of Edgartown*, 151 and 260; Dukes County Jail Records, Edgartown, 15 April 1811, MVH.

[31] Lists of Prize Shares of Seamen Who Served on the *BonHomme Richard*, *Pallas*, and *Vengeance* in 1779, RG 45, NAW.

[32] In the prisoner list his name is given as "Joseph Days." Dandridge, *American Prisoners*, 459.

[33] "Account Book of Brigantine Yankee," ca. 1813, De Wolf Family Papers, MSS 382, Ser. 2, S-s 2, RIHS; Maclay, *History of American Privateers*, 274; Munro, "Most Successful American Privateer," 12-62. In his *History of Martha's Vineyard*, Charles Banks states that during the war Dias, like his father before him, was captured by the British and sent to Dartmoor Prison in England. But he offers no evidence for this. (Banks, *History of Martha's Vineyard*, 1:423.

[34] "Account Book of Brigantine Yankee."

[35] Pilot Boat Favorite Account Book, 1837-42, Ships' Papers Coll., MSS 79, Ser. F, S-s 2, ODHS.

[36] Chyet, *Lopez of Newport*, 3.

[37] Ibid., 44-45.

[38] Ibid., 125. Edouard Stackpole states that "during his merchantile [*sic*] life Lopez owned thirty-nine ships and owned parts of sixty-one others." (Stackpole, *Sea-Hunters*, 71.)

[39] Chyet, *Lopez of Newport*, 149.

[40] See for example the log of the brig *Hannah*, 1768 (Aaron Lopez Papers, MSS 541, f. 3, RIHS).

[41] No indication of the home port is given. "Guide to the Papers of Aaron Lopez (1731-1782), MSS P-11, AJHS.

[42] "Instructions to Capt. from Aaron Lopez," brig *Leviathan*, Lothrop, Newport, R.I., voyage of 1773, s-g 13 Misc. Ships, Ser. 8-0, S-s 1, f. 1, Nicholson Whaling Coll., PPL.

[43] Alden, "Yankee Sperm Whalers," 269-72.

[44] The idea of a Brazilian whaling fleet that would hunt sperm whales along the coasts of Brazil and Argentina was formally suggested a few years later, but it was evidently never put into practice. (Bonifácio, 'Memória,' 2:283-84.)

[45] Alden, "Yankee Sperm Whalers," 297-80.

[46] Ibid., 280.

[47] Stackpole, *Sea-Hunters*, 74.

[48] Stackpole, *Sea-Hunters*, 71-80; Stackpole, *Whales & Destiny*, 7-9; McDevitt, *House of Rotch*, 223-26.

[49] Chyet, *Lopez of Newport*, 193-94.

[50] Ellis, *History of New Bedford*, 57-58.

[51] Thomas Jefferson, "State of the Whale Fishery in Massachusetts, from 1771 to 1775," *Report of the Secretary of State, on the Subject of the Cod and Whale Fisheries* (Wash., D.C: U.S. Dept. of State, 1791), 12:25.

[52] Starbuck, *History*, 177.

[53] Ibid., 78.

[54] Ibid., 79.

[55] Barbosa, "Hipólito José da Costa," 41.

[56] Alfred Hower, "A Proposal to Transfer the New England Whale Fisheries to Brazil," Edouard A. Stackpole Coll., MSS 335, f. 933, NHA.

[57] Barbosa, "Hipólito José da Costa," 50-51. Translated by the author.

[58] "Insert," Edouard A. Stackpole Coll., MSS 335, f. 933, NHA.

[59] Starbuck, *History*, 85.

[60] Cf. Malloy, *"Boston Men" on the Northwest Coast.*

[61] *New Bedford Journal*, 28 Jun 1799; Mawer, *Ahab's Trade*, 94.

[62] Starbuck, *History*, 213-14.

[63] *Vital Records of New Bedford*, 2:163. The *Polly* sailed from Nantucket in 1793-94 and later from New Bedford in 1797. It is uncertain, however, what activities it was engaged in during 1796. (Starbuck, *History*, 188 and 194.); ship *Maria*, Coffin, New Bedford, Mass., 1808-10, Register for Seamen's Names and Descriptions, New Bedford Port Society, ODHS (hereafter referred to as the Seamen's Register, ODHS); ship *Sally*, Clark, New Bedford, Mass., 1808-10, Seamen's Register, ODHS; *Vital Records of Nantucket*, 2: 438 and 4:338.

[64] Ship *Manilla*, Barnard, Nantucket, Mass., Crew List and Protection Paper, Edouard A. Stackpole Coll., MSS 335, f. 976, NHA; ship *John Adams*, Folger, Nantucket, Mass., Whalemen's Shipping Paper, Edouard A. Stackpole Coll., MSS 335, f. 1014, NHA; ship *Independence*, Swain, and schooner *Industry*, Macy, Nantucket, Mass., Whalemen's Shipping Papers, Coll. 335, f. 994, NHA. See also p. 39.

[65] Kartunnen, *Other Islanders*, 124.

[66] Ship *John and James*, Clasby, Nantucket, Mass., Crew List and Protection Paper, Manuscript Vertical Coll. MSS 384, NHA.

[67] Kartunnen, *Other Islanders*, 124; *Vital Records of Nantucket*, 2:468 and 4:459.

[68] Starbuck, *History*, 216-18.

[69] *Moby Dick*, Chap. 73.

[70] Joseph Silver Indenture, 11 Oct 1802 , MSS, Box 3S, Env. 9, MVH. Cf. "Still, somehow, I could read, write, and cipher to the Rule of Three." Richardson, *Abraham Lincoln's Autobiography*, 10.

[71] Joseph Antone Indenture, William H. Tripp Coll., MSS 58, box 1, f. 12, MSM.

[72] Grover, *Fugitive's Gibraltar*, 52.

[73] Seamen's Register, ODHS.

[74] Ibid.

[75] Lund, *Whaling Masters*, 724.

[76] Crew Lists, U.S. Customs Service, RG 36, NAB.

[77] Ibid. It was common to list men from neighboring islands who signed on at Faial as being natives of that port. Folger was from the island of Pico, even though he was listed here as being from Faial.

[78] Schmitt, "James Fennimore Cooper," 4.

[79] Ship *Hannibal*, Green, Sag Harbor, N.Y., 9 Aug 1827, Log 204, MSM.

[80] Harry D. Sleight, "Catholic Church on Long Island," *The Corrector (Sag Harbor, N.Y.)*, 24 Jul 1869.

[81] Ship *Timoleon*, Starbuck, New Bedford, Mass., 1824-25, Seamen's Register, ODHS; Census of the Town of Nantucket, 1820, MSS, NAL.

[82] *New Bedford Weekly Mercury*, 3 Jun 1825.

[83] Ship *Hidalgo*, Randall, New Bedford, Mass., 1826-27; bark *Lydia*, Frazier, New Bedford, Mass., Seamen's Register, ODHS.

[84] Ship *Ann Alexander*, Bowen, New Bedford, Mass., 1824-25; ship *Pacific*, Potter, New Bedford, Mass., 1824-27, Seamen's Register, ODHS.

[85] Ship *George and Martha*, Chase, New Bedford, Mass., 1823-24, Seamen's Register, ODHS. Much later, in 1857, three seamen named John Adams, Thomas Jefferson, and George Washington, sailed out of New Bedford aboard the whaleship *Eliza Adams*. From the crew list it is evident that they are foreigners, but their nationality is not stated. (Seamen's Register, ODHS.)

[86] Ship *Maine*, Worth, Fairhaven, Mass., 1834-35, Seamen's Register, ODHS.

[87] See also pp. 172 and 195-96.

[88] Afonso, "Baleação pelos Açores," 1282.

[89] Anonymous, Journal, brig *Aurora*, Macy, New York, N.Y., 30 Jul 1820, Log 334, Ships' Logs Coll., MSS 220, NHA.

[90] Ship *Swift*, Arthur, Nantucket, Mass., 24 Jun 1823, Log 246, Ships' Logs Coll. MSS 220, NHA.

[91] Ibid., 3-6 Jul 1823.

[92] *New Bedford Weekly Mercury*, 12 Sep 1828.

[93] Ship *Galatea*, Russell, New Bedford, Mass., Seamen's Register, ODHS.

[94] Protest of Thomas Hickling, São Miguel, 12 Sep 1829, Doc. 3, Despatches from U.S. Consuls in Fayal, 1795-1897, vol. 1, March 4, 1795-November 28, 1832, RG 84, NAW.

[95] Protest of Thomas Hickling.

[96] Despatches from U.S. Consuls in Fayal, Doc. 6, 27 Jul 1829 and Doc. 16, 8 Aug 1829.

[97] I. P. Hutchinson to Thomas L. L. Brent (Chargé d'affaires of the U.S. consulate, Lisbon), 30 Dec 1829; Brent to Martin Van Buren, Despatches from U.S. Consuls in Fayal, Doc. 121, 31 Dec 1829; I. P. Hutchinson, Lisbon, to Thomas Hickling, U.S. Vice-Consul, São Miguel, Despatches from U.S. Consuls in Fayal, Doc. 26, 21 Aug 1829.

[98] *New Bedford Daily Mercury*, 2 Aug 1831; Milton O. Gustafson, NAW, to Philip F. Purrington, Old Dartmouth Historical Society, 6 Jul 1973, MSS, vertical file folder, *Galatea* (ship), ODHS.

[99] Ship *Commodore Perry*, Hobron, New London, Conn., Log MS540, Whaling Logs Collection, YUL.

[100] For a discussion of a possible sixteenth-century discovery by an Italian, Portuguese, or Spanish navigator (Giovani/ Gaetani/Caetano/Gaitán) see Knowlton, *Portuguese in Hawaii*, 3-13.

[101] Ebenezer Townsend, Jr., Journal, ship *Neptune*, Greene, New Haven, Conn., 27 Aug 1798, Log 869, KWM.

[102] Knowlton, *Portuguese in Hawaii*, 25.

[103] A contemporary describes him as "of mixed English and Portuguese ancestry." Chamisso, *A Voyage around the World*, 111.

[104] Knowlton, *Portuguese in Hawaii*, 26; Bancroft, *History of California*, 274.

[105] Stackpole, *Sea-Hunters*, 267.

[106] Kuykendall, *Hawaiian Kingdom*, 1:306.

[107] Ibid., 307.

[108] Quoted in Morgan, *Hawaii*, 77.

[109] Kuykendall, *Hawaiian Kingdom*, 2:136

[110] Knowlton, *Portuguese in Hawaii*, 31.

CHAPTER 3

Heyday of American Whaling, 1835-1860

> Nowhere in all America will you find more patrician-like houses; parks and gardens more opulent, than in New Bedford.... Yes; all these brave houses and flowery gardens came from the Atlantic, Pacific, and Indian oceans. One and all, they were harpooned and dragged up hither from the bottom of the sea. (Herman Melville, *Moby Dick*, Chap. 6)

As the authors of a recent study of the American whaling industry remind us, "One hundred and fifty years ago, the world was different. Whaling was a major economic activity, and it was centered in the United States. In value of output, whaling was fifth among U.S. industries; it provided raw materials for the chief lighting and lubricating products of the day. During this period some seventy percent of the world's whaling activity took place on American vessels. New Bedford, Massachusetts, the leading whaling port, was said to be the richest town, per capita, in the country."[1] And in the Commonwealth itself whaling had become the third most important industry, after cotton and shoes.

The year of 1835 marks the beginning of a period of extraordinary growth in the American whaling industry. Between 1835 and 1860 the whaling fleet averaged 620 voyages annually, the greatest number of those taking place in the 1850s. For 1835 Starbuck lists seventy-two whaleships sailing from New Bedford, along with twenty-one others from nearby ports; twenty-four from the New London, Connecticut, area; twenty each from Nantucket and Long Island; fourteen from Rhode Island ports; ten from the Boston area; from up the Hudson River came six vessels, another six from New York City; and four from Martha's Vineyard. Newark, New Jersey, and Wilmington, Delaware—in their second year of whaling—sent out one vessel each. During its relatively brief whaling cycle the latter port sent out a total of sixteen voyages, and employed quite a number of Azoreans and Cape Verdeans. Midway through this period, in 1846, the American whaling fleet reached its greatest size, with a total of 735 vessels; but production would increase into the next decade before beginning to diminish.[2]

By this time whaleships had begun to move beyond the central Pacific. Despite the distance now to the whaling grounds, voyages to the Pacific still lasted only two to two and a half years—significantly less than what they would soon become. Nevertheless, the Atlantic—and particularly its islands—would continue to play an important role in American whaling.

As Jacob Hazen pointed out in *Five Years before the Mast*, the bank crisis of 1837 affected much of the American economy and encouraged a good number of young men to try out the thriving whaling industry. The author signed aboard the ship *Hudson* of Sag Harbor that very year. Upon arriving at Faial he noted the cat-and-mouse game of tobacco smuggling: "On reaching the landing, we were accosted by the custom house officer, who examined us very closely, and thrust his hands, in an unceremonious manner, into our pockets, in search of contraband tobacco. If a whole plug was found, the owner was permitted to twist off what he deemed sufficient for the day, and the balance was retained in the hands of the officer, to be delivered over on his return to the ship."[3]

Unfortunately, at the end of the day, as the crewmen were headed back to the ship, neither officer nor tobacco were to be seen. But the real exchange took place at sea a few days later as the *Hudson* was departing. After cruising some twelve miles out until sunset the vessel turned around and "at nine o'clock, a light was hung under the bow, and half an hour afterwards a boat full of men appeared looming through the darkness, within hailing distance of the ship." Then tobacco was unloaded in exchange for casks of wine.[4]

One of the interesting observations made by Hazen was of the women of Faial and the contrast between those of the town and those from the country. "The costume of the female portion of the inhabitants presents rather a peculiar appearance to an American eye. The ladies of the city usually appear in the streets, either veiled, or covered with large black hoods, contrasting strangely with those from the country, who appear to discard the cowl-like cloaks of their town friends, and in lieu thereof, decorate their persons with short gowns, and broad brimmed straw hats."[5]

Traditional cloak and hood, Azores. Courtesy of the New Bedford Whaling Museum.

The Ascendance of the Port of New Bedford

O land of Betfet,
From where we seek the whale,
From there I have come, there I'll return
As the sea touches the sand.
(Old folksong from the island of Flores)[6]

New Bedford was now supreme, easily outdistancing Nantucket in number of voyages. While the ports of Nantucket, New London, and Sag Harbor were in decline, New Bedford would not reach its peak until 1857, at which point it would control one-half of the nation's whaling business. In 1834, New Bedford whaling vessels carried nearly 100 Portuguese crewmen, the majority of them from the western Azores. In 1835, over forty Portuguese sailed from the northern Connecticut ports.[7]

Azorean women with baskets, island of Faial.
Courtesy of the New Bedford Whaling Museum.

This is also the period of great growth in Azorean and Cape Verdean participation in the industry, at all levels. It is often assumed that their ascension to the quarterdeck was a phenomenon of the late nineteenth century; but already by 1824 Joseph Folger, an Azorean, had assumed command of a New Bedford whaleship as replacement master.[8] By the 1840s and 50s a number of Portuguese were participating as whaling captains. One of the earliest was Frederick Joseph, from the island of Faial, who took the bark *Peri* of New Bedford on a cruise to the Indian Ocean from 1843 to 1845.

By 1850 the Portuguese were firmly established in New Bedford and environs. The census of that year shows a total of ninety-four residents born in Portugal. The number breaks down as follows: Azores, sixty-six; Cape Verde, ten; Madeira, three; Lisbon, one; and "Portugal," fourteen.

Almost all resided in the city itself, with a few settled in neighboring Dartmouth. Of this total, seventeen were female, with sixteen from the Azores and one from Cape Verde. Virtually all were residing with their husbands. While the majority of the men were sailors, others had settled down to shoreside occupations such as shipkeeper, rigger, grocer, farmer, etc.[9] But the real growth was yet to come, as the cotton mills began to attract immigrant workers toward the end of the century.

Among the many sailors who came and went along the New Bedford waterfront about this time, Rev. Moses How of the Seamen's Bethel was one their few steadfast supporters, taking up the cause for many a distressed mariner. At times, accompanied by an aggrieved sailor, he would visit a ship's agent to help settle the man's voyage. But, as he lamented in 1855, "I am more and more convinced that any man who will be a true friend to the Sailor, and look out honestly for his temporal welfare will be unpopular with most of this trading community."[10]

Rev. How was also a pioneer in encouraging the spread of Protestantism in the Azores. He noted in 1857 that "A Portuguese who has lately returned from sea & been converted on board Bark Bell of Fairhaven, spoke and prayed [at the Bethel]. He appears to be a very excellent man. He is a native of Pika [Pico], and wants to go home and tell the people what the Lord has done for his soul, and try to do good among children by learning them to read their Bible."[11]

Life in the Islands

In the Portuguese islands universal conscription, allied with overpopulation, tradition, and a sense of adventurousness were all factors in encouraging island youths to depart, as were the blights that wreaked havoc with the Azorean citrus crops and viniculture by the mid-nineteenth century. In the Cape Verdes, with their fragile agricultural base and periodic severe droughts, signing on to an extended whaling voyage became one of the principal means of escaping the poverty of the islands.

In his poem "Irmão" [brother] Cape Verdean poet Jorge Barbosa evokes the experience of his compatriots:

> You crossed Seas
> In your whaling adventures

on those voyages to America
from which at times the ships did not return.

Your hands are calloused from pulling
the lines of the boats on the stormy seas;
you lived hours of cruel hopes
in combat with the storms;
you were bored with the maritime tedium
of the long interminable calms.[12]

In "Romance," the Azorean poet Pedro da Silveira tells the story of one of the old men he knew as a youth on the island of Flores who had one day been picked up by an American whaler and had spent his adult years at sea before finally retiring to his native isle:

> In the long hours on watch Antonho Cristove [António Cristóvão] climbed the masts. He harpooned whales in all the seas—in the cold waters of the Arctic, in the warm waters of the Pacific. And he remembers the right whale that one day overturned the boat and killed the boy from Corvo. (Blood of the whale, blood of the dead sailor, coloring the sea. Tears of the sailors join with the sea.) Antonho Cristove, crewman on all the sailing vessels, crossed all the oceans, knows all the ports, knows the fury of the cyclones and the calms of the Gulf. And the marks on his body remain from the fits of American captains, of Azorean captains, of captains from Cape Verde. The island always within him and the hope of returning. An escape in each port and another ship to sail away on. Women-of-the-street in the four corners of the Earth. His companions in adventure lost so long ago: the water of the sea has carried them off, dirt of the earth has covered them. Uncle Antonho Cristove recounts his life and his tired eyes still look to the sea as if he were yet sailing upon it.[13]

Departure, for Azoreans and Cape Verdeans, became a natural part of the maturation process. As a native of the island of Pico recalled, "One of these stories of such an adventure, told by my father, I still remember. A boy of fourteen was hopping from rock to rock along the coastline… where a codfisher was whistling its presence [to take him on as a crew member on its way to the Grand Banks]. The father followed after him as best he could and shouted, before seeing him disappear, 'leave your sandals for Antonio!'" She goes on to point out that in her village in the mid-nineteenth century a particularly large number of boys left the island at the age of twelve.[14]

In the process of obtaining provisions (referred to at the time as "recruits"), crewmen would be added, often illicitly at night from some remote point. As mate John B. Wood of the bark *George and Mary* of Westport, Massachusetts, noted, as the vessel was lying at the island of Flores in September, 1852, "our recruits came off consisting of 60 bushels of potatoes 11,000 onions one Calf one Hog a lot of... Grapes Eggs Cucumbers and so forth Two men took French leave (Richerd Quin of England and William Williams of Ireland) shipped 2 Portugues emanuel and Sylvestra."[15]

In Cape Verde it was not just aversion to military duty or the desire to go to America that propelled young men of the archipelago, but also the periodic severe droughts. Writing a few decades later, a government official on Brava reported that "the majority of the whaleships do not anchor in our ports but remain under sail along the coast, taking on those they need; and not even the police are a guarantee as agents of authority in these matters, because it has happened that they have been the first to embark, leaving their swords on the beach as a testimonial."[16]

Horatio Greene departed New Bedford aboard the whaleship *Two Brothers* as a passenger bound for the Azores and then on to a journey through Europe. When the ship called at Flores Island in 1851, Greene summed up what brought whaling masters to ship so many of these young men: "Here may always be found enough who are willing to undertake a sea voyage even for their cloathes and a trifle more and they are found to be willing, orderly and hard working men, who will seldom if ever desert a ship when there is a dollar due them, we procured several, that is five or six, having Come out from home short handed for the purpose."[17]

While it became regular practice to fill out a crew in the Azores or Cape Verde, whaling masters had to take care at the time that crew members did not desert there, as in the case of the *George and Mary* above. So it was that Manuel Silveira Andrade got a glimpse, and no more, of his home on Faial:

> Flores, August 11th. We sailed off and on until the next day. Then we squared away to Fayal at 4:00 o'clock p.m. under a full sailed strong N.W., and we made a good progress, got in Fayal bay in 22 hours, to be in the light sails, and laid there off wondering the night; one boat was dispatched ashore at 2 o'clock, and came aboard right off; I was in her, but I could not obtain permission to go ashore on my own native land! Without doubt I've been very sorry for it, but I couldn't help myself! I only could get a previlige to have the spy glass on my hand for a moment, to look upon my own dwelling and some others in my neighbourwood! How pretty they looked to me!![18]

However, most often the arrival at any of the several Portuguese Atlantic islands was a time momentarily to forget the drudgery of shipboard life. As the keeper of the log of the ship *Lucy Ann* of Wilmington, Delaware, noted in January of 1842 after the ship arrived in the harbor of Praia on the island of Santiago, Cape Verde:

> The Captains boat was cleared away & lowered, the boats crew were ordered to get themselves in readiness to go on shore which they very quick did, their rig was white duck trousers, red flannel shirts, pumps tarpaulins & black silk neck handkerchiefs with bandannas in their pockets tied to the button hole to keep (as they said) the young ladies on shore from walking off with the slack of them, in the evening they returned to the ship in middling kind, the description of St. Jago [Santiago] & its inhabitants put the rest of the crew in their tantrums & all the wonder was who should go on shore the next day.[19]

The following day the keeper and several of his shipmates went ashore with "tobaco, side combs, & handkerchiefs to trade for fruit."[20] Contrary to what they might usually expect, especially in the Azores, they found there was no need to conceal the items, as the officials were not searching anyone. The group traveled into the countryside. There they found coconuts, cabbages, dates, bananas, watermelons, grapes, figs, and especially oranges in abundance. "The appearance of the soil is as if it never felt rain but the plantations looked flourishing. we saw plenty of Orange groves Pine Apple bushes vineyards Sugar cane plantations etc."[21] The next day, back aboard ship and ready to sail, "all hands

[were] in the downs at leaving St Jago So soon. some swearing they could live there all the days of their lives & a spell longer...."[22]

To the green hand who has left the American shores for the first time the sight of any land greatly brightened his spirit. Nathaniel Taylor was a physician who had signed on in 1851, quite uniquely, for a voyage on the ship *Julius Caesar* of New London with the intention of improving his health. His description of the first sighting of the Azores from the vessel, on its way to the Kerguelen Islands in the South Indian Ocean, is particularly poignant:

> At five o'clock on the morning of the sixth of September, 'Land ho' resounded in clear and joyous tones throughout the ship. In an hour, the island of Corvo was distinctly visible. Soon Flores appeared beyond, reflecting from its mountain sides the rays of the rising sun, and its brown and furrowed surface, unadorned with flowers, affording no indication of the origin of its name. A nearer approach, however, exhibited a luxuriant landscape of vineyards, orange and lemon groves, with green fields bounded by extensive woodlands.[23]

A green hand aboard the ship *Dartmouth* of New Bedford related in a letter to his mother his experience upon visiting the Portuguese islands in 1845:

> [After riding out a severe gale the vessel] arrived safe at the weston Islands in 21 days we anchored at port prayer [Praia] on Island of tesarh hee [Terceira] we had a days liberty each watch of us I was glad to get on shore once more the people seemed very kind and tried all they could to entertain us they gave us greate quanterteys of fruit grapes figs appels pairs peaches and so [on] one of the men took us in to the church and shewed us their Imeges and the high alter with the concecrated lamp burning befour it and many other things to numeras to mention when we sailed from heer we went to the Cape Devirde Islands and lay 4 days we then put to sea and bent our corse towards Cape Horn....[24]

In September 1851, young green hand Enoch Cloud of the ship *Henry Kneeland* of New Bedford noted in his journal his first day in a foreign land, on the island of Flores: "At 8 in the morning cleared away

Street scene, island of Brava, Cape Verde, 1904. Courtesy of the New Bedford Whaling Museum.

& lowered the Starboard Boat & pulled for the land, distant abut 3 leagues; an ugly 'chop' sea was running and 'pulling' was anything but easy! We finally reached the shore. Pulling around a bluff we found ourselves in a small smooth cove at the foot of an abrupt mountain where we were met by the Custom House Officer — an old Portuguese wearing a 'hat, that had been a hat,' barefooted and graced with a mouthful of an old cigar — who inquired 'who we were, where from, where bound & if we had any sickness on board.'" Two days later they "made the port of Fayal. A beautiful place, and quite a City. Immediately opposite Fayal, towering far up in the skies, with its side for some distance up dotted with houses and its top lost to view in a cloud of mist, looms the 'Peak of Pico.'"[25]

When green hand Ben-Ezra Stiles Ely was given a short leave on the island of Brava, from the whaling bark *Emigrant* of Bristol, Rhode Island, in the mid-1840s, he was most impressed with the inhabitants:

> [There he met] a very beautiful young girl; and with her broken English and bright glances made quite an impression on a sailor's heart. She did not seem to be bent on conquest however; for she took it for granted that I had a sweetheart at home. –She presented me with a string of

Street scene, Ribeira Brava, São Nicolau, Cape Verde, 1904. Courtesy of the New Bedford Whaling Museum.

> green beads, and told me if I wore them, I should be safely returned to my lady in America.
>
> The females in this Island smoke a great deal; and having motioned to me to sit beside her, she took my hand in hers, and my pipe from my mouth and made free use of it.
>
> I told her as well as I could, how badly we were treated on board ship, and how I longed for land; at which she seemed to be much affected, and gently pressing my hand endeavoured to persuade me to leave my vessel and live with her, until I could go home. Indeed, I should have taken her advice, had a good opportunity offered, but it was ordered otherwise.[26]

Women smokers at Cape Verde were often remarked on. When John Wilson, surgeon aboard the whaleship *Gipsy* of London, England, was strolling around the island of Brava with the captain of the vessel in December of 1839, they "saw numbers of dark-looking, giggling damsels, who flocked to the windows & doors, & appeared to make merry at our expense. Walking along, we were stopped by the people of a house & desired to enter & partake of any refreshments they could afford; we thanked them & walked on; the Skipper observing, that so much hospitality did not exist in England. The women wear a loose dress, with a scarf thrown over the shoulder & brought across the body: we met some of gigantic stature: almost all had long pipes in their mouths."[27]

The feeling of strangeness that overwhelms the first-time traveler in a foreign land is well evoked in the prose of Thomas Higginson, who spent a winter and spring on the island of Faial in the late 1850s, not as a whaleman but with a small group of tourists. It gives a sense of how disoriented the green hand from a whaler might feel in the Portuguese islands, as well as the initial impressions that a young islander might have upon landing in New Bedford:

> One general truth came over us instantly, and it was strange to think that no one had happened to speak of it before. The essence of the surprise was this. We had always been left to suppose that in a foreign country one would immediately begin to look about and observe the foreign things, —these novel details having of course that groundwork of ordinary human life, the same all the world over. To our amazement, we found that it was the ground-work that was foreign; we were shifted off our feet; not the details, but the basis itself was wholly new and bewildering; and, instead of noting down, like intelligent travellers, the objects which were new, we found ourselves stupidly staring about to find something which was old, —a square inch of surface anywhere which looked like anything ever seen before, —that we might take our departure from that, and then begin to improve our minds. Perhaps this is difficult for the first hours in any foreign country; certainly the untraveled American finds it utterly impossible in Fayal. Consider the incongruities. The beach beneath your feet, instead of being white or yellow, is black; the cliffs beside you are white or red, instead of black or gray. The houses are of white plaster on the outside, with wood-work, often painted in gay stripes, within. There are no chimneys to the buildings, but sometimes there is a building to the chimney; the latter being a picturesque tower with smoke coming from the top and a house appended to the base. One half the women go about bareheaded, save a handkerchief, and with a good deal of bareness at the other extremity, —while the other half wear vast conical hoods attached to voluminous cloth cloaks which sweep the ground. The men cover their heads with all sorts of burdens, and their feet with nothing, or else with raw-hide slippers, hair outside.[28]

Sixteen-year-old Thomas Milligan, a green hand aboard the bark *Noble*, of Sag Harbor, was particularly impressed as the vessel approached the island of Flores in the fall of 1846. "The day was lovely, and, as soon as we got near enough, the island offered to our view a charming sight—slopes thickly wooded, mountains clothed luxuriantly with vines and orange trees, some of which sprang even from the interstices of the rocks, peaks shod in heath and bounded with light, cascades joyfully tumbling down on all sides into the sea, and the whole scene of fertility and beauty as if freshly bathed in morning dew." Upon landing the sailors "were greeted by a number of natives, who reminded me of the scattered Italians I had seen in New York, the men wearing the [incomprehensible], calico shirts, dungaree (a Kind of coarse jeans) Trousers, slight in stature, swarthy, barefooted, Their feet looking like cowhorns, and all, men and women, jabbering Portuguese."[29]

Milligan spent the day taking in the sights of a foreign land for the first time when in the afternoon he was,

> surprised, on turning my eyes towards the heights by which my companions had disappeared, to see a company of ragged, rakish-looking fellows coming down, with baskets on their heads (the unique Portuguese baskets), full of grapes, oranges, jackass cheeses, and so forth, and, as the ragamuffins drew near, I was amused, as well as surprised, to find that they were my own shipmates, who, in the true spirit of a sailor's lark, had traded their go-ashore, Sunday clothes for the old toggery of the natives, and the fruit to boot. All hands, well-pleased with their adventures, now returned to the ship; and, the result of the trips having been the purchase of ample ship's stores, in the shape of large, mild onions, potatoes, poultry (the Flores poultry was then esteemed the finest in the world), and the like, we squared away for Fayal, about one hundred miles Southeast.[30]

Robert Weir was a well-educated, somewhat effete, young man who fled home after some domestic difficulties and made his way to New Bedford, which, incidentally, he considered detestable. His first impressions aboard the bark *Clara Bell* of Mattapoisett, in 1855, were equally negative and probably driven in large part from conversations with some of his fellow crew members or with other sailors during his two week stay in a New Bedford boardinghouse prior to sailing. As he noted soon after departure, "So far the crew have deported themselves very peacefully. Nearly half of them Are Portuguese don't like them – though I can make out to Live peaceably – Our Mate is a villain. I can see it but too plain, but he must be careful – the second mate is a boy – 3rd Mate – a Gee."[31]

But stereotypes and first impressions have a way of mellowing after long contact. Evidently Weir's nearly three years at sea changed some of the negative opinions he had of his crewmates; for, at the end of his journal—written sometime, perhaps years, after the end of the voyage—he takes up again the first days of his voyage, rewriting his experiences and dropping many of his negative impressions. Now, on the day of sailing, there are no more "Gees," nor villainous mates. When the latter sent him aloft, "a sailor named Manuel, a potugese kindly told me what to do, after getting the order 'you there' lay aloft & slush the to'gallant & topmasts. I went up in fear and trembling and followed kind Manuels Instructions, carrying a slush bucket & plastering the masts with the contents melted fat from the cooks galley." A few days later, seasick and contemplating feeding himself to the accompanying sharks, he was given his turn at the wheel, "being two hours under the eye of another portugese named Mike Miguel. He did not have much to teach me as he said I knew all about it after I handled the wheel five minutes."[32]

While at mid-century the majority of whaling vessels headed for the Indian or Pacific oceans, a goodly number remained in the Atlantic, taking advantage of the various islands to refresh their supplies, and crew if necessary, as well as perhaps to transship their oil and bone. The bark *George Washington*, under the command of Capt. William O. Harps, followed such a trajectory on her two-year voyage in the years 1851 and 1852. The vessel sailed from New Bedford on November 6, 1851, and a month later made its first land at Brava. After picking up a few men it then headed across the equator to the River Plate Grounds, to the east of Buenos Aires and Montevideo. In April it was at St. Helena, in May at Ascension, and then on to the Azores, around which it spent all summer whaling. During that time it called there eleven times for provisions and men and to land oil. Then, in October, it was off again for the island of Santiago, Cape Verde—followed by the island of Santa Catarina in November. Known as St. Catherine's by American whalemen, it hugs the coast of the Brazilian state of the same name. Besides being a popular port of call for provisions, the island was an important

Road leading down to the port of Furna, Brava Island, 1904. Courtesy of the New Bedford Whaling Museum.

center of the Brazilian shore whaling industry and a locus of Azorean immigration dating from the eighteenth century.[33] Curiously, however, and in contrast to the Portuguese islands, the inhabitants apparently showed little interest in signing aboard the American whalers. It was now November again, and the same itinerary was repeated, taking the vessel to the River Plate Grounds during the southern summer before it headed to the Azores for the northern season, this time calling on the way at the Brazilian island of Fernando de Noronha, and finally arriving home in December.[34]

Upon leaving the Azores, the *Julius Caesar*, with Dr. Nathaniel Taylor on board, headed for the Cape Verde Islands, where he again afforded a lively, albeit bigoted description:

> Porto Praya, a town of St. Iago [Praia, Santiago], alone has a good harbor which unfavorable winds prevented us from reaching, and we bore away for Brava, where our ship lay off and on during the twenty-ninth, taking on board hogs, poultry and provisions. This island is rugged and mountainous in its aspect, the land rising abruptly into high konical peaks with a melancholy and barren appearance, much in unison with the character

> of the inhabitants, who are principally negroes, speaking the Portuguese language, but very low and degraded in the scale of civilization.
>
> The bay itself was full of life and activity; nineteen vessels were sailing back and forth while taking on fresh supplies. On board of the *Enterprise* [whaleship of New Bedford] was a brother of our cooper whom he had not met for seven years. Unfortunately he was unable to communicate with him, although he recognized him with the spy-glass. Of course there was no boat at his service, and "duty before decency" is a fixed law on board ship. The island is remarkable for a kind of sour wine called *aguardiente* by the natives. The sailors smuggled some of this on board in calabashes or cocoanuts from which the milk had been extracted. Others more cunning procured it by means of the long intestine of the hog, which, being properly cleaned and prepared, was filled and worn around the body coiled like a rope.[35]

The New England Whaling Fleet

By 1846 the American whaling fleet had reached its peak. According to Starbuck, that year it consisted of 680 ships and barks, thirty-four brigs, and twenty-two schooners. Whaling was now the third largest industry in Massachusetts after shoes and cotton. During the second half of the decade the fleet averaged 638 vessels sailing from nearly fifty different ports in southern New England and New York and employing some 70,000 persons.[36]

By mid-century Nantucket, while past its heyday, was still a major participant, seeking especially the sperm whale in all the oceans of the world. For the Nantucketer there had always been an unrealistically romantic tinge to the whale fishery, and to a lesser extent in other ports. As a Nantucketer noted in 1838, with a bit of hyperbole: "In the present day, every energy, every thought, and every wish of the Nantucketman is engrossed by Sperm Oil and Candles. No man is entitled to respect among them, who has not struck a whale; or at least, killed a porpoise: and it is as necessary for a young man who would be a successful lover, to go a voyage round Cape Horn, as it was for a young knight, in the days of chivalry, to go on a tour of adventures and soil his maiden arms with blood, before he could aspire to the snowy hand of his mistress, and enjoy the delights of 'ladye love.'"[37]

One of the early Nantucket masters of Portuguese American descent was William H. Swazey, born in Nantucket in 1811, the son of Jonathan Swazey and grandson of Anthony.[38] As mate on the ship *John Adams* of Nantucket (1841-45) he took over command upon the death of Captain Isaac Stockman in April 1842. Two years later, in the beginning of 1844, he in turn gave over command at the port of Paita, Peru, upon becoming ill. Unfortunately, Swazey was serving on a schooner out of New Bedford when it was lost with all hands in April of 1851.[39]

But several factors, both economic and physical, combined to diminish Nantucket's participation. The first was the Great Fire of 1846 that devastated the business section and the wharves. As whaling vessels grew in size they had increasing difficulty in crossing the sandbar at the entrance to Nantucket harbor; and so the island became more and more dependent on Martha's Vineyard for the clearance and entry of its whaling vessels. Another problem involved obtaining supplies and distributing oil, given the island's distance from the coast. Still another was the California Gold Rush which siphoned off the best of the entrepreneurs and veteran whalemen. During 1849 alone, more than 500 Nantucket men headed for California. At the same time more than forty-two whaleships were withdrawn from the industry to carry passengers and goods to the Gold Country.[40] And finally there was the Civil War, which virtually put an end to Nantucket as a whaling port.

Portuguese on the Quarterdeck

The first well-documented voyage of an American vessel under Portuguese command was that of the ship *Liverpool* of New Bedford (1838-40) under Captain Joseph Thomas, born on the island of Terceira, Azores. Again under his command in 1840-42, the *Liverpool* whaled in the Indian Ocean. Thomas's final voyage, to the Northwest Coast in the ship *St. George*, was relatively brief (eighteen months) but successful.[41] In fact, although the Portuguese still formed a relatively minor component of the American whaling fleet at this time, a number of natives of the Azores and Cape Verde had begun to ascend to the highest levels of shipboard responsibility.[42] Severino D. Pierce (Pires) was born on the island of Brava about 1817. His first recorded service is aboard the ship *Magnolia* of New Bedford (Jun 1845-Oct 1848). He then served as second mate of the ship

Gratitude (Dec 1848-May 1851). In November 1848, between these two voyages, Pierce managed to travel to Stonington, Connecticut, where he married Harriet H. Nichols.[43] Within a fortnight he was off on the *Gratitude*. A month after his return, Pierce sailed on a three-year voyage as mate of the ship *America* of New Bedford (Jun 1851-Jun 1854).[44]

Upon his return Pierce had four months at home before embarking as mate of the ship *Magnolia* of New Bedford (Oct 1854-May 1858).[45] Following this voyage he had a scant two months at home before departing in July 1858 as master of the same vessel. Although the *Magnolia* was condemned at Sydney four years later, its voyage was a success, yielding 330 barrels of sperm oil, 2400 barrels of whale oil, and 31,675 pounds of whalebone.

Azoreans were particularly active on Provincetown whalers, mostly schooners. Francis Joseph was the first to command a vessel out of that port. In 1851, he took the schooner *Rienzi* for a four-month cruise. He was followed by Joseph M. Caton in the same vessel in 1852 and Manuel Costa on the *Antarctic* in 1854.[46] During this period, a significant portion of some of the whaling crews was Portuguese (virtually all Azorean). In 1846, the schooner *Samuel Cook* sailed with five, and the following year the bark *Fairy* carried seven. Ten years later, in 1857, when the schooner *John A. Lewis* left port, all but the captain and mate were Portuguese (Azoreans except for one Madeiran).[47]

The Connecticut Ports

One of the more interesting Azorean seamen of this period was Nicholas C. Anthony, born between 1806 and 1810 on the island of Flores, who early settled in New London, Connecticut. Anthony served on a series of Connecticut vessels before moving up to his own command. He was with the ship *Wabash* of New London from 1829-30; second mate on the ships *Caledonia*, of nearby Stonington, 1835-37; and *Friends* of New London, 1839-41. He was mate on the bark *Cervantes* (New London, 1841-43), and then returned to the *Friends* (New London, 1843-45) as second mate.[48]

Anthony had married a local woman, Lydia Ann Spencer, in New London, in 1830, and for many years the family resided there on Tilley Street.[49] During several voyages Anthony maintained a very insightful journal. On the ship *William Lee* of Newport, Rhode Island, he described a harrowing voyage on a vessel full of worm holes that needed progressively more pumping as it made its way north from the South Atlantic. In October 1851, the ship made 7000 strokes in twenty-four hours and 8000, the next, while meeting heavy gales. "And if we don in on Leak ship may God have mercy on Souls," he lamented. Fortunately the captain decided to head for the nearest port, where they made New York harbor a few days later. "The Lord as been with us these day," Anthony noted.[50]

Since at least 1835, he had been almost constantly at sea, and yet managed to maintain a family. But the distance had begun to affect him. In February of 1850, during the voyage of the *William Lee*, Anthony recounted a nightmare whose images haunted him. Upon arriving home from a voyage, he was unable to locate his house or family. He finally found his wife living in a hut. She had had a stroke and had been bedridden for many months. Their daughter Ellen was also very sick. Moreover, he was informed that their house had burned down and their son had died.[51]

In 1852, Anthony took command of the brig *Topaz* of New London for a whaling and possible sealing voyage. Unfortunately, in April of the following year, as the vessel was working into the harbor of Beaver Island in the Falklands, it struck on a rock. After three days the crew was able to get the brig afloat; but with the main keel broken, some twenty feet of the garboard strake stove in, and five feet of water in the hold, the *Topaz* was hardly seaworthy. Eventually making it into the harbor at Stanley, the vessel was discharged and the cargo sold at auction. At that point, the *Topaz* was condemned and the captain, officers, and crew taken into Montevideo.[52]

In October 1853, Anthony sailed as mate on the recently launched ship *Petrel* of New Bedford under Capt. Moses G. Tucker. It must have been a difficult voyage, for the keeper of the log (not Anthony) notes that after coming down from the Okhotsk Sea to Tahiti in early November of 1854, by the first of January virtually the entire crew had run away. The captain later retrieved them from the "Calebuse," where they had been on bread and water for about twenty-five days.[53]

Perhaps it was after having arrived home from this voyage that Anthony had another nightmare, and upon awakening noted in the journal he had

Antone DeSant, ca. 1880-86. Courtesy of Mystic Seaport Museum.

kept for years, "I tryed to go to slip a gain. I did not know but I my dream the same dream again but been near day light. I could not Geet my yees to clear a gain I fild so vey bad that I fonde my silfe sobing very bad when wake I all see my wife sobeing very bad when she seeing me and she sade a. Nichalas a Nicholas o my God a Long you have been gon."[54]

And so ended the whaling career of Nicholas C. Anthony. In 1860, he was apparently captain of a vessel, perhaps in the coasting trade and living with his wife in New Haven, Connecticut. In 1870, he was still sailing, but by 1880 he had retired.[55]

Another well-known whaleman from New London was Antone DeSant, born in Cape Verde, ca. 1815. DeSant sailed on at least seven whaling voyages out of New London, beginning in 1832. In 1850, he evidently tried his luck in the California goldfields, sailing as an officer on board the bark *Portland* from Boston to San Francisco. Later he established himself as a merchant in the waterfront district of New London.[56]

The 1850 federal census reveals fifty-three natives of Portugal in New London, several of whom were aboard whalers. There was only one Portuguese woman residing in the town. Most of those who stated their provenance were Azoreans (thirty-five), while eight were from Cape Verde. As did his compatriots in other whaling ports, thirty-eight-year-old Manuel Enos (not to be confused with the whaling captain from Cold Spring Harbor, New York) was running the Portuguese Seaman's Boarding House.[57]

Stonington, Connecticut, was an active whaling and sealing port between 1818 and 1892. In 1845, twenty-seven ships and brigs were operating out of the port.[58] As they had in neighboring New London, Portuguese began shipping out of Stonington around 1835. Some were tempted, early on, to settle there. Frank G. Sylvia was born in Faial in 1824 and arrived in Stonington on a whaleship about 1840. Sylvia worked as a successful builder and farmer; and in 1874 established, with his son-in-law, Joseph A. Vargas, also from Faial, a partnership known as Sylvia & Vargas Ice Dealers. In the winter, ice was harvested from two local ponds, called Sylvia's and Vargas. The firm, later known as the Vargas Brothers Ice Company, continued to harvest ice from these ponds until 1946.[59]

The U.S. census of 1850 lists nine men and one woman from Portugal residing in Stonington, including a clergyman, Emanuel Goncalves. In the census there are sixty sailors from the Azores and Cape Verde serving on Stonington vessels. In the 1840s and 50s, there were twelve marriages of Portuguese in the borough. In 1851, the Portuguese helped establish St. Mary's Roman Catholic church, whose worshippers in the early years were mainly Irish. By the 1870s, a significant number of Portuguese settled in Stonington after having left the local whalers and sealers. By 1914, a local Holy Ghost Society had been founded. The annual blessing of the fleet and the Holy Ghost festival and parade are legacies of the Portuguese presence that continue today. As the activity in whaling and sealing diminished toward the end of the nineteenth century, many mariners took up commercial fishing. Today the Portuguese are still a significant element in the local fishing industry.[60]

The Long Island Whalers

The Sag Harbor whaling fleet continued to grow along with the industry as a whole, reaching its high point in 1847 with sixty vessels and rivaling in importance at the time Nantucket and New London. In spite of the presence of Azoreans and Cape Verdeans on the majority of voyages out of Sag Harbor, it was rare to see them leave the forecastle except as boatsteerers. For example, in fifty-six voyages between 1828 and 1855 belonging to one of Sag Harbor's principal owners there were more than 200 Portuguese on fifty-two of these sailings, an average of four per voyage. Among them were a good number of boatsteerers but only one officer, Caton Joseph (or perhaps Joseph Caton), who served as second mate on the bark *Nimrod* between 1853 and 1855.[61] Since these fifty-six voyages represent only seven percent of the whaling voyages out of Sag Harbor, this affords a good approximation of the total number of Portuguese who sailed from this Long Island port. But very few remained there.

A look at some of the records of the Sag Harbor whaling fleet affords us an idea of the prevailing attitudes there concerning the Portuguese. At the bottom of the list of foremast hands of the ship *Ontario* (1850-54) one reads:

Antoine Portugee
Antoine doRun Away
Lewis do
George do
Emanuel do
"Got at Bravo one of the Cape De Verde Islands":
Antone Portugese
Peter do
Joseph do[62]

In 1850, there were twenty Azoreans, two Cape Verdeans, and one "Portuguese" residing in or near Sag Harbor, most of them sailors and several laborers, but also a caulker, a barber, a confectioner, and a farmer, plus three wives and two daughters, all born in the Azores.[63] Also that year a cotton mill began operation in the village. In 1870, Azorean Thomas Sylvia was working there as an engineer, along with his three children: Emily, 16; Thomas, 12; and Willie, 9.[64] Unfortunately, the mill burned down in 1879; but in any case Sag Harbor was never destined to be a great industrial center, nor a mecca for Portuguese immigration.

Of the various Portuguese who did remain in Sag Harbor, the De Castros are probably the best known. Like so many of the successful settlers from the Azores, Joseph De Castro left his native island of Pico at an early age, serving as a cabin boy at the age of fourteen on a Sag Harbor whaler.[65] He evidently served as mate on the Sag Harbor schooner *Susan* on a whaling and sealing voyage (Dec 1857-Feb 1860). In April 1858, as the schooner was lying at Bahía Unión on the central coast of Argentina, De Castro took time to compose a long letter to his wife. They had seen many whales off the River Plate, but the weather had been bad and the whales so swift that they could not fasten. So they had only thirty barrels on board. The terrain around Bahía Unión, at the mouth of the Río Colorado, was very enticing—much game and no settlers—so much so that Captain Edwin Smith had applied for a grant to settle there. De Castro was asked to bring down his family as well, and he mentioned to his wife that "I wish you and the children were here now and I would not go back again the land is very rich and a man could make his fortune in a few years horses and cattle are very plenty all you want is a house and farming utinsils Mr Payne says this is the most beautiful land he ever see he has been most every where he says this beats all he ever see there is a good market for everything close by I think the easiest way a man could make his living would be to come out here."[66]

In his letter De Castro stated that this was both a whaling and sealing voyage, but he added that if the whales did not appear later during that southern winter, they would be off to the Falklands in search of penguins. He hoped to be home the following winter (i.e. early in 1859) or no later than that spring. Evidently luck was not with them as they spent almost two more years before returning with only sixty-five barrels of sperm oil and 285 of "whale" oil, a generic term in this case for whatever sea mammal they might have come across in their hunt.[67]

De Castro vowed that this would be his last extended absence from his wife and family; and so it was as he later became captain of the coastal schooner *Louis Walsh* of Sag Harbor.[68] In 1860, his brother John De Castro owned a boardinghouse in town and ten years later ran a carriage house alongside the long wharf.[69]

That Sag Harbor was less welcoming to the Portuguese than the whaling communities of New England can be surmised from the words of early twentieth-century local historian Harry D. Sleight: "Portuguese sailors, Kanakas, Figians, Malays and Montauk and Shinnecock Indians, all good whalemen, walked the streets, showing scrimshaw work and gimcracks, flush with money, spending and giving away lavishly as they went. Drunk, but good-natured, they rolled along the main street promenade, laboring like ships in a heavy sea, literally 'half-seas over.'"[70] He also suggests a degree of social hierarchy that is not reflected in descriptions of contemporary New Bedford, for example: "The agents and owners were aristocrats. They did not deign to bump elbows with the bibulous whalemen, but dressed in fine linen and rode in their private chaises."[71]

The Portuguese fair no better in Frederick P. Schmitt's description of a day in the life of the minor whaling port of Cold Spring Harbor. Again they are conflated with South Sea islanders in an unflattering portrayal of both groups. The author evokes the excitement engendered whenever one of the port's whaling vessels returned from one of its extended voyages: "Just as soon as the sailors hit the tiny port, Cold Spring came to life. Numerous peculiar looking characters, speaking in a babel of languages, swaggered along Main Street, which was soon dubbed 'Bedlam Street,' rather appropriately, by the disgusted villagers. A number of Portuguese sailors and Pacific Island Kanakas stayed in the village for a time, living at the Stone Jug, a combination inn and saloon, where the proprietress, it is said, had the pleasant faculty of getting along with them."[72]

Cold Spring Harbor is of particular interest because of one of its longtime residents, Manuel Enos. At the height of its activities, between 1836 and 1862, Cold Spring Harbor could count nine vessels at sea. But, even more than Sag Harbor, it did not encourage foreigners. Except for two or three Portuguese and several Irish, the local population consisted of native-born New Yorkers.[73] Nevertheless, one outlander stood out above all the local whalemen. In the words of local historian Frederick Schmitt, Manuel Enos was "probably the best known whaleman of Cold Spring Harbor."[74] And without a doubt the most enigmatic. "Just enough is known of Enos to give him almost mythic proportions," wrote Robert D. Farwell in 1981.[75]

Manuel Enos was born in Lajes, on the island of Pico, on May 22, 1826, one of eleven children of José Inácio Macedo and Maria Carmo. He is supposed to have arrived in the U.S. at an early age and presumably began his whaling career as a young man.[76] His common name, however, makes it makes it difficult to ascertain his early voyages. He first appears at the port of Cold Spring Harbor as a crew member on the whaling ship *Huntsville*, which sailed from there in October of 1849.

After a relatively short voyage of one and a half years, the vessel returned in March of 1851. During his stay in town awaiting the next sailing of the *Huntsville*, Enos must have met his future bride, Susan Brush, the daughter of a well-established local family. The *Huntsville* sailed that December with Enos as third mate. Three weeks after his return in April 1854, the couple married.[77] Once again, five months after the nuptials, Enos embarked on another whaler, the ship *Sheffield*, on which he would serve almost five years, returning in May of 1859. One can imagine the dialogue at the time between the young whaleman, his wife and in-laws for the couple had been married five years, yet Enos had been present just five months. Moreover, with the approaching end of whaling activities at the port, continued participation in the industry would have necessitated his sailing from some other location. So, perhaps with the assistance of Susan's family, Enos opened a store. And he witnessed the birth of his daughter Melna.

A significant number of whaling voyages (126) were conducted out of the port of Greenport, on the North Fork of Long Island, from 1824 to 1860. There is little record of the Portuguese participation except for one man. Manuel Claudio first arrived in Greenport in 1854 aboard the whaleship *Neva*, and continued to sail until 1870. He then decided to settle down and found Claudio's Tavern, now Claudio's Restaurant—recognized by the National Restaurant Association as the "the oldest same family run restaurant in the United States."[78]

Numerous smaller ports existed along the Northeast seaboard, but Portuguese did not gravitate to many of them in significant numbers. The port of Lynn, Massachusetts, made thirty-eight voyages between 1830 and 1857.[79] One of the captains was Antonio Da Costa, who commanded the schooner *George* on its voyage of 1835.[80] San Francisco's time was yet to come, as it only sent out its first vessel in 1851.

The Port of Fall River

New Bedford's close neighbor, Fall River, caught the whaling fever quite late, in 1832. Ten years later, among its three sailings was that of the ship *Holder Borden*, on its ill-fated maiden voyage under the command of Capt. Jabez J. Pell. In the crew were fifteen Portuguese. The vessel arrived at Honolulu in March of 1844 with 800 barrels of sperm oil and

700 of blackfish and departed the following month. Antonio Dean, of Flores, was second officer; and two boatsteerers were also Portuguese. In fact, the makeup of the crew resembled those at the end of the century: eleven Portuguese seamen in addition to the second mate and boatsteerers, and less-experienced American seamen from the Northeast states. Less than a week after leaving port the *Holder Borden* went aground upon a shoal on tiny Lisianski Island, subsequently known for a time as Pell's Island, some 900 miles northwest of Honolulu. There they found water, seals, turtles, and ducks and began to construct a schooner that would carry the bulk of the crew back to Honolulu that September. After purchasing the brig *Delaware* there, Captain Pell sailed it back to the island to pick up the remaining crew members and the oil, finally arriving in Fall River in July of the following year.[81]

The Sealers

As has been mentioned, nineteenth-century voyages out of New London area ports such as Bridgeport, Groton, Mystic, Norwich, Stonington, and New London, often had a dual purpose of whaling and sealing. In fact, it was quite common for whaling to be only ancillary to the principal object of the voyage: obtaining skins from fur seals and oil from elephant seals in the southern Atlantic and Indian oceans. Historian Briton Cooper Busch has estimated that one-fourth of so-called "whaling vessels" leaving New London from 1840-90 went in search of elephant seals.[82] Another port that often sent out sealers was Sag Harbor, inspiring James Fennimore Cooper's early maritime novel, *The Sea Lions*. And some of these voyages sought not only seals, as can be seen from "Capt. Edwin Smith's instructions and directions in regard to taking Penguin and Penguin Oil."[83]

Sealers traded the relative hardships of life and activity aboard the whaleship for that of spending time on remote, frigid islands somewhere north of Antarctica. Tierra del Fuego, South Georgia, and the South Shetland Islands formed one preferential region for the sealers. The other, in the South Indian Ocean, was comprised of the Crozet and Kerguelen islands, the principal island of the latter commonly known to whalemen and sealers as Desolation. Another common destination was Heard Island, south of Desolation.

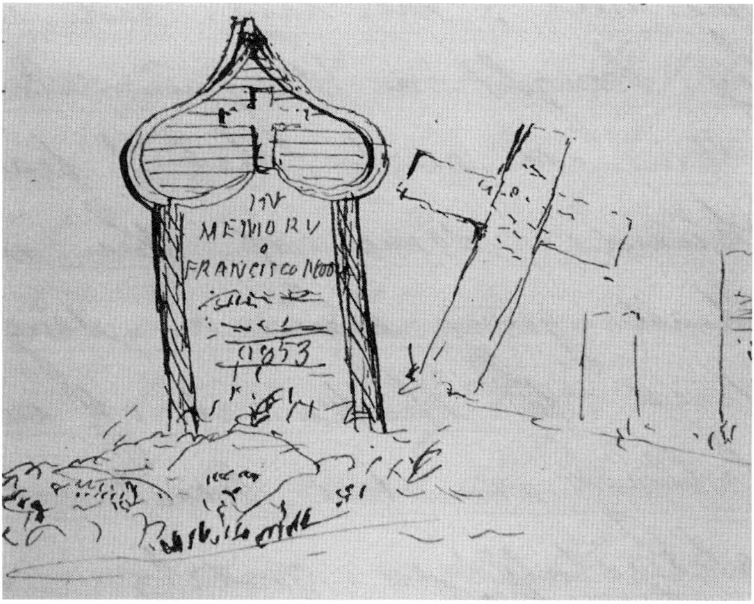

Grave marker for Francisco ——, Heard Island. Sketch by Charles Kennon, ship *Laurens*, New London, 1855. Courtesy of the New London County Historical Society.

Crewmen would be dropped off at these remote spots with provisions and then, if all went well, retrieved a month or perhaps a season later. Schooners were preferred, as they could run closer into shore on

Pot Harbor, Desolation Island, ca. 1852, from Taylor, *Life on a Whaler.* Courtesy of the New London County Historical Society.

these rocky islets. Theodore Lewis, the cooper aboard the ship *Atlantic* of Bridgeport, Connecticut, in 1835-37, described in vivid detail in his journal what it was like "to be left alone upon a solitary rock for weeks together with provisions enough to last half of the time," with waves from the frequent gales breaking over and threatening to wash men and provisions into the sea.[84]

After calling at the Azores and Cape Verde, the *Julius Caesar* finally arrived at Desolation Island. The climate there, north of Antarctica, is harsh. Rain or snow falls most days of the year. Lying in the path of the "Furious Fifties," its winds blow constantly, while waves may reach forty or fifty feet in height. However, the seas around the island are ice-free year around, and its many fjords offer protection from extreme weather.

Nathaniel Taylor describes the activities there:

> The object of the present assemblage of the tenders in Pot Harbor was to make arrangements for working the bay and to place men upon the beaches for assaults upon the herds of sea elephant already beginning to leave the water to shed their coats upon the land. From these animals we expected to obtain twelve or fourteen hundred barrels of oil. Hell's Gates, upon the windward side of the island one hundred fifty miles from the ship's anchorage, was fixed upon as the destination of the tender, *John E. Smith,* and four months were allotted to her as sufficient to procure seven hundred barrels of elephant blubber.
>
> The side of the island to which she was assigned is represented to be so savage and repulsive as to make the stoutest heart quail and despair. Its beaches, washed by the unbroken waves of the South Pole and pounded by unceasing storms, are strewn with large fragments of icebergs and masses of volcanic rocks....
>
> Portions of the crew, furnished with provisions, are to be landed on such shores as can with safety be reached. There, by clambering rocks and wading mountain torrents, they are to penetrate to these last haunts of the elephants. They are to drive them day after day and week after week from the rocky fastnesses, and then, after killing them, retrace their steps, heavily laden with blubber. Others of the crew are to be stationed at intervals of fifteen miles, where, living alone in huts of turf, they can work the beaches seven or eight miles in each direction.[85]

José, one of the six Azoreans picked up by the *Julius Caesar* at Faial, became a favorite of the crew while at Pot's Harbor on Desolation. But a moment of carelessness almost cost him his life. To Nathaniel Taylor, he was "a merry laughing, bright-eyed boy sixteen years old who tends the cows and does other chores."[86] José was very content with his new-found responsibilities on shore which included assisting the cooper. Evidently the similarity to the life to which he was accustomed on Faial was especially enticing. As Taylor observed, "if we may judge from an expression of his own, 'Fayal no star bon; Desolation ver gude', we should pronounce him the most contented person at the island. He follows the cows around, and with the dog, Fanny, at his heels, he whistles and whiles away the hours as cheerfully as if on his native hills...."[87]

One day, while the weather was perfectly calm, and ignorant, perhaps, of the fact that conditions in these latitudes could change in a moment, José took a little skiff and set out in search of fodder for his cows. No sooner had he made it out of the harbor than a violent squall came down on the island. And so, in the clouds of snow blown off the mountains and the spray raised up from the growing waves, little José disappeared from view—presumably never to be seen again. After five days the storm abated; and that morning one of the men on deck shouted out that he had seen the young Azorean on shore, waving a handkerchief. Upon hearing this the men rushed on deck, but there was no José to be seen. Another crewman summed up the generalized feelings: "You mistook a king penguin for little Jose's ghost. Now the poor boy is gone, I move we drop the subject."[88]

Later that day, however, as the crew went ashore to assess the damage, there in the cooper's hut was José, "pale and wan from hunger and damp."[89] As all conjectured as to how he had survived at sea and made his way back to the harbor, José kept repeating, "Desola-shn no gude; Fayal mucher better. Kelp 'sta bon.... De kelp, oh, yes de kelp much-er gude."[90] And so it turned out that José's little boat had gotten caught fast in a free-floating bed of kelp that sustained him and calmed the waters, allowing him to reach land twelve miles away and where he subsequently hid in a cave to protect himself from the elements. As Taylor reminisced, "Jose soon recovered from this adventure; and notwithstanding his assertions, 'Desola-sh-n no gude,' his attachment to the cows had so increased that when we sailed for home, having sold them to a ship which then arrived, he insisted upon remaining behind as their companion and keeper. The elf-like glance of his brilliant black eyes still sparkles in my memory."[91]

The tale of Little José and its fortuitous ending is touching and probably more common than the literature would suggest, for it is the stories of bravado and even brutality that tend to dominate the canon of whaling literature. More representative of this tendency is Taylor's narration of an incident that occurred on the schooner *Marcia* of New London—Simeon Church, master—which was also at Desolation at the same time, along with eight other vessels from the same port, including Taylor's own ship:

> A favorite expression of Captain Church during his outward-bound voyage had been: "Hurry up your cakes, or I'll pitch you overboard." Now this threat contained nothing of fearful import to those of the crew who knew the humor of the man, but it happened, when at the Azores, that several Portuguese had shipped in the *Marcia*, one of whom, a fellow of some education, had a smattering of the English language. Evidently he understood English better than he did human nature, for he was so frightened on one occasion when our worthy captain made use of this expression to him that he took the opportunity, as soon as evening came, to anticipate the threatened plunge by jumping overboard. The poor fellow was never seen afterwards, and must have perished, a victim to his fearful forebodings, a warning to that class of mankind who are ever borrowing trouble. After this occurrence Captain Church was in the habit of advising the men in a tone of mock politeness to jump overboard if they could not work more lively.[92]

New London sealers also prosecuted their business in the North Pacific, especially in the Sea of Okhotsk and at the Pribilof Islands. The *Whalemen's Shipping List* reported in November of 1854 that three New London sealers had returned from the former with respectively, for the schooner *E. L. Frost*, "7000 fur seal skins and sixty bbls. of seal oil, worth nearly $30,000" (ca. $750,000 in today's dollars); for the brig *Zoe*, "10,000 skins and a proportionate quantity of oil, worth about $40,000" (ca. $1,000,000); and for the bark *N. S. Perkins*, "13,000 skins, and 600 bbls oil—a part of the latter whale, worth $60,000" (ca. $1,500,000).[93]

The paper went on to observe that "with such immense slaughter, of 30,000 seals upon a single island [Robin Island, thirty-five miles in circumference], the stock has, of course, been greatly diminished." Yet it confidently predicted, "it is believed there are other places in the Northern Ocean where the seals resort in large numbers."[94] And so the slaughter continued.

Rebellion among the Crew

> At sunrise he summoned all hands; and separating those who had rebelled from those who had taken no part in the mutiny, he told the former that he had a good mind to flog them all round—thought, upon the whole, he would do so—he ought to—justice demanded it; but for the present, considering their timely surrender, he would let them go with a reprimand, which he accordingly administered in the vernacular.
>
> 'But as for you, ye carrion rogues,' turning to the three men in the rigging—'for you, I mean to mince ye up for the try-pots;' and, seizing a rope, he applied it with all his might to the backs of the two traitors, till they yelled no more, but lifelessly hung their heads sideways, as the two crucified thieves are drawn.
>
> (Herman Melville, *Moby Dick*, Chap. 54)

Dissatisfaction among the crew might arise from several causes, two of the most common being mistreatment by the mates and displeasure with the fare on board. Another was the deep sense of exploitation, as can be seen in the following episode. Its most common manifestation was one of intransigence on the part of an individual crew member. More serious was the open sharing of these concerns, leading to a common refusal to obey orders, technically a mutiny. The most serious crime aboard a vessel was the concerted attempt by members of the crew to take over command.

By mid-century voyages were extending to four and even five years. While this was due in part to the difficulty in filling the ship, there were other factors involved. One was the growing practice of transshipping oil and whalebone from Hawaii on merchant ships or whaling vessels headed home with less than a full hold. Another was the tendency of whalers to engage in inter-island trade, especially in coconut oil.

While in the port of Honolulu in May 1854 the bark *Belle* of Fairhaven, under the command of Ichabod Handy, discharged almost 8,000 gallons of coconut oil, as well as 1631 gallons of sperm oil. Presumably in preparation for more trading, the captain also took on board 170 boxes of tobacco.[95]

But thirteen of the crew, several of them Portuguese, refused duty, presumably because of the change in mission of the vessel and the resultant doubts about who would share the profits from the trading. With the skipper ill, the mate, who had been ashore, related:

> When I Came on Board I Called all hands aft and asked them what the Mater was Some said they wanted to leave the ship some one thing and Som an other.... I told them thay must turn too or I would put them in prison on Shore and they said they would go to prison and I Sent Mr. Stowell [the second mate] on Shore after proper athoroty & two poleasman Came on board and they asked them if thay was again to turn too or go on Shore and thay Said on Shore and thay went on Shore.[96]

The mate at that point was forced to hire a group of natives to break out the coconut and sperm oil. Almost three weeks later, nine of the men who had been held in the fort were returned to the ship, still in a mood of rebellion. Among them was Joseph Silva. When the mate inquired if they were ready to go to work, they replied in the negative, whereupon the mate "put Them all in hand Cuffs and put them Down in the Bluber room." Two days later the captain ordered several of the men to be fastened to a midship stanchion. At that point Silva asked to speak with the captain, explaining that he had been sick and was sent to the fort mistakenly. The captain had little sympathy for Silva's plight: "Manifesting a Mutinous Dispersition the Capt. ordered him below again in irons." On May 31, Captain Handy had had enough. Calling the prisoners on deck he placed them in the rigging and commenced to lash each one until he acquiesced. As the mate explained, "the Capt. Said if I cannot bring You into good subordination by good treetment and kind words I will thrash You till you will be good and quiet Men." Without enumerating their punishment the mate noted that "Andrew Bond and Joseph Silva also acknowledged there faults and promised good behaviour for the future—and they were turned to Duty."[97]

No mention is made in the log of the type of lash used; but since 1850, at least, the cat-o'-ninetails had been outlawed. Flogging among the whaling fleet continued, in fact, at least until the end of the Civil War. Even apparently mild-mannered Joseph Dias Jr. did not hesitate to use the lash. While master of the ship *St. George* of New Bedford in 1854, he was awakened at four o'clock in the morning "by a noise on deck went up and found the mate and one of the hands clenched in a fighting found that the mate had asked him why he did not turn out when the watch was called began some Insolence and declared he would do as he pleased about it parted them and put Monegan in the riging at 7 gave him two Dozen and Sent him to the mast head."[98]

Dias, a resident of Martha's Vineyard, was born on Christmas Day 1822. He was the son and grandson of the two Joseph Diases before him. Like these two he seemed destined for the sea, shipping first as cabin boy on the ship *Clifford Wayne* of Fairhaven in December of 1836, just as he turned fourteen.[99]

A strong motive for the master of any whaling vessel to exert his authority was the possibility of a concerted action against him—in other words, a mutiny. Almost all of these were hatched in private by the conspirators, but they could take place spontaneously, as Captain George N. Macy and the first mate of the bark *Plover* of New Bedford feared, after the mate ordered the watch to work one evening:

> one Porteguis by the name of John Forth did not move. I then went forward to see the reason wil he said in an insolant manner there was a nuf man gon after that I orderd him to go & help he refused & I tried to make him go he then struck at me severel times Capt Macy there upon came to my assistence were upon the said forth got old of Capt macy by the wisker & scrached his face in avery bad maner. John forth then cald out for Manuel forth to come to his assistence Manuel forth cauld out at the gangway for all the Pirtigues to come up & pitch in & give the oficers & Captain a trashan but no body came but Manuel forth were upon [whereupon] Mr. Craft [3rd mate] stopt him & they ad to back down Capt Macy orderd them to be put in the ryun [run] & he floged them & put them in irons. They then bagged off & said they would not resest a gain – were upon he (Capt) orderd them out of Irons & go to duty so ends the day.[100]

In February 1847, a Portuguese by the name of Manuel Lorenzo sued Captain Henry Manter of the ship *Pocahontas* of Holmes Hole, Martha's Vineyard, and two of its mates. After a confrontation between boatsteerer Lorenzo and the third mate during the voyage of 1844-46, Manter had Lorenzo seized up in the rigging and struck him five or six times with a rope's end. After some five or six hours had passed, the captain again gave him five to eight more blows with the rope before cutting him down. After hearing the arguments the judge of the Court of Common Pleas in Boston ruled for the plaintiff and ordered Manter to pay him $20. The *Boston Star* conjectured that the judgment was "Just enough to carry costs—probably on the ground that for the offence of insolence to the mate and fighting with him—although a flogging might be proper enough, yet that it savored of cruelty to leave the sailor tied up so long."[101]

At times having recourse to the courts might be utterly futile, as one of the boatsteerers on the ship *Callao* of New Bedford during its voyage of 1855-58 discovered. Antonio Machado found himself ill, and wishing to escape the oppressive atmosphere below decks, came above for some fresh air. Upon being ordered to go below by the mate, he refused, whereupon the mate began to beat him severely about the head and body with a bung driver, as he was being held by the second mate. All the while Captain Alden Howland stood by shouting words of encouragement to the mates, "Give it to him, kill him." Machado filed a libel in the U.S. District Court of Boston immediately upon the vessel's return, asking for damages of $500. Unfortunately nothing further materialized, and in 1862 the case was dismissed for want of prosecution.[102] Much later, another Machado, with the help of a Portuguese American attorney, was able to collect damages for mistreatment aboard the *Canton II*, under Nicholas R. Vieira. Evidently threatening to go to court, Manoel Machado da Silva reached an agreement with the vessel's agents in which he acknowledged having "received full satisfaction from an alleged assault by Captain Nicholas R. Vieria [*sic*], on the whaling cruise after leaving Fayal Azore Islands in Fall of year 1905 and arrival at Dominica, in year 1906."[103]

One of the more egregious examples of punishment took place on the ship *Canada* of New Bedford during its voyage of 1846-49, as reported in the journal of the cooper, Sylvanus C. Tallman. The problems began on March 5, 1847, as the vessel was cruising off the Galapagos Islands, when Captain William H. Reynard seized the steward, William Persons, in the rigging for dirtiness. The next day it was found that the cutting stage had been cut away, and a green hand by the name of William Freeman was put in the rigging on suspicion. The next day more damage was done; this time the jib and flying jibstay were cut, and another man was put in the rigging. The following day the mainstay was found cut. This time the captain called all hands aft and questioned them. Suspicions fell again on the steward, who denied any complicity. But after being put in the mizzen rigging and flogged, he confessed the plan for a mutiny. Following this he was put in irons. But the damage continued. A week later the jib and other stays were found cut. The steward, "after persuasion," confessed, but stated that he had been assisted by seaman Manuel Silva. At this point the captain evidently surmised that examples were needed in order to avoid an outright mutiny. And so Silva was put in the rigging and given 18 lashes. But the real punishment was meted out again to the steward, who was placed in the rigging and given "about 250 lashes." Subsequently the captain "took him down and hauled him up in a cask, and stowed down." Eleven days later, on July 29, "in the morning took the steward out of the Cask for an airing at 7 PM put him back again and took in sail." Subsequently, on the third of September, he was again taken out for an "airing." A week later, while the vessel was anchored at Lahaina, steward Persons was taken ashore.[104]

Resistance of the crew members to a situation they judged intolerable at times took a more violent form, challenging the captain's authority and leading to mutiny. Such was the event aboard the ship *Rousseau*, under the command of John C. Brayton, of New Bedford, in November 1844. Having sailed in April 1841, the vessel at the time was in the Pacific bound home toward Cape Horn when one of the foremast hands, a twenty-year-old Portuguese named William Swain, who had been signed on at Maui for the return voyage, became delusional and "very unruly," demanding to know why he could not be master of the vessel. On the orders of the captain the mate put the man in irons in the ship's run. Later, in checking on the prisoner it was discovered that he had crawled into the forecastle, from which the crew—evidently aware of the young man's disturbed mental state—refused to give him up. Informed of this, Captain Brayton headed down with a loaded musket and two pistols and told the men that he would shoot the first one who attempted to stop the officers from removing Swain. As these attempted to take the prisoner out of the forecastle, Frank Sylva, one of the foremast hands, grabbed the musket and pulled it out of the captain's hands. It was then a free-for-all: Edward Gillman, a Yankee from New Hampshire, leaped on Brayton's back, while an Englishman, George Baker, pulled at his hair. Then they hauled him down on the windlass, beating him in the face.

It should be noted that this resistance to Capt. Brayton and the subsequent treatment of him were most egregious offenses under maritime law. Finally, some nine seamen hauled him into the forecastle and kept him there for a couple of hours before releasing him, presumably after he had agreed to forgive the entire episode—which, most uniquely, he did—for it is not brought up again in the log.[105]

This was not the only bizarre episode on the voyage of the *Rousseau*. Earlier that year, while the vessel was lying off Nuku Hiva in the Marquesas, the crew was busy wooding during the day and going on liberty at night. A reader of Melville's *Typee* would understand that doing business on the beach with these Polynesians at the time was one thing, but socializing with them quite another. And so it was that a scuffle ensued one January evening, resulting in the serious wounding of one of the crewmen and the abduction of the Portuguese cooper, Antone Francis. For two days nothing was heard of his fate, and then the keeper noted, in the typical succinct prose of a logbook, "The sad news has come That thay have eat the Cooper."[106]

Reflecting on some of these episodes at the completion of the voyage, the keeper penned these verses before turning over the log:

> Misfortunes comes from Heaven
> And why should we complain
> For we have done our best endeavor
> A cargo to obtain.[107]

The solidarity of the men in the forecastle, despite their disparate backgrounds, is patent in the following episode from the voyage of the ship *Daniel Webster* of Sag Harbor in 1848. In January of that year, one of the crew members wrote the following letter, evidently intended for an American consul, describing a scene of violence to which a number of ascribers took exception:

> This morning while setting sail the mainsail was loosed by a Portuguese named Manoel Joaquin who through mistake neglected to fasten one of the gaskets for which he was again sent aloft by the Sec. mate who told him at the same time that he had ought to have the end of it, whipped up about his back. To this he replied that it was only a mistake and the officer told him to shut up which not being immediately obeyed when he came down he was caught by the hairs (by the afforesaid officer) snatched down on his knees struck several times in the face (which set him bleeding at the nose) and he also kicked him several times
>
> Written by Charles F. Chester W the request of Manoel Joaquin
> This affair was witnessed by individuals the names of whom are here inserted.
> [signed]
> Charles F. Chester
> Selah Oakley
> Conkline Raynar
> George Hess x Jason Purdue x Steward
> Sidney Smith Henry M. Bennett
> Allan Wilkie
> Samuel P. Field Third mate
> Charles H. Hardon Boatsteerer
> This officer's name is George S. Tooker[108]

In order to avoid such complications owners might advise the avoidance of ports where there was consular representation. When he sailed from St. Helena in March of 1858 aboard the brig *Leonidas* of Westport, Captain Rescom Borden wrote to his owners complaining of problems with the American consul there. Two months later they wrote, counseling "if possible to keep away from all consuls, as they will make trouble & expense and your crew will be more likely to trouble you, if they have a Consul to run to. I would advise you to go to such places for recruits & Water, and what you need, where there is no American Consul."[109] While it is true that American crewmen might run to the American consul, it was more common, unfortunately, for these poorly paid officials in remote corners of the globe to consider first their purse strings and side with those who might help them monetarily, i.e. the masters and owners.

Portuguese and Cape Verdean seamen were not necessarily victims of unwarranted punishment, as the following extract from the log of the bark *Three Brothers* of Nantucket in 1848 suggests:

> At 4 PM. Nat Currier, 1st officer, ordered Antonio Puito (a Portuguese seaman) to grease the masts and he refused to obey. When Mr C— took hold of him and repeated the order, and the man struck him a frightful blow in the face cutting and bruising his eye badly I then interfered and saw Mr C.s order obeyed, nothing more occurred till 6 P.M. when Mr C— gave the order accordingly, he again refused when I ordered him placed in irons—A scuffle then ensued, where all hands came aft, the said Antonio Puito drawing his knife and making several desperate plunges at Mr C— and myself both of us narrowly escaping with our lives by flying to the cabin I then armed myself with a loaded pistol, called all hands aft and told them the man must be placed in confinement Accordingly without further resistance placed him in irons not considering the lives of myself and officers in safety with such a Desperado at liberty.[110]

Several days later the log keeper added, "The Portuguese seaman still in confinement, he appears very sulky and not inclined to speak He is without a doubt a *most desperate villain*."[111] Nothing further is mentioned concerning Puito, suggesting that some quiet time in irons was in itself sufficiently persuasive to control this Portuguese seaman. One notes that this more benign approach was more typical than the previous egregious examples.

An Odiferous Cargo

Just as whaling vessels in the Pacific might take time out for a trading cruise, carrying coconut oil to Honolulu, for instance, others might pick up a less pleasant cargo, guano, highly valued in Europe and America as a fertilizer. In fact the U.S. Congress passed in 1856 "An act to authorize protection to be given to citizens of the United States who may discover deposits of guano," referring in particular to dozens of islands in the South Pacific. Within three years, according to the *New Bedford Daily Mercury*, nearly fifty islands, mainly in the Pacific, were "discovered" and reported—and essentially acquired by the U.S.[112] The extent to which whalers were engaged in the trade, apart from the aspect of discovery, is unclear. But those passing along the west coast of Africa, particularly New London vessels returning from a whaling/sealing voyage to the Crozets or Kerguelen, were often tempted to fill any vacant space with guano.

The topsail schooner *Emeline* of Mystic, returning from an elephant-seal hunt in the Crozets in early 1844, sprang a leak and barely made it into Cape Town, South Africa. After selling his oil and repairing the vessel, Captain William Eldridge was given sealed orders by the U.S. Consul, to be opened after leaving port. The vessel was directed to sail for tiny Ichaboe island off the coast of Namibia to load guano, a location unknown to American vessels at the time. There they found seventeen British ships loading guano. Later the *Emeline*'s cargo was sold in the West Indies on the return home. To give some idea of what it must have been like to carry this bird waste across the seas, a later notation in the schooner's log observes that in 1912 this same document "was still permeated with the fumes of guano."[113] The following year, the *Emeline*, again on a sealing voyage, returned to Mystic with 500 fur-seal skins and 75 tons of guano.[114]

In December of 1844 an article appeared in the *People's Advocate* of New London, relating an experiment comparing the productivity of ashes as a fertilizer with that of guano, the latter producing two and one-half times as many bushels of turnips.[115] From then on it was common for the sealers and an occasional whaler to return with a partial cargo of guano.

A curious incident took place on Ichaboe Island around this time. The crew of the British ship *Colchester*, while scraping guano there, discovered, about six feet down, a mummified body. Accompanying it was a stave from a flour barrel identifying the man as Christopher Delano, a Portuguese sailor aboard an American whaler, who was buried there in 1791. His tongue was "as thin as a wafer… and the nose is slightly flattened, but, in all other respects, the body is perfect." It was subsequently carried back to England. A few years later the hair of this "petrified man of the Island of Ichaboe" was displayed in New Bedford, "in a state of perfect preservation."[116]

The California Gold Rush

> I write to say that the fever for California "Gold digging" is all the rage now, and thinking it would derange you people in the same way. I think it would be well to keep clear of the coast, recruit at the Marquesas till the rage is over.
> (Whaling agent to ship's master, 1849)[117]

Since at least the 1820s, American whaling vessels had been touching at Mexican California. Already in 1833 it was announced that the port of Monterey, on the central coast of California, had been opened as a free port for whalers.[118] And by the early 1840s, a good number of American whaleships began to anchor off Sausalito, inside the Golden Gate, in what is now known as Richardson Bay. There they were welcome to supply themselves with water from the springs of former English whaling captain, William A. Richardson.

The discovery of gold in California in January of 1848 soon brought an influx of gold seekers, principally beginning the following year—thus the term "49er." It was no longer a time for American whaling vessels to call at San Francisco, for they risked quickly losing their crews. The apprehension of whaling captains can be gleaned from the diary of Joseph Dias Jr. of the bark *Ocmulgee*. The *Ocmulgee* was one of the first whaling vessels to enter the Arctic Ocean. It had left Holmes Hole on the second of September, 1847, with Joseph Dias as mate. Dias was a man given to reflection and irony, and his journal reflects sundry feelings of nostalgia, hope, and regret. Two days after sailing, he wrote, "Some light hearts and Some heavy ones and about 20 green hands to train which is a quite a job for a poor devil that is as home Sick as I am." And, as he points out, "this is not ment for an every day journal but mearly for the owner."[119]

One of the curiosities of nineteenth-century whaling was the way in which sailors kept in touch with relatives and friends at home. At times there was a mail box located on a remote island of the Pacific in which letters were deposited or retrieved. Whaling vessels often carried mail intended for others; at times the letters actually arrived at their destination. Writing in 1843, the Rev. John Abbott of Nantucket noted, "I asked a lady the other day, how many letters she wrote to her husband during his last voyage. 'One hundred,' was the answer. 'And how many of them did he receive?' 'Six.' The invariable rule is to write by every ship that leaves this port, or any other port that can be heard from for the Pacific Ocean."[120] It was in this manner that mate Dias received a letter from his sweetheart at home. On the sixteenth of December, some three and a half months out of Holmes Hole, the *Ocmulgee* spoke the *Liverpool II* out of New Bedford. There he was handed a letter from Mary, unfortunately without "much news in it."[121]

As the New Year found the ship near the Cape of Good Hope, the crew informed mate Dias that they had about fifteen signatures on a temperance pledge and another seven on a pledge to avoid swearing.[122] And, of course, there was the inevitable disciplining. On the twentieth of February in the "after noon [I] boxed a thing's ears for not minding his business while tacking Ship in a Strong breese."[123] Nevertheless, it was an especially tranquil voyage. Upon departing their first port of call, Mangonui, in northern New Zealand, which they had not reached until almost six months after first sailing, Dias marveled that "Every man that left home in the Ship and not one has even made an attempt to leave a very uncommon thing for the first poart."[124]

But 1848 was destined to be a notable year in the history of the United States; and, as the news of the discovery of gold in California began to spread far and wide, all aboard the *Ocmulgee* must have entertained thoughts of leaving the ship for the goldfields. The *Ocmulgee* had just departed from the Sandwich Islands, amid the rumor that the ship was headed to San Francisco. However, even the mate himself had not been informed of the vessel's destination. "But the Lord only knows whare we are bound," he wrote.[125] After sailing for nine days in the general direction of California, the captain ordered the ship turned around; and once again they were headed for Oahu. "So good bye *California* and the Gold mines. You can Se by this that we had Some notion of going to California but Something was droped by Some of the crew which made the captain abandon it and perhaps it is all for the best but it is rather hard to think So."[126]

As it was for California and the settling of the American West, 1848 was also pivotal for the nation's whaling industry. The California Gold Rush, which drew scores of vessels and thousands of men to a new adventure, was the first of several factors leading to the industry's decline. In the frenzy to reach the goldfields whaling vessels were chartered or purchased in their home ports and carried off much of the local elite and more adventurous. When the whaling bark *Russell* sailed from Nantucket in March of 1849, carrying over forty residents of the island, it included among the passengers John Praro and William Sylvia, both Azoreans.[127]

The *Whalemen's Shipping List* reported that during the year of 1849 almost fifty vessels had sailed from New Bedford for San Francisco, virtually all—presumably—loaded with eager gold seekers. It has been estimated that some 800 individuals sailed from Sag Harbor at the height of the Gold Rush. Typical of the frenzy at the time was a letter from a seaman aboard the ship *Massachusetts* of Nantucket, dated 16 June 1849. According to the writer, Captain Seth Nickerson had decided to go to the mines instead of proceeding on a whaling voyage. He would take all hands with him and give the crew two-thirds of the gold they found. But evidently cooler heads eventually prevailed, for by October of that year the vessel was cruising for whales in the central Pacific.[128] Many whaling vessels did, however, end their days in San Francisco Bay, as entire crews deserted for the goldfields.

One of them was the famed ship *Niantic* of Warren, Rhode Island, rediscovered in landfill during construction in the waterfront district of San Francisco in 1978. When Captain Henry L. Cleaveland arrived at Paita, Peru, early in 1849, he heard of the discovery of gold in California and also received instructions from his owners to delay whaling and proceed to Panama to pick up passengers for San Francisco who had managed to cross the Isthmus and were waiting for transportation. Cleaveland managed to accommodate 246 men eager to reach the goldfields, charging them $150 for steerage and, for a few, the relative luxury of cabin space at $250 (some $4000 and $7000 respectively in today's currency). One of the steerage passengers, F. D. Almeida, was evidently either Portuguese or Brazilian.[129] When the *Niantic* reached San Francisco after a sixty-six-day passage not only did the passengers depart, but the crew began to melt away as well. So the *Niantic* was soon beached and ended its days being utilized variously as a storehouse, a hotel, and offices, until it was destroyed in the San Francisco fire of 1851.

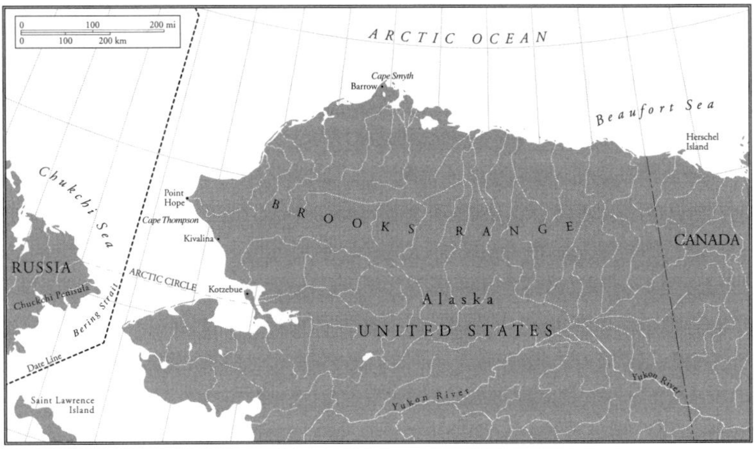

Western Arctic

Into the Western Arctic

Over-hunted and increasingly shy, whales were becoming much harder to find. It was the daring exploit of one man, Captain Thomas Welcome Roys, of Sag Harbor, that brought another half-century of success to the industry. In July of 1848, after an unprofitable cruise about the Crozets and Desolation, and then north around the South Seas, Roys—against the wishes of his owners and the fears of his crew—took the bark *Superior* where no other whaler had ever dared, into the Arctic Ocean.[130] As author John Bockstoce has observed, "Roys's cruise was not only the most important whaling discovery of the nineteenth century, it was also one of the most important events in the history of the Pacific."[131] Roys's account of his success with the bowhead whales he had discovered in the Arctic was published that fall in the Honolulu newspaper *The Friend*. Consequently, the next spring there was a race to the new grounds, and the *Ocmulgee* was one of the first to arrive. There, as Joseph Dias noted, they saw "whales in every direction" and were soon "blubber logged."[132]

It was now September of 1849 and the gold fever was reaching a crescendo. In the *Ocmulgee*'s passage south into the North Pacific, after a very successful season in the Arctic, Capt. Frederick Manter gammed with the master of an English schooner bound from Hawaii to San Francisco. From him they "heard grand news from *California* wich brought the Feaver out afresh. I Say afresh because it has not been broke intirely at any rate it has bothered me Some for the last 9 months."[133] October found the *Ocmulgee* in the port of Lahaina on the island of Maui, and again Dias contemplated the goldfields: "As I have had some notion of leaving the Ship to make or rather try to make my fortune in *California* I have been ashore to make Some Inquerys but I find no one to give me any incouragement that has been thair but those that have not been are all very anxious to get thair. The Feaver rages So high that they are almost leaving Ships destitute of Officers and men but I am inclined to think Some of them will be Sorry but not till it is to late."[134] In the end, probably due to a somewhat conservative nature, as well as loyalty to his captain, Dias held back from abandoning the *Ocmulgee* and seeking his fortune in the goldfields of the Sierra Nevada.

Within a few decades San Francisco was destined to become the principal port of the American whaling industry. But little activity took place during the height of the Gold Rush. The first two vessels to sail from San Francisco Bay destined for a whaling voyage were the barks *Popmunnett* of Fairhaven and the *Sarah* of Edgartown. The *Popmunnett* was on its way to whaling grounds in the Pacific when it called at a South American port for provisions. Soon the crew had deserted; but Captain Charles L. Eldridge did not have any trouble once he announced that the vessel, now loaded with merchandise, would be headed for El Dorado. It was easy to lose a crew in San Francisco, but not impossible to find disillusioned mariners who had found that "striking it rich" was not as simple as they had imagined. And so with a new crew of whalemen, the *Popmunnett*, after unloading its passengers and freight, once again set off whaling. After depositing the gold seekers it had carried from the East Coast, the bark *Sarah* of Edgartown soon followed.[135]

But neither of these whaling voyages truly originated in San Francisco. That would be the role of the bark *Russell* and the ship *Nile*. They sailed in April and June 1851, respectively.[136] As Lloyd Hare has observed, "The eastern shipowner perceived no challenge in either the *Russell* or the *Nile*, two old tubs sailing out of the Golden Gate into the setting sun bound for the Arctic's icy horizon. And yet little San Francisco, 'knee britches' in the whaling business, was one day to supersede the Colossus on Buzzards Bay" — at least briefly.[137] In spite of this early activity, very quickly it became apparent to whaling captains in the Pacific who wanted to pursue their trade that San Francisco must be avoided at all costs. And so, during the next two decades Honolulu and Lahaina became the major ports of call for American whalers in the Pacific.

But another industry soon developed in the region: shore whaling. In 1854, ex-whaleman John Pope Davenport gathered up a crew of Portuguese at Monterey, California, and commenced shore whaling. Within a short time stations were established along the entire length of the California coast, from Crescent City in the north to San Diego. Shore whaling along the California coast lasted until the end of the century and was dominated by the Portuguese. David Bertão, in his comprehensive study of this industry, has identified twenty-six stations that operated during the second half of the nineteenth century.[138]

The Wide Pacific

During the 1840s and 50s, the great majority of American whaling vessels depended on the Hawaiian Islands as their field base in the Pacific. In that first decade, the most popular port for whalers was that of Lahaina. In 1846, 429 whaling vessels refitted there, while 167 preferred Honolulu. During the 1850s the numbers reversed. In 1853, for example, 246 called at Honolulu, 177 at Lahaina, and 110 at other Hawaiian ports, principally Hilo.[139]

While the ship *London Packet II* of New Bedford was whaling off the coast of Australia in March of 1839, several men deserted at Kupang, Portuguese East Timor. One of the deserters was given thirty-six lashes by order of the governor, upon which the log keeper noted, "his sentence was from the Govenor 100 but Capn Samson begged him off with 36—After getting him on board he had 20 fits in the course of an hour owing chiefly to his eating raw corn which he had lived upon while in the bush—We done all in our power to releive him but all to no purpose."[140] A few days later the log notes, "Four hands off duty—Sent the boat on shore last night on duty and one Portuguese Antonio deserted from the boat which will detain us another day if not more." And the following day the mate noted in frustration, "And oh that Portuguese If I had a towline round his neck—I'd I'd I'd him." And so the ship stood out to sea, "Leaving the portuguese behind to lookout for himself," which was probably just fine for the crewman

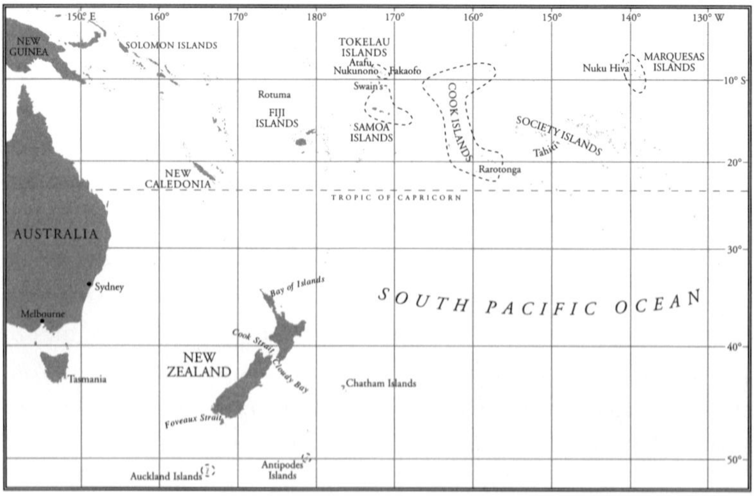

South Pacific

in this far-flung Portuguese colony—if he could just avoid the Governor's justice.[141]

In ports that were commonly frequented by American whaleships, such as Talcahuano, Chile; Paita, Peru; Mangonui, New Zealand, and many others, it was common to find Portuguese sailors awaiting a voyage out or even, perhaps, thinking about settling down. For example, the bark *Alice* of Cold Spring Harbor, while at Tasmania (then known as Van Diemen's Land) in 1852, took on Manuel and Charles Sylva as seamen, and while at Hilo in March of the following year, signed on Francis and Philip Silva as ordinary seamen for the season, discharging them in Honolulu that November.[142]

American whaleships had begun to frequent New Zealand in the 1820s, preferring the growing white settlements at the Bay of Islands on the North Island. By the mid-1830s it was a favorite provisioning spot. But Portuguese had begun to arrive in the region much earlier. Four or five were among the crew of an Australian sealer, the fifty-eight-ton *Antipode*, that touched at the Antipodes Islands and possibly the Chatham and Auckland Islands in 1809-10. In succeeding decades many whalers engaged in bay whaling. While anchored in a protected bay a vessel would send out its boats in search of whales, which were then towed back to the ship or to shore to be tried out. Cloudy Bay, at the north of the South Island, was a favorite of American whalers in the mid-

Farm of Manuel Goomes (center), Bravo Island, New Zealand. Christopher Aubrey, 1879. Courtesy of the National Library of New Zealand.

Azorean Joe Flores (right), resident of the Chatham Islands, ca. 1872-75. Courtesy of the Canterbury Museum, New Zealand.

1830s. Shore whaling on the South Island also attracted Portuguese at an early date.[143]

Many of the Portuguese who settled in New Zealand and the neighboring Chatham Islands were from Cape Verde. In 1826, Joseph Antonio, from São Nicolau, settled on Codfish Island in Foveaux Strait and later married a Maori woman. Another Cape Verdean settler was Manuel Goomes (Gomes), whose father is said to have originated in continental Portugal and owned a cotton plantation with slaves and overseers on the island of Brava. Manuel sailed on a whaler bound for New Zealand; and there, on Stewart Island at the tip of the South Island, in early 1858, he left the vessel. At some point young Goomes, accompanied by a fellow deserter known as "Yankee" Smith, rowed out to a small island, on which they settled, naming it Bravo. In October of that year the two friends married the daughters of Joseph Antonio. Manuel's daughter Esther later recalled the lingering influence of maritime life: "Father had endured a very hard time on shipboard where he had learned very early rising and the law and order of a ship, and he kept us all going, and expected us to fall into the routine and work as if he was the captain and we the crew."[144]

The principal area that attracted American whalers, however, was the Bay of Islands, at the tip of the North Island. During the 1830s it was a rather lawless place, "composed of natives, half castes, missionaries, honest and dishonest traders, respectable artisans, escaped convicts, and runaway sailors." By 1840, there were estimated to be over 1,000 white settlers.[145] That same year the British took command of the lawless colony, imposing duties that, along with the overfishing of right whales, led to a decline in the number of whaling vessels visiting the region.

A curious notation appears in the account of the French expedition of 1837 to the South Seas under Capt. Jules Dumont d'Urville. Arriving in New Zealand in March of 1840, d'Urville met up with a Portuguese whaleship that he noted as being the *Speculacao* (Especulação), under the command of an Englishman named Robinson, which had recently been sealing. When he gammed with the whaler, d'Urville was informed "that the Portuguese Government had given a subsidy to develop the whaling trade, and that this vessel was the first to be fitted out at Lisbon…." The following month, the expedition encountered another Portuguese whaler called the *Adventeur* (Aventura). It had been out for nineteen months and had taken 3500 barrels of sperm oil.[146]

For the great majority of crewmen aboard American whalers, the South Seas afforded quite a cultural shock. When the ship *Elbe*, from the Hudson River port of Poughkeepsie, anchored in the harbor at Nuku Hiva in the Marquesas, crew member Joseph B. Bryant observed, "at 10 am Came too in 11 fathoms of water at 11 the Ship was full of natives fore and aft male & female all crying out —— for Showing us their Pritty little Legs and a little higher up…."[147] But, as often occurred, the attraction was too tempting and the captain was forced to abandon the island after a few days. "This day had to leave all those Pritty Charmers and Proceed to sea on account of The crew for two had made their exit and all the others had it in contemplation but we were two quick for them but had to Send all Those man Eater on Shore to have a tuck out on those two that had left the Ship it gave me Some unpleasant feelings to have to leave all those butyfull creatures Behinde…."[148]

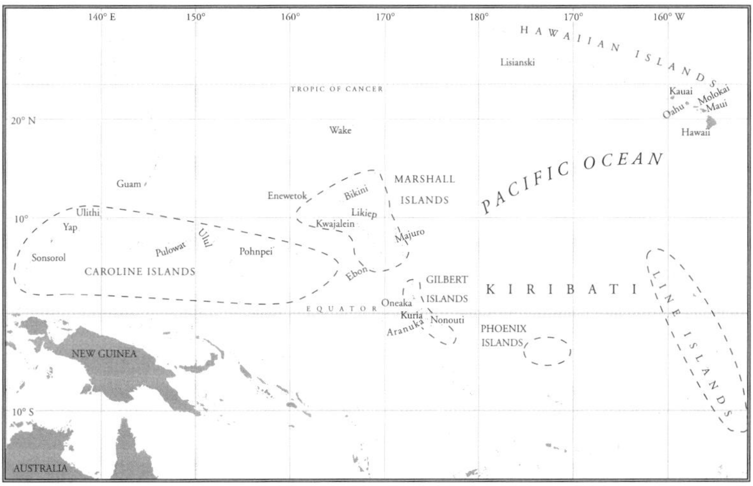

Central Pacific

Trade was a big attraction, often involving the exchange of much-coveted tobacco for food, clothing or sex. The keeper on the ship *Ann Alexander* of New Bedford recounted the scene of encounter in February of 1848 as they dropped anchor at Kuria in the Republic of Kiribati (Gilbert Islands). "At 10 O'clock the Canoes began to come along side to trade some with Couconuts some with fowlz fish Shellz cooked Tarro (the largest I ever saw) mats Ladies and ladies dresses all of which could be bought for a few pounds of tobacco except the Ladies who could not be bought but for a short time and then they were very reasonable only a head and sometimes half a head of tobacco a piece at [most] They all left us for the shore and to all appearances very well satisfied and I think both parties were equally satisfied."[149]

The Legacy of the Beachcombers

By the 1840s it was common for American whaleships in the Pacific to spend the brief summer up north, then to refit in the Hawaiian Islands and spend the remainder of their time cruising in more equatorial climes. The result of much cruising of American whaleships among the temperate and tropical islands of the Pacific Ocean, with the attendant desertions, shipwrecks, etc., was a legacy of beachcombers spread throughout this vast region.[150]

One of the regions frequented by American whalers was Micronesia. These islands were not new to the Portuguese, for Sonsorol, one of the western Carolines, was sighted by Magellan in 1522. And three years later, Diogo da Rocha, having been blown off course to the east, reached either Yap or Ulithi, which he termed the "Islands of Sequeira," and where he and his crew remained for almost four months.[151] But of course it was now a quite different era. More recent contacts had created a sense of wariness and antagonism on the part of both natives and outsiders, not infrequently resulting in violence.

Beachcombers would soon discover, of course, that they were living in a highly complex culture, inscrutable at times with its taboos and savagery. In his list of beachcombers in the Caroline and Marshall islands, Fr. Francis X. Hezel identifies some twenty Portuguese and Cape Verdeans, several of whom spent considerable time there. They range from the notorious to the successful. John Goliah arrived at Pohnpei in 1839 and worked as a pilot in the harbor. As described by a contemporary, "Goliah is a very large man, his tones are hoarse and peremptory. He treats the king as little more than his equal, and gives him one-half of all that he gets for pilotage in return for which he receives all the native food he needs for himself and his three wives. He wears a dirty shirt and trousers while piloting ships, but otherwise dresses only in a Pohnpeian skirt."[152]

Pohnpei, capital of the Federated States of Micronesia, and known in the whaling days as Ascension Island, was a favorite port for provisioning in the western Pacific. Its tall peak and lush terrain attracted not only the whaleships, but also led to a flurry of desertions. In the mid-nineteenth century, quite a few Portuguese whalemen settled there and on neighboring islands. When John Wilson passed by in 1841 aboard the British whaler *Gipsy*, he commented, "I understand there are upwards of *eighty white-men* on the Island! *English, Portugese, American*, &c."[153] These men could make a living practicing a trade, whose knowledge they brought with them, such as carpentry, blacksmithing, or coopering; or like Goliah they could become harbor pilots. Above all, they could become traders, functioning between the natives and the masters of visiting vessels. Antone Pantone was a seaman on board the ship *George and Mary* of New London when he deserted at Pohnpei along with four fellow crew members in December of 1851. A few weeks before he had had an altercation with the mate while in the harbor that would have acted as encouragement to leave the vessel: "Begins with fine weather all hand emploed gittin Redy to tak the for mast out I sent Mr Rogers aloft to Send down for topsail yard Antone pantone Got into words & Mr Scovel told Antone to cum down he told me that he wold not I went aloft to Bring him down Antone Drawed A Sheaf knive and and swore that he wold Stick it in to me the first mate… [when] capt Green cum of from Shore I told him the Curcumstance capt Green told him to go down below that ther was no more duty on board the ship."[154]

The Portuguese presence in the Pacific was extensive enough that it even at times entered the world of fiction. In "Deschard of Oneaka," Australian author Louis Becke—himself a bit of a rogue and adventurer—included as one of four beachcombers on the island of Kuria, in the Gilbert Islands, Pedro, "a Portuguese deserter from a whaler." In this fictional account the four, "hardened and merciless," with "blood-stained" pasts, control the natives of the island and induce them to seize a brigantine and its crew lying offshore.[155]

Becke's fictional account may differ in the details, but it is similar in its theme to the story of the massacre on board the whaleship *Triton* of New Bedford in 1848 at Nonouti (Sydenham's Island) in the same island chain. In his narrative of the incident, Capt. Thomas Spencer relates how, upon approaching the island he was met with several canoes filled with natives, and "in one of these canoes I found a Portugese by the name of Manuel, whom I allowed to come on board, who spoke very good English," and who had been discharged from a French whaler some ten or eleven months before. The captain came ashore to purchase a fluke-chain and other items offered by the Portuguese, but then was kept from returning to his ship. That night he "found Manuel with about one hundred of the wildest looking natives armed with muskets, pistols, cutlasses, lances, harpoons and spades." Reluctant to do away with the captain in cold blood, Manuel explained that "the James Stewart of St. Johns, New Brunswick, had been there about three weeks since, and had purchased articles to the amount of about $300; that upon his going on board to receive the money, they refused to pay him, and drove him overboard without giving him a farthing, and now he intended to have his pay." Spencer warned him that he would be killed in the attempt, and so he was, but not after a bloody fight on board in which most of the crew were murdered. Meanwhile, Capt. Spencer, a boatsteerer by the name of Peet, and the carpenter, John Gomes, managed to survive on shore and eventually escape the island.[156]

In the maritime folklore it was easy to build upon this story and somehow equate "Portuguese" with "unruly beachcomber," as did the mate of the ship *Ann Alexander* of New Bedford, when his vessel arrived nearby the next month:

> There was one white man came on board with some of the natives and reported the Ship Triton Spencer New Bedford with 700 bbls Sperm taken by the natives at Henderville [Aranuka] Island and 16 of her crew massacred it appears to me by what he said that there were some Portagues on the Island that were concerned in taking the Ship The Ship Columbia of Sag Harbor was wrecked on said Island some time previous no doubt had some articles on board which the natives fancied and Thinking they could get a like prize was a great inducement to them besides being urged on by the portaguse [*sic*] Beach Combers that are far beneath the poor natives in principles or in fact anything else (he the white man said).[157]

Actually, the Portuguese might just as well be the victims, as was almost John Gomes, above, of the *Triton*. Or they might take on the role of hero, as did Manuel José dos Reis aboard the ill-fated ship *Sharon* of Fairhaven in 1842. Reis was a youngster from Faial who signed on in the vessel's home port in September of 1841 as an ordinary seaman at a lay of 1/175.[158] It was a difficult voyage with a veteran captain, Howes N. Norris, who was out of control and fixated on punishing the black steward, Babcock, whom he pummeled and tortured mercilessly, both before and after promoting him to cook and then relieving him of that responsibility, before the man finally succumbed to the captain's reign of terror.[159]

In February of 1842, with the vessel in deteriorating condition, Captain Norris brought her into the harbor of Rotuma, one of the Fiji islands lying southwest of Samoa. There nine of the crew soon deserted. And in October of that year, while at Pohnpei, twelve more fled. Norris then sailed away with only seventeen men. Reis had been promoted to steward when the late Babcock became cook. He was soon destined to play a central role in one of the most infamous maritime incidents of the century. Being so shorthanded, one day in the open sea when whales were sighted the captain could order only two boats off. Remaining on board with him were the steward, Reis, and three South Sea natives.

Around five o'clock that afternoon, as the two boat's crews were preoccupied with chasing a pod of whales, they descried the ship's flag flying at half-mast, a signal that they must return to the vessel with all haste. Fortunately Manuel had been in the rigging when he heard the cry of Capt. Norris as he was attacked by the three islanders. There he remained for some two hours before the flag that he had run down was noticed. While the natives were hurling various missiles at the approaching boats, the officers instructed the youngster "to cut the top-gallant, top-sail and jib halyards, to let the sails run down and then clear them from the yards and let them hang by the rigging, so as to stop the ship's way—in doing which he displayed great activity and courage, passing from one masthead to the other by the rigging, and deserves the highest praise."[160]

Credit for retaking the ship falls largely to the bravery of third mate Benjamin Clough, who swam to the *Sharon*, climbed the rudder, and entered the after-cabin window, whereupon he killed one of the three mutineers and disabled another, allowing his fellow crewmen eventually to board the vessel and take control. After a court hearing at Sydney on which he was questioned about the incident, and attempts were even made to implicate him, followed by a return to the U.S., Manuel José dos Reis evidently had had enough of whaling voyages, for he was never heard from again.

The Atlantic Slave Trade

> Mock whaling vessels occasionally entered the trade. Because they had a right to be double-decked, and to have small boats, extra men, casks, furnaces, and bulging larders, dummy whalers were in some ways nicely suited to the trade.
> (Warren S. Howard, *American Slavers and the Federal Law, 1837-1862*)[161]

Slavery, in one form or another, is as old as civilization itself; but its modern form, that of stuffing hundreds of hapless Africans into a vessel to transport them to the New World, was especially horrific. With scant food and water, and the utter lack of sanitation, those who perished en route—and they were not an insignificant number—were jettisoned overboard, while those who survived were sold at a handsome profit into the various slave economies of the Americas.

By 1808, the slave trade had become illegal in both the British Empire and the U.S. Spain officially banned the trade not long after, followed by Portugal in 1820. Brazil initially declared the trade illegal in 1831; nevertheless, enforcement was lax, and slave voyages there continued for another two decades until the British finally forced an end at mid-century. At that point Cuba became the principal destination.

In 1808, the British had established a West African Squadron to interrupt the trafficking of slaves across the Atlantic. And in order to help curtail the traffic, in 1843 the U.S. government placed Commodore Matthew Perry in charge of the newly formed African Squadron. Based in Praia on the island of Santiago, Cape Verde, the Squadron consisted of only a handful of vessels and was largely ineffectual.

The increased vigilance of British and American naval vessels operating along the coast of Africa demanded in turn greater subterfuge on the part of the slave runners. One ruse was to pretend to be on a whaling cruise. This could explain an empty hold and plentiful provisions. And the trypots made perfect containers for preparing meals for the hundreds of captives who would be squeezed into the vessel.

The bark *Fame* sailed out of New London on June 18, 1844, headed for the Pacific whaling grounds. Captain Joseph B. Mitchell died at sea as had the mate before him. And so at Honolulu Anthony Marks, a native of the Azores who had signed on originally as second mate, was appointed captain by the U.S. Consul. Marks brought the *Fame* into Rio in December of 1846, declaring her unseaworthy; but soon he had her repaired and departed, ostensibly to continue whaling but really to obtain slaves on the coast of Africa. After depositing 527 slaves at Cabo Frio, Brazil, out of an original cargo of 530, Marks left the vessel in the care of the mate and disappeared with some $40,000 in profits (ca. $1,000,000 in today's dollars).[162] He was apparently never apprehended.

During his voyage on the *Cynosure* of Stonington, Nicholas C. Anthony witnessed the secondary effects of the slave trade on whaling. The vessel, under Captain Walter Simonds, called on its way out at Faial in September of 1845 and at São Vicente the next month. By the time the vessel arrived at the Falklands in December Anthony must have had a good idea that this was no ordinary voyage, for the captain showed little interest in obtaining oil. Instead, he was shooting geese; and Anthony began to keep track of lost opportunities, noting the days lost to the whale hunt. In January 1846, at Talcahuano, Simonds spent seventeen days ashore with his wife. The next month Anthony noted another sixteen days "lost" at that port. And so the lack of interest in chasing whales persisted. In March of 1847, now back in the Falklands, a frustrated Anthony wrote in his journal, "this day we Anchor at port stephens—Falk Island for to get oil but the Capt. got about 500 geeses and did not car a dam ffor the oil 5 days Lost."[163]

Since any remuneration at the end of the voyage depended on the vessel's catch, it is easy to understand Anthony's frustrations. His journal does not reveal the moment when the motives for Captain Simonds's lack of interest in whaling became clear, but Anthony certainly understood when he penned these words as the *Cynosure* anchored at Salvador, Bahia, Brazil, in July of 1847: "This day we Anchor in the harbor of Bahia and her Capt. will make her oners fill sory."[164] The vessel was soon sold, and it was up to Anthony to make his way home. The following month the *Cynosure* is reported to have landed slaves on the coast of Brazil.[165]

By the late 1850s Cuba had become the last major destination for the slave trade. It was inconvenient, however, for the traders to establish their headquarters in Havana, since the focus of their activities would be too obvious. And so, as the business died out in Brazil, several Portuguese who had been based there or in Portuguese West Africa, moved to New York City, with its active legal African trade. Thus in 1852 Manuel Basílio da Cunha Reis founded the so-called "Portuguese Company" with the Portuguese Consul General, César de la Figanière and his brother William, a naturalized American citizen. The Portuguese Company was an amorphous operation whose members, it is said, "ran their business on gangster lines."[166] Ironically, the brothers' father had worked for the Portuguese government both in Africa and Brazil ostensibly to abolish the trade.

By the end of the decade the Portuguese Company was taken over by the Spanish Company, but many of the same figures continued in the trade. During the years immediately preceding the Civil War, there was great growth in the industry. In 1859, in spite of the minor U.S. and significant British patrols in the Caribbean, there were more vessels in the trade than at any time since 1820. That same year President Buchanan ordered more ships into the Africa squadron and also moved it from Cape Verde to Angola, to be closer to the center of slave-trading activity. It has been estimated that "out of 170 slave-trading expeditions, presumably to Cuba,

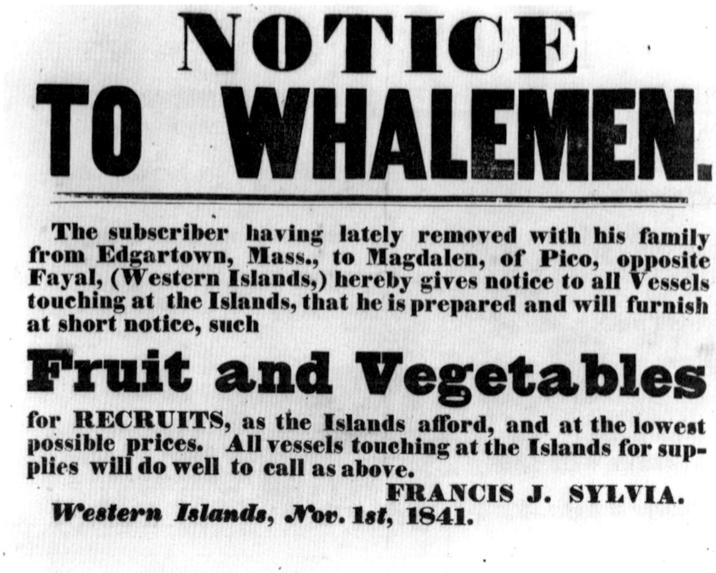

NOTICE
TO WHALEMEN.

The subscriber having lately removed with his family from Edgartown, Mass., to Magdalen, of Pico, opposite Fayal, (Western Islands,) hereby gives notice to all Vessels touching at the Islands, that he is prepared and will furnish at short notice, such

Fruit and Vegetables

for RECRUITS, as the Islands afford, and at the lowest possible prices. All vessels touching at the Islands for supplies will do well to call as above.

FRANCIS J. SYLVIA.

Western Islands, Nov. 1st, 1841.

Courtesy of the New Bedford Whaling Museum.

fitted out in the three years preceding 1862, seventy-four were known or believed to have sailed from New York." Of all the vessels involved in the slave trade at the time, nine-tenths were apparently American.[167]

In June of 1859, the *New Bedford Evening Standard* noted an apparent resurgence of interest in whaling, stating that, "although there has been considerable doubt expressed as regards the propriety of investing money in the whaling business, there are some who do not believe in a failure of this enterprise. Two ships have been purchased within a day or two—one in New York and one in Boston by Portuguese gentlemen, to be fitted at this port."[168] What the paper did not know—or did not state—was that these Portuguese gentlemen were about to embark on a plan to employ their whaling vessels, the ships *Atlantic* and *Comora*, in a subterfuge to carry slaves from Africa to the Americas. In truth it was not the first time that such a ruse had taken place.

The two men who brought the *Atlantic* and the *Comora* into New Bedford in July of 1859 were indeed respected Azorean whalemen. Francis J. Sylvia was born on the island of Pico in 1810. In 1838, he married a woman from Edgartown, and soon after the two moved to Pico, where Sylvia established a provisioning station.[169] His is the well-known "Notice to Whalemen" that has appeared in many publications on the history of American whaling and its connection with the Azores. Sylvia's Azorean enterprise apparently was not a great success, for 1843 finds the family once again in Edgartown. In July 1844, Sylvia sailed as mate of the ship *Governor Troup* of New Bedford. The following spring the captain died at Honolulu and Sylvia took over as master for the remainder of the voyage, but not without a bit of trouble, just as happened to mate Joseph Roderick of the *Charles W. Morgan* some sixty decades later, when the crew refused to sail with the new master.[170]

As William A. Allen noted in his journal aboard the ship *Samuel Robertson* of New Bedford as it lay at Oahu:

> I could tell you of a good many scenes and scrapes that I have seen but I will not—it only tires me to write over the rascality of some people. One instance and I am done. The Gov. Troup came in with the capt sick and they brought him on shore. The crew did not want to go to sea with the mate who was a portugese and would not weigh their anchor.

> Wm Hooper the Asn. Consul went of with a boat load of irons to flog them till they would go. Put a man in the rigging and flogged him till they consented to go![171]

When the vessel returned to New Bedford two years later it brought in 120 barrels of sperm oil, 3400 of whale-oil, and 14,000 pounds of whalebone.

Of José C. S. Maciel less is known. Just before taking the *Comora* out of Boston, while in the city of New York, and being a naturalized citizen, he took time to obtain an American passport. He was just thirty-one years old at the time.[172] Undoubtedly he had connections with his compatriot slave traders in Manhattan. Officially he was both owner and master of the ship *Comora*. Maciel also evidently sold the ship *Atlantic*, 699 tons, to Sylvia, who brought her into New Bedford from New York to be fitted out. Presumably the financing for these two vessels originated with the New York slave traders. It was not long after the appearance of the *Evening Standard* article, however, that suspicions were raised around town concerning the intended purpose, and the *Atlantic* was briefly seized by the government. Some of the items that aroused concern were the 1200 bushels of corn on board, supposedly destined for Faial; the casks in the lower hold that were filled with fresh, instead of salt water; and a cracked trypot. It was claimed additionally that $20,000 had been sent from Havana to the firm of Ameses, Ambroges and Co., in New York, to purchase these two vessels. But ultimately both vessels were allowed to sail.[173]

The *Atlantic*, after being seized on July 21, 1859, was finally released on September 2 and sailed six days later. It is uncertain just how far Francis Sylvia took the vessel, for early the next year it was brought back into New York under Edward S. Devoll. In the meantime it had landed at least one cargo of slaves in Cuba.[174]

The 439-ton *Comora* cleared from New Bedford on July 31, 1859, and made at least one successful trip across the Atlantic with its hold full of slaves, landing them also in Cuba. Subsequently, in July 1860 the vessel was reported in Bahia, Brazil. In March of 1861 the Australian *Sydney Herald* announced the arrival of the *Comora* in a leaking condition, having taken so far only fifty barrels of oil. In November of that

"Chawed Boat."
George Albert Gale, 1937.
Courtesy of the New
Bedford Whaling Museum.

year, it was reported that the vessel had been sold and her name changed to *Lady Young*. Captain Maciel was not heard from again.[175]

Stove by a Whale

There are very few confirmed incidents of whaleships being stove by an angry whale. The sinking of the *Essex* (1819), the *Ann Alexander* (1851), and the *Kathleen* (1902), are three famous encounters. But such an incident occurred during Joseph Dias Jr.'s first command. The *Ocmulgee* had returned to Holmes Hole on 21 April 1850; and less than two months later, on 10 June, Dias, then only twenty-seven years old, took command of the ship *Pocahontas*, in which he eventually would make a successful three-year voyage in the Pacific. But six months after sailing, while the vessel was in the South Atlantic, Dias made the mistake—which could have resulted in tragedy—of toying with an enraged and wounded whale. Early in the morning of December 12, 1850, several sperm whales were sighted and two boats lowered. An hour later "the Larboard boat struck a very large whale the whale sounded a little came up and appeared to be quiet and resigned to his fate and Mr. B was in the act of going up to lance him when he got within a few fathoms of him the whale milled for the boat and rooled jaw up and as quick as thought he bit here in too… but he was so intent upon his business that we could scarcely pick up the crew of the stoven boat two of which was hurt quite seriously," noted Dias in his journal. As the wounded men were taken back to the ship the whale continued about his business, grinding up the boat. "It is useless for me to try to describe how fine he used her up even breaking the oars mast and everything that was large enough for him to get at."[176]

But the young master was determined to take up the challenge, confronting the leviathan directly from the ship. "Finally after letting him chew the boat for nearly 2 hours I concluded to let him take a bite at the Ship. So we got every body stationed ready to dart some kind of craft in him." At the first pass, the whale evaded the weapons that were directed at it. The *Pocahontas* then tacked back, planning to pass alongside the whale, now motionless among the pieces of the stove boat. But the whale had other ideas, and as the ship approached it started, "to all appearances with all his might," head-on for the bend in the bow and struck just below the water mark, breaking a plank and three timbers and causing the ship to leak. Fortunately, the sheathing remained, and thus the leak was contained until the *Pocahontas* could reach the port of Rio de Janeiro, some 750 miles away.[177]

Unfortunately, Dias's luck with the weather was no better. As they slowly made their way to the Brazilian coast, through alternating calm and heavy headwinds, Dias had one of his many moments of anguished reflection. Rather than being thankful for the close call that could have seen them attempting to survive the voyage in a few hastily provisioned whaleboats, he cursed his luck: "Well this is getting along indeed. I don't know what to do with myself. Talk about trouble I never knew what trouble was before and I should like to know what I have done to deserve this heep it takes away my appatite and in fact kills me right out and I believe if it holds calm much longer I shall loose what little reason I have got. Now is the time I suppose people want Christian fortitude to bear up under such afflictions. But I suppose I must bear it without the help of such things."[178]

The passage to Brazil was agonizingly slow; and when, nine days after the incident, he was forced to take in sails due to a strong headwind, Dias again cursed his luck. "If this is not enough to make a man curse," he lamented, "I don't know what is but I Suppose Christians would say it is all for the best and I hope it may be So."[179] Finally, two weeks after being stove, the *Pocahontas* arrived at Rio, and after repairs was soon off again.

The *Pocahontas* returned to Holmes Hole on May 7, 1853. Its voyage was relatively successful, but not before its master spent half of it in anguished anticipation of failure and humiliation. Putting into words what other whaling captains might only think, in May of 1851 Dias lamented, "at 1 Saw a Sperm Whale close to the Ship lowerd two boats one boat would have had a chance to Strik the whale but a d---d Ship ran down and frigten the whale off. I don't know what will come next to prevent us from getting a whale but I suppose Something that was never heard of before this is the 14th time we have seen Sperm whales and got 140 bbls. And I don't hardly know what to do with myself."[180] Two months later one of the crew members became ill, and Dias noted the necessity of making port to find him treatment. "But that is not what troubles me," he noted, "for I expect the rest of them will run away or a part of them and I shall be bothered to get men and I don't know if I can get them Short of the Society Islands but to port I must go Soon. What will people think at home to hear from me 13 mos. out with nothing."[181]

Summer came and went still with no additional oil, as Dias fretted, "I am in hopes for a whale for I do not despair yet altogether if others do, but it is rather hard for a person to keep a Smiling face on when everything looks so dark but I Suppose I must Sacrafise my own feelings for the Sake of encouraging others."[182] But their luck finally turned around, and the *Pocahontas* arrived home with 320 barrels of sperm, 1720 of whale-oil, and 1,000 pounds of whalebone, having previously sent home an additional 16,998 pounds of bone.

But as was typical with whalemen, Dias spent little time ashore. After a brief few months at home in Tisbury, Joseph Dias Jr. headed for New Bedford to take command of the 408-ton ship *St. George*. As the vessel set sail that September he penned in his journal some thoughts on the life of a mariner. "See if you can think Strange," he ruminated, "that Sailors live fast what little time they Stay on Shore." He, himself, must "forget all the flirtations with the girls," and get on with "the Stern realities of a whaling voyage whos trials and troubles no one that has not tried it can form the least Idea of. Just think of it to be pen[n]ed up and tossed about on the wide ocean for I Suppose the Space of three years."[183]

Later on the same doubts about the catch continued to assail him, as well as concerns about the loyalty of the crew. At Point Hope in the Beaufort Sea in September of 1854, now a year out, Dias noted that he "had an interesting conversation with C & J on the Propriety of my Still continuing master they did not like to Say that they were opposed to it." Three days later he remarked that "at 10 Saw right whales and lowered and the Larboard boat went on and the boat-steeerer missed him as I Suppose intentially as I am fully convinced that there is a conciderable dissatisfaction, they are all afraid that they will [not] get money enough to pay their debts."[184] But again, fortune was with him; after a voyage of nearly four years that ended in May 1857, the catch of the *St. George* totaled seventy-eight barrels of sperm oil, 2158 of whale-oil, and 19,350 pounds of whalebone.

The Adventure of the Western Arctic

> The whaling industry introduced the outside world to the natives and, by nearly eliminating the whales and walruses, it change the natives' world as well.
> (John R. Bockstoce, *Whales, Ice, and Men*)[185]

The question of timing for the bowhead migration must have come up quite early in the development of the Arctic fishery. The bowheads move south along the Chukchi Sea in late October to early December and then begin their return north in April. By wintering in the Arctic, a vessel could take advantage of these movements. John E. Simmons (1819-1908), born in Faial, made a pioneering attempt at this on his second voyage as master of the bark *Cleone* of New Bedford in October of 1858.[186] By the summer of 1859, having had poor luck whaling in the Chukchi Sea, he began taking walruses for their oil. His was one of the first American whaling vessels to engage seriously in walrusing. A decade later almost everyone was taking walruses on a regular basis, leading almost to the decimation of the Arctic walrus upon which much of the native population in the Bering Strait was dependent.

That September, accompanied by the brig *Wailua* of Honolulu, Simmons headed into Plover Bay with the pioneering intention of wintering over. Unfortunately this first attempt was a disaster. As the bowheads swam south that fall, the young ice impeded the approach of the whaleboats; and during the return of the whales the following spring the whalemen found themselves too far from open water to effect a hunt. They had, as well, been unprepared for the extreme cold and the growing scourge of scurvy. Ultimately fifteen men aboard the *Wailua* died of the disease as well as two from the *Cleone*.[187]

As it was in other regions of the vast Pacific, the presence of native women was a temptation to men long absent from family and lovers. The fourth mate of the *Cleone* fell for a native by the name Penneneur, who eventually moved into the after cabin with the skipper and in the spring presented him with a little girl whom she claimed was his. If so, John Simmons can be seen as a pioneer of a different sort; for over the next decades it was not unusual for whalemen, including the captains themselves, to form more than fleeting relationships with native women of the Arctic.[188]

American whaling was, indeed, a complex industry. One of the most overlooked aspects were the peripheral commercial activities in which individual whaleships participated, from sealing, to smuggling tobacco, trafficking in coconut oil and other commodities, carrying home a load of guano, or even entering the slave trade. Extended cruises throughout the wide Pacific Ocean left their mark, as well, in the form of settlers in the most far-reaching locations.

As the most productive period of the American whaling industry approached its climax, the situation of the Portuguese segment had changed dramatically from three decades before. Azoreans and Cape Verdeans were now serving at the highest levels aboard ship. Excluding replacement masters, the number of sailings had grown from only five prior to 1850 to twenty-one during this latter decade: nine out of New Bedford, ten from Provincetown, and one each from Holmes Hole (Vineyard Haven) and New London.[189] And while the number of mates and captains was still small relative to the industry overall, it was destined to grow dramatically in the years to come, as was the Portuguese role in ownership.

Notes

[1] Davis, Gallman, and Gleiter, *In Pursuit of Leviathan*, 4, 19. See also *New York Times*, 23 Aug 1853.

[2] Starbuck, *History*, 98, 314-22. Cf. "Wilmington whaleships had an unusually large number of educated young men aboard; yet the crews, like those of all whaleships, were cosmopolitan, seasoned with European seamen, Polynesians, and especially Azores and Cape Verde islanders." Martin, *Delaware Goes Whaling*, 35.

[3] Hazen, *Five Years before the Mast*, 67.

[4] Ibid., 69-70.

[5] Ibid., 67-68.

[6] "Ó terra de Betefete [New Bedford], Donde se vai prà baleia, De lá vim, lá tornarei, Como o mar bate na areia." Dias, "Baleeiros Portugueses na América," 181. Translated by the author.

[7] Seamen's Register, ODHS, 1834; Crew Lists, New London area, 1835, NAB.

[8] See p. 87.

[9] U.S. Census, Bristol Co., Mass., 1850.

[10] Rev. Moses How (b. 1789 – d. 1881), Journals, 1813-1881, 8 Dec 1855, NBL. On July 17, 1853, How noted that he had "settled 104 voyages amounting to 25,605.81."

[11] Ibid., 22 Aug 1857.

[12] Jorge Barbosa, "Irmãos," *Obra Poética*, 61. Translated by the author.

[13] Silveira, "Romance," *A Ilha e o Mundo*, 44-46. Prose adaptation and translation by the author.

[14] Amorim, "Mortalidade e Emigração," 905. Translated by the author.

[15] Bark *George and Mary*, Manchester, Westport, Mass., 8 Sep 1852, Log 604, KWM.

[16] "Report of 2 September 1880 (205) of the Administrator of the District of Brava," in Carreira, *Migrações das Ilhas de Cabo Verde*, 253. Translated by the author.

[17] Horatio N. Greene, Diary, ship *Two Brothers*, Nichols, New Bedford, Mass., Log 240, Ships' Logs Coll., MSS 220, NHA.

[18] Manuel Silveira Andrade, "The True Remarks of a Whaling Voyage from North Atlantic to North Pacific Ocean around Cape Horn in 1859 on Board of the Bark Pacific (396 tons) of New Bedford, Capt. Jacob A. Howland," MSS, JAF. (This journal, written originally in Portuguese, was later translated by Andrade into English. The original is not extant.)

[19] Ship *Lucy Ann*, King, Wilmington, Del., 15 Jan 1842, Log 434, KWM.

[20] Ibid., 16 Jan 1842.

[21] Ibid.

[22] Ibid., 17 Jan 1842.

[23] Taylor, *Life on a Whaler*, 21.

[24] James Webb, letter to his mother, 4 Oct 1845, Business Records Coll., MSS 56, Ser. W, S-s 6, ODHS.

[25] Cloud, *Enoch's Voyage*, 23-24, 26.

[26] Ely, *"There She Blows,"* 44-45.

[27] Forster, *Cruise of the "Gipsy,"* 15.

[28] Higginson, *Atlantic Essays*, 230-31.

[29] Thomas Milligan, "A Sea-Yarn without a Strand of Fiction," Carlton D. Morse Whaling Coll., MSS 79.22, ser.5, box 9, f. 11, JHL.

[30] Ibid.

[31] Robert Weir, Journal, bark *Clara Bell*, Robbins, Mattapoisett, Mass., 31 Aug 1855, Log 164, MSM. The third mate, Franklin Perry, was from the island of São Jorge, Azores (Seamen's Register, ODHS and Whalemen's Shipping Paper, bark *Clara Bell*, mic 180, NBL). Whereas "Portagee" was a common and hardly deprecatory term in the old whaling days, "Gee" was much more disparaging. Whereas used here against an Azorean, it was most often used to refer to Cape Verdeans, viz. Melville's "'Gees."

[32] Robert Weir, Supplement to his Journal (but dated 19 and 22 Aug 1855).

[33] See pp. 76.

[34] Bark *George Washington*, Harps, New Bedford, Mass., 1851-52, Log 930, ODHS.

[35] Taylor, *Life on a Whaler*, 30.

[36] Starbuck, *History*, 67; Bockstoce, *Whales, Ice, and Men*, 94.

[37] Comstock, *A Voyage to the Pacific*, 4-5.

[38] Eliza S. Barney, "Barney Genealogical Record," http://12.46.127.86/bgr/BGR-o/index.htm (accessed 19 Jul 2008), NHA. See also pp. 38, 73, and 82.

[39] *WSL*, 6 Aug 1844; Account Book, ship *John Adams*, Stockman, Nantucket, Mass., 1841-45, Account Books Coll., MSS 10, AB68, NHA; Memorial plaque, Seamen's Bethel, New Bedford, Mass.

[40] Downey, "Consequences of California Mania," 25.

[41] "Marriages Recorded in the *New Bedford Mercury*, 1807-1845," typescript, NBL; U.S. Census, Newport Co., R.I., 1850-70.

[42] V. Joseph Folger, Nicholas C. Anthony, and William H. Swazey.

[43] Manifests of Whaling Voyages, Card File, Genealogy Room, NBL. Referred to hereafter as Whaling crew list card file, NBL; White, *Connecticut Town Vital Records*, 5:215.

[44] Whaling crew list card file, NBL.

[45] Ibid.

[46] Caton was born in Massachusetts, the son of Azorean Manuel Caton. His surname appears variously as Caton or Katon, sometimes with the middle initial and at times without. (U.S. Census, Barnstable County, Mass., 1850-80.)

[47] Crew Lists, U.S. Customs Service, RG 36, NAB; It has not been possible to identify the ethnicity of two other Catons. Francis Caton served as master of the schooner *Emporium* during three voyages in the late 1850s to the early 1860s. Additionally, Joseph H. Caton commanded the schooner *Life Boat* out of Newburyport, Massachusetts, between 1867 and 1870. Nevertheless, it has been assumed that both these men were either Portuguese or Portuguese American.

[48] U.S. Census, New London County, Conn., 1850; U.S. Census, New Haven County, Conn., 1860; New London County, Conn., 1870; and New Haven County, Conn., 1880; Crew Lists, U.S. Customs Service, RG 36, NAB. 163

[49] White, *Barbour Collection*, 4:41.

[50] Nicholas C. Anthony, Journal, 28 Oct and 6 Nov 1851, Log 37, MSM. Anthony maintained the same journal during multiple voyages.

[51] Ibid., Feb 1850.

[52] *Republican Standard (New Bedford)*, 11 Aug 1853.

[53] Ship *Petrel*, Tucker, New Bedford, Mass., 26 Jan 1855, Log 188, Ships' Logs Coll., MSS 220, NHA.

[54] Anthony, Journal, 26 Jan 1855.

[55] U.S. Census, New Haven County, Conn., 1860 and 1880; and New London County, Conn., 1870.

[56] Calabretta, "Picture of Antoine DeSant," 94; Brown and Rose, *Black Roots in Southeastern Connecticut*, 109-10. This latter identifies DeSant as hailing from the island of São Vicente, but his marriage record in 1842 shows him to have been born in Boa Vista (White, *Barbour Collection*, 5:101).

[57] U.S. Census, New London County, Conn., 1850.

[58] Haynes, *Stonington Chronology*, 64.

[59] Vargas, *Stonington Ice*, xvi-xvii, 51, and 74-75; Lewis, "Stonington's Portuguese Fishermen," 1-2.

[60] U.S. Census, New London County, Conn., 1850; Henrietta Mello Mayer, "1846 to 1948: Marriages of Portuguese People in Stonington, Ct.," MSS, SHS; Haynes, *Stonington Chronology*, 77; Boylan, "Festival of the Holy Ghost," 1-8. Cf. Calabretta, *Fishing out of Stonington*.

[61] "Crew Lists for Vessels Owned by Charles T. Dering of Sag Harbor, 1828-1855," MSS, EHL. (Due to the difficulty in identifying nationalities, this is only an approximation.)

[62] Ibid.

[63] U.S. Census, Suffolk County, N.Y., 1850.

[64] William R. Palmer, "The Whaling Port of Sag Harbor," (PhD dissertation, Columbia University, 1959), 306; U.S. Census, Suffolk County, N.Y., 1860.

[65] Capt. Basil Gordon De Castro, interview by Donald Warrin, Bridgehampton, N.Y., 17 Sep 2001.

[66] Joseph De Castro to Hannah De Castro, 26 Apr 1858, De Castro family document.

[67] Starbuck, *History*, 558-59.

[68] Custom House Record, Sag Harbor, 15 Apr 1874, MSS, De Castro family document.

[69] De Castro interview.

[70] Sleight, *Whale Fishery on Long Island*, 16.

[71] Ibid.

[72] Schmitt, *Mark Well the Whale*, 120.

[73] U.S. Census, Suffolk County, N.Y., 1850-1880.

[74] Schmitt, *Mark Well the Whale*, 99.

[75] Farwell, "Manuel Enos."

[76] Rafael Enos Aguirre, "Orígenes e historia de la familia Enos," private document; Rafael Enos

Aguirre to Donald Warrin, email, 18 Mar 2009. http://sarmento.eng.uminho.pt/cgi-bin/geneweb?b=Lajes;lang=pt;p=jose+inacio;n=macedo (accessed 10 Jul 2009).

77 Stevens, *Index to the Long Islander*, 1974.

78 "Claudio's, Our History," http://www.claudios.com/(accessed 8 January 2009).

79 Lund, *Whaling Masters*, 723.

80 It is not certain that this was a whaling voyage. *Abstract of Vessels Registered (Foreign Trade) at Marblehead Custom House, Essex Institute Historical Collections*, 56 (January 1930), 112, PEM.

81 *New Bedford Weekly Mercury*, 21 Mar 1845; *The Friend (Honolulu)*, 1 Nov 1844 and 15 Aug 1846 (in *American Activities in the Central Pacific*, 4:39-40 and 60).

82 Busch, *War Against the Seals*, 175.

83 Abstracts from Whaling Voyages, MSS, SHWM.

84 Theodore Lewis, Journal, ship *Atlantic*, Cooper, Bridgeport, Conn., Log 822-23, MSM, 35.

85 Taylor, *Life on a Whaler*, 140-41.

86 Ibid., 96.

87 Ibid., 96-97.

88 Ibid., 136-37.

89 Ibid., 137.

90 Ibid.

91 Ibid., 138.

92 Ibid., 140.

93 *WSL*, 12 Dec 1854.

94 Ibid.

95 Bark *Belle*, Handy, Fairhaven, Mass., 9 May 1854, MSS, "Abstracts from Whaling Voyages," SHWM.

96 Ibid.

97 Ibid., 28-31 May 1854.

98 Ship *St. George*, Dias, New Bedford, Mass., 8 Aug 1854, Log 202/203, ODHS.

99 Joseph Dias Jr., Journal, ship *Ocmulgee*, Holmes Hole, Mass., 25 Dec 1847, Log 204, ODHS; ship *Clifford Wayne*, Downes, Fairhaven, Mass., 1836-37, Seamen's Register, ODHS, 1836. On the Vineyard the family name is pronounced 'die-us.' See also pp. 130-31.

100 Bark *Plover*, Macy, New Bedford, 21 Aug 1863, Log Wh P7315 1862L, Nicholson Whaling Coll., PPL.

101 *New Bedford Weekly Mercury*, 19 Feb 1847.

102 Antonio Machado vs. Alden Howland etal., USDC Special Court, Dec 1861, no. 37, U.S. District Court, Mass., Docket 1861-62, NAB.

103 Manoel Machado da Silva, legal release from J. and W. R. Wing Co, J. and W. R. Wing and Co. Papers, MSS 35, Ser. B, S-s 8, f. 5, ODHS.

104 Sylvanus C. Tallman, Journal, ship *Canada*, Reynard, New Bedford, Mass., 5 Mar-10 Sep 1847.

105 Ship *Rousseau*, Brayton, New Bedford, Mass., 16 Nov 1844, Log 4174, NBL.

106 Ibid., 15-17 Jan 1844; Statement of Captain Brayton, New Bedford, Mass., 18 Feb 1845, Incoming Crew List, NAB.

107 Ship *Rousseau*, Log, n.d.

108 Ship *Daniel Webster*, Log, Misc. Ships Coll., MSS s-g 13, Ser. 4A, ss 1, f. 1, Nicholson Whaling Coll., PPL.

109 C. A. Church to Rescom Borden, 4 May 1858, Cory Family Coll., MSS 80, s-g 3, Ser. K, S-s 4, f. 3, ODHS.

110 Bark *Three Brothers*, Mitchell, Nantucket, Mass., 16 Apr 1848, Log 309, Ships' Logs Coll., MSS 220, NHA.

111 Ibid., 22 Apr 1848.

112 *New Bedford Daily Mercury*, 25 Mar 1859.

113 Schooner *Emeline*, Eldridge, Mystic, 1843-44, Log 147, ODHS.

114 *New Bedford Weekly Mercury*, 16 May 1845.

115 *The People's Advocate (New London)*, 4 Dec 1844.

116 *New Bedford Weekly Mercury*, 28 Feb 1845; Perry, *A Treatise on the Human Hair*, 43.

117 Coleman Sanford, agent, Field and Sanford to David Bunker, master, ship *Lexington*, Nantucket, Mass., 16 Jan 1849. Quoted in Downey, "Consequences of California Mania," 26.

118 *New Bedford Mercury*, 12 Jul 1833.

[119] Joseph Dias Jr., Journal, ship *Ocmulgee*, Manter, Holmes Hole, Mass., 4 Sep 1847, Log 204, ODHS.

[120] *New Bedford Weekly Mercury*, 31 Mar 1843.

[121] Dias, Journal, ship *Ocmulgee*, 16 Dec 1847.

[122] Ibid., 1 January 1848. In the early years of the industry liquor was rationed out to crew members, as it always had been in the merchant service. Beginning in the 1830s the temperance movement began to have some influence aboard whaling vessels, and it grew in importance in the succeeding decades. At that point liquor was no longer given out to crew members on a regular basis; and at times whole vessels became temperance ships.

[123] Ibid., 20 February 1848.

[124] Ibid., 3 Mar 1848.

[125] Ibid., 9 Nov 1848.

[126] Ibid., 18 Nov 1848.

[127] Bark Russell, Crew List, *WSL*, 13 Mar 1849. The former was either John P. Praro Sr. (see p. 196) or John Prairo, a mariner and laborer from the island of Pico who was married on Nantucket in 1840. William Sylvia was born in the Azores ca. 1825, and by September of 1850 was back on Nantucket (*Vital Records of Nantucket*, 2:393 and 4:298; Pat Amaral, *They Ploughed the Seas*, 107; U.S. Census, Nantucket County, Mass., 1850.)

[128] *WSL*, 7 Aug 1849 and 14 Jan 1851; Schmitt, *Mark Well the Whale*, 29; ship *Massachusetts*, Nickerson, Nantucket, Mass., 1848-51, Log 287, ODHS.

[129] James F. Cleaveland, Journal, ship *Niantic*, Warren, R.I., March and April 1849, SFM.

[130] Tønnessen and Johnsen suggest that there was at least one much earlier penetration. In any event it evidently did not lead to further exploration at the time. (Tønnessen and Johnsen, *History of Modern Whaling*, 16.)

[131] Bockstoce, *Whales, Ice, and Men*, 24.

[132] Dias, Journal, ship *Ocmulgee*, 29 Jul 1849.

[133] Ibid., 20 Sep 1849.

[134] Ibid., 17 Oct 1849.

[135] Hare, *Salted Tories*, 11-12.

[136] Ibid.

[137] Ibid., 12.

[138] Bertão, *Portuguese Shore Whalers of California*, 63.

[139] Kuykendall, *Hawaiian Kingdom*, 1:307, 309.

[140] Ship *London Packet II*, Sampson, New Bedford, Mass., 27 Mar 1839, Log 760, ODHS. There is no mention in the log that the lashing played any role in the man's illness, although it would seem to suggest otherwise. Within a few hours his seizures had moderated, so presumably he recovered (op. cit.). The identity of the man was not given.

[141] *London Packet II*, Log, 31 Mar-1 Apr 1839.

[142] U.S. Consular Documents in the log of the bark *Alice*, White, Cold Spring Harbor, N.Y., 1851-54: Hobart Town, Van Diemen's Land [Hobart, Tasmania], 10 Mar 1852; Hilo, Hawaii, 19 Mar 1853; Honolulu, Hawaii, 29 Nov 1853, CSHM.

[143] Richards, *Whaling and Sealing*, 16-19. Cf. McNab, *Old Whaling Days*, 187 ff.

[144] Richards, *Foveaux Whaling Yarns*, 103; "Murihiku in the Making," *Southland Daily News (Invercargill, N. Z.)*, 31 Dec 1937; Peter Goomes and Rosemary Taiaroa, personal communication. Apparently some of the descendants today use the name Goomes, while others prefer Gomez. Originally it would have been Gomes.

[145] Kenny, "Yankee Whalers," 30.

[146] NcNab, *Old Whaling Days*, 288, 327-28.

[147] Joseph B. Bryant, Journal, ship *Elbe*, Whipple, Poughkeepsie, N.Y., 4 Apr 1835, Log 891, Ships' Logs Coll., MSS 220, NHA.

[148] Ibid., 12 Apr 1835.

[149] Ship *Ann Alexander*, Sawtell, New Bedford, Mass., 17 Feb 1848, Log 394, ODHS.

[150] A literal reading of the term "beachcomber" suggests a person who wanders the shoreline of some South Pacific isle while seeking flotsam and jetsam and surviving on fish and coconuts. But the term was actually much broader. Once these men obtained a functional level of communication skills in the local dialects, they could serve an important function as intermediaries between the locals and visiting whalers and other vessels. As they grew in importance within the commerce of the local society they became known as "traders."

[151] Lessa, "Early Descriptions of Carolinian Culture," 314-15.

[152] Hezel, "Beachcombers, Traders & Castaways in Micronesia," http://www.micsem.org/pubs/articles/historical/bcomber/index.htm(accessed 29 Mar 2009).

[153] Wilson, *Cruise of the Gipsy*, 160. In *The First Taint of Civilization*, Francis X. Hezel noted that by mid-century, "Beside the veteran Englishmen on the island, the foreign colony was composed of Americans and Portuguese, largely from New England whaleships, and natives of other Pacific islands: Rotumans, Gilbertese, Maoris, and Tahitians." (p. 124.)

[154] Ship *George and Mary*, Greene, New London, Conn., Nov-Dec 1851, Log 84, MSM.

[155] Becke, "Deschard of Oneaka," *Ebbing of the Tide*, 160.

[156] Thomas Spencer, "Narrative of the Events Attending the Massacre of Part of the Crew Belonging to the Whaleship Triton of New-Bedford," *The Friend*, Sep 1848.

[157] Ship *Ann Alexander*, Log, 17 Feb 1848.

[158] He has been referred to as an "apprentice boy," so he was probably quite young. (Thomas H. Smith, mate, and Nathan S. Smith, second mate, *New Bedford Weekly Mercury*, 14 Mar 1845).

[159] Druett, in *In the Wake of Madness*, 147-230, examines the history of the voyage and its attending circumstances in great detail.

[160] Smith and Smith.

[161] Howard, *American Slavers*, 19.

[162] Gorham Parks, U.S. Consul, Rio, to James Buchanan, Secretary of State, 24 Jul 1847, HR Ex. Doc. 61, 30th Congress, 2nd Session, 5-7; Decker, *Whaling City*, 86.

[163] Anthony, Journal, Sep 1845-Mar 1847, passim.

[164] Ibid., 13 Jul 1847.

[165] Reilly, "Slavers in Disguise," 188.

[166] Clarence-Smith, *Third Portuguese Empire*, 71.

[167] Thomas, *Slave Trade*, 770 and 773.

[168] *New Bedford Evening Standard*, 30 Jun 1859.

[169] Railton, *History of Martha's Vineyard*, 175-76.

[170] See p. 216.

[171] William A. Allen, Journal, ship *Samuelson Robertson*, Jenney, New Bedford, Mass., 14 May 1845.

[172] Jose C. S. Maciel, U.S. Passport Application, 22 Jun 1859, U.S. Passport Applications, 1795-1905, General Records of the Department of State, RG 59, NAW.

[173] *New Bedford Evening Standard*, 21 Jul 1859; Charles W. Agard, "Live Ivory Trade I," Charles Walter Agard Papers, MSS 23, Ser. A, f. 2, no. 2, ODHS; *WSL*, 13 Sep 1859.

[174] Reilly, "Slavers in Disguise," 188.

[175] Ibid., 188 (Reilly mistakenly reports the *Comora* back in New Bedford in 1861); *Sydney (Australia) Herald*, 11 Mar 1861, quoted in *WSL*, 11 Jun 1861.

[176] Joseph Dias Jr., Journal, ship *Pocahontas*, Holmes Hole, Mass., 12 Dec 1850, Log 202/203, ODHS.

[177] Ibid., 12-24 Dec 1850.

[178] Ibid., 20 Dec 1850.

[179] Ibid., 21 Dec 1850.

[180] Ibid., 23 May 1851.

[181] Ibid., 15 Jul 1851.

[182] Ibid., 24 Oct 1851.

[183] Ship *St. George*, log, 11 Sep 1853.

[184] Ibid., 27 Sep 1854. C and J appear to be the third mate, C. H. Canfield, and either cooper John C. Belvie or blacksmith Joseph Turner (Seamen's Register, ODHS).

[185] Bockstoce, *Whales, Ice, and Men*, 13.

[186] His first voyage as master of the *Cleone* (Aug 1855-Apr 1858) had been quite successful, with a total catch of 367 barrels of sperm oil, 2563 of whale-oil, and 15,384 pounds of whalebone.

[187] Bockstoce, *Whales, Ice, and Men*, 185-86.

[188] Robert B. Strout, "A sketch of performance on board Bark Clione [*sic*], on her third season in the Arctic," Robert Bruce Strout Coll., MSS 210, MSM.

[189] See table, p. xx. It should be noted that these and succeeding voyage statistics represent ports in which the vessels were registered. Many whaling vessels, while registered in ports such as Provincetown, actually fitted out and departed from New Bedford, as well as discharging their crew and product there upon their return.

CHAPTER 4

Decline of American Whaling, 1860-1900

> For many years past the whale-ship has been the pioneer in ferreting out the remotest and least known parts of the earth. She has explored seas and archipelagoes, which had no chart, where no Cook or Vancouver had ever sailed.
>
> (Herman Melville, *Moby Dick*, Chap. 24)

The previous decade ended on a portentous note for the whaling industry. On August 24, 1859, the greatest conflagration in the history of New Bedford took place along Water Street, destroying many of the support industries of the whaling fleet, along with Wilcox's wharf, and 8,000 barrels of oil.[1] Of more universal impact was the drilling, only a few days later, of the first commercial oil well at Titusville, Pennsylvania. And soon the Union would begin to split apart, carrying ominous implications for the maritime industry.

The Civil War had a profound effect on the whaling fleet, from which it never really recovered. In the fall of 1861 the Union government attempted to block maritime commerce through Charleston, South Carolina, by sinking a number of vessels in the harbor. A total of forty-five were purchased, all but seven being whaleships. Many were aged craft, and their owners welcomed the opportunity to reinvest in more profitable enterprises. The Stone Fleet, as it was called, did impede maritime traffic for a short while, but it essentially had little impact. Nevertheless, it was an important influence in the decline of the whaling industry.

More destructive yet were the activities of the Confederate privateers. With the assistance of the British, privateers in the service of the Confederacy began to inflict heavy damage on the whaling fleet. The *Alabama*, under the command of Capt. Raphael Semmes, began her plunder around the Azores in late 1862. There she met and burned nine vessels of the American whaling fleet, including the *Ocmulgee*, sailing at the time out of Edgartown.

With the advent of war in April of 1861, New Bedford cautiously sent out fewer than half the whaling vessels that had sailed in the prior year (twenty-seven vs. sixty-eight). On the other hand Provincetown almost doubled its fleet—from sixteen sailings in 1860 to twenty-seven in 1861. And it paid a price. In May of 1861, with the Civil War now in full progress, Joseph M. Caton was sailing the schooner *John Adams* in company with the small ship *Mermaid*, and the brig *Panama*, some ninety miles south of Belize, when they were overtaken by the Confederate steam tugboat and privateer *Calhoun* with 120 men. The crews, including Joseph's brother, Manuel, mate of the *Mermaid*, were carried to the *Calhoun* and taken to New Orleans. As was the practice the three vessels, with prize crews aboard, were taken there also. Finding themselves deep in rebel territory and without funds, some of the men were forced to join the Confederate forces, while others were able to ship to Liverpool in merchant vessels in order eventually to reach home. Some of the officers were more fortunate. Captains Robert Soper Jr. of the *Mermaid*, George Powe of the *Panama*, and Joseph M. Caton of the *John Adams*, being allowed to travel cross-country to New England, along with mate Manuel Caton of the *Mermaid* and a couple of others.[2]

Most destructive were the activities of the Confederate warship *Shenendoah* under Capt. James I. Waddell. Operating late in the war and continuing, in fact, after the war had ended, Waddell managed to decimate the western Arctic fleet, burning twenty vessels and—because he did not have sufficient room for prisoners—bonding four more to carry them to port. By the summer of 1865 the depredations of Waddell and the *Shenendoah* were well known among the Arctic whalemen. On July 11, the bark *Cleone* of New Bedford, that John E. Simmons had commanded a few years before, was, as described by the log keeper, "working to the NE along the ice at 7 oclock the Capt came on board reported the Shenandoah Capt Wardello was seen a steeming out the Straits to the southward we made all sail working to the NW," and so the bark escaped the fate of so many others that season.[3]

Joseph F. Francis, a seaman on the ship *Milo* of New Bedford, and a future whaling master himself, recalled many years later the day that

Captain Waddell interrupted a pleasant gam between three New Bedford masters and the ensuing destruction. The *Milo* was cruising at the time in the Bering Sea, some twenty-five miles eastward of the Siberian coast. Nearby were the bark *Jireh Swift*, Captain Thomas W. Williams, and the ship *Sophia Thornton*, under the command of veteran master, Moses G. Tucker—both sailing out of New Bedford. The latter vessel had just come up from Honolulu loaded with fresh provisions, and the captains were aboard her sharing the latest news when a cry from the masthead alerted them to an approaching steamer. Fearing the worst the *Sophia Thornton* and the *Jireh Swift* made all sail in an attempt to escape. Both were no match, however, for the *Shenendoah*; and soon they were run down—not without a cry of protest from Captain Williams that could have come from the mouth of any of the masters of the twenty-four whaleships bonded or burned by the Confederate privateer.

Francis recalled that, as the *Shenendoah* approached the *Jireh Swift* Captain Williams vented to Waddell, calling him a coward and suggesting he should be somewhere south in the Pacific taking on the Union's men-of-war instead of destroying these defenseless whalers. Williams at that point shouted to Waddell to come on to his quarterdeck and fight him. If Waddell thrashed him, then he could burn the *Jireh Swift*, but if Williams won, then he and his vessel should be left to go in peace: "but not award was said from the Privateer, she sent her prize crew on board to cut away the mast[s] and then set fire to her and in short time the Bark Jerry Swift was all in a blaze."[4] The *Milo*, which had been boiling at the time and was on short sail, could not even make an attempt to escape. It was bonded by Waddell and given the task of returning the *Shenendoah*'s prisoners to San Francisco.

But other more permanent factors had been affecting the profitability of the whaling industry. Significant was the revolution in illuminants. By the 1840s, lard oil and camphene, as alternatives to whale oil, were diminishing the latter's dominance. By the early 1850s New Bedford began to move to coal gas in its street lamps. As Rev. Moses How of the Seaman's Bethel remarked on February 5, 1853, "This evening our Streets are lit up with Gass for the first time. Is it not strange that New Bedford which is emphatically the Oil Factory of the world, should abandon Oil which is for their own interest to burn, and follow the example of other Cities in the Use of Gass? but so it is. The cost of light[ing] our City and stores with Gass will, no doubt, be considerable more than the cost of Oil, but yet, the light, it is hoped, will be enough better to pay for it, or justify the change. This is an age of improvement in almost every thing, but Religion."[5]

Kerosene, in particular, first distilled from coal tar and later from petroleum, provided an efficient and inexpensive alternative to whale oil for illumination in the home. By 1865, it was selling for a third to a quarter of the price of sperm or whale-oil per unit of energy.[6]

Risk was another factor in the decline of whaling. Compared with shore-based industries it was considerably more of a gamble. A returning whaling vessel might possibly bring in a bonanza or, on the other hand, it might never come back. Out of 787 New Bedford whaling vessels 272 were lost. Since most sailed multiple times this nevertheless translated into a loss percentage of over six percent of all New Bedford whaling voyages.[7]

Another cause of decline in the industry was the rise of the cotton cloth industry in New England, much of it centered in the same region as the whaling industry. Thus capital in New Bedford began to flow to a safer and surer investment in the local mills that would ultimately become, for many years, the basis of the local economy. By the 1820s, nearby Fall River was becoming an important center for the manufacture of cotton. In 1846, two cotton mills were established in New Bedford, the New Bedford Steam Company and the more important Wamsutta Mills. In 1871, the Potomska Mills began operation; and then in the 1880s seven more. What is noticeable from the beginning is the siphoning off of capital that previously would have been directed to the whaling industry. The names of the owners and directors of these companies are a litany of old maritime names: Rotch, Grinnell, Rodman, Howland, Knowles, Bourne, etc. New Bedford's mills grew to seventy in number by 1920, before beginning a rapid decline.[8]

By the middle of the nineteenth century several Portuguese were part-owners of whaling vessels. The earliest owner on record was Antone Joseph of New Bedford, born in São Jorge, Azores in 1805. His portion of ownership was generally a modest one-sixteenth; but he continued as partial owner of a considerable number of New Bedford whaling vessels, beginning with the bark *Edward II* in 1846, and ending with the bark *Laconia* in 1871.[9] Many minor owners of whaling vessels had close maritime business connections and might act as well as recruiting agents. Joseph was no

Portuguese dry-goods dealer Frank Lewis with whaling agent W. R. Wing and daughters, New Bedford, 1868. Courtesy of the New Bedford Whaling Museum.

exception, for he ran a clothing and outfitting store at 173½ South Water Street, in a district that early on had become known as Little Fayal, for the large number of island Portuguese who had settled there.[10]

Antone L. Sylvia, born on the island of São Jorge in 1840, operated a similar establishment at 159 South Water. He was a more serious investor in the New Bedford whaling fleet, beginning with a one-eighth ownership of the bark *Tekoa* in 1865. Over the next several decades he held at least a one-quarter ownership in the barks *Crowninshield*, *Lydia*, *Peru*, *Minerva II*, and the *Greyhound;* and the schooners *Mary E. Simmons*, *Adelia Chase*, *Pedro Varela*, and *Carleton Bell*.[11] Toward the end of the century, Sylvia participated in other aspects of mainstream New Bedford capitalism, as a director of the Bennett Manufacturing Corporation, manufacturers of cotton yarn, and as a director of the New Bedford Board of Trade.[12]

Antone Thomas was born on the island of Madeira about 1815, and in 1849 was the owner of a grocery store at 111 South Water Street. Later he became owner of both a seamen's boardinghouse and outfitting store at the same address. Thomas continued in the business into the 1880s. And, like many of his contemporaries, he began to invest in whaling vessels, beginning with a small ownership in the barks *Hope* and *R. L. Barstow* of New Bedford in 1856. By the 1870s, he was the major owner of the schooners *Galena*, *George J. Jones*, and *Pedro Varela* and the bark *Edward Everett*, all of New Bedford, as well as the bark *Janet* of Westport.[13]

Joseph Vera (José Vieira Dias) was born in Lajes, Pico Island, Azores, in 1826 and his brother Frank (Francisco Vieira Dias), ten years later. Both sailed initially on American whalers out of Horta, Faial. After returning from a very successful three-year voyage on the ship *Montreal* of New Bedford (Jul 1850-Mar 1853), Joseph received as his share of earnings $3760.56 (worth some $100,000 in today's dollars), which he invested in a sailors' boardinghouse on South Water Street. Joseph continued to sail for at least another decade, and eventually opened an outfitting store that catered to Azorean whalemen. He was part owner, as well, in several New Bedford whaling vessels, unfortunately including the ship *Sophia Thornton*, which was burned by the privateer *Shenendoah* in June of 1865, more than two months after the end of the Civil War. Brother Frank whaled for a time in the western Arctic, then settled on a ranch in California, before finally returning to New Bedford and taking over some of his brothers' business interests.[14]

Another important ship owner was Henry Clay, who was also a whaling master for a number of years. But well before his first command he had invested his earnings in a half ownership of the schooner *Franklin* in 1862 and a quarter ownership in the brig *Star Castle* in 1867. Later he had major investments in the schooners *Golden City*, *Eleanor B. Conwell*, *Pearl Nelson*, and *Union*, and the barks *Tropic Bird*, *Palmetto*, *Gay Head II*, *George and Mary*, and *Bertha*.[15]

One of the most successful Portuguese involved in shoreside activities was Thomas Luce. Born on the island of Flores in 1827 or 1828, he came to the U.S. in August of 1843 aboard the bark *Brunette* of Falmouth. Captain Edward M. Luce took a liking to the youngster; his own young son, Thomas R. Luce, having recently died. Becoming a father figure to the young Azorean, Luce saw that he adopted the name of his late son.[16]

Luce sailed the next year as an ordinary seaman on the ship *Roman II* of New Bedford at a 1/170 lay, returning in July of 1847. As he had signed with an X for the name Thomas Luiz, it is apparent that at this date young Luce was, like so many of his compatriots, illiterate.[17] In 1849, he tried his hand, evidently somewhat successfully, at mining in

California during the Gold Rush. Upon his return he established himself in the cooperage business and, in 1852, married Hannah, the daughter of Captain and Mrs. Luce. In later years Luce was active as a whaling agent. Starting with a small-percentage ownership in 1867 of the New Bedford whaler *Robert Edwards*, he eventually became a major or sole owner of such vessels as the schooners *Antarctic*, *Francis Allyn*, *Charles H. Hodgdon*, *Clara L. Sparks*, *Era*, *Mary E. Simmons*, *Pearl Nelson*, *Sarah W. Hunt*, *Star King*, and the barks *Desdemona*, and *Gay Head II*.[18]

Thomas Luce evidently had a close relationship with the well-known Arctic whaling captain George Comer. Luce purchased the *Era* in 1895 with the intent of having Comer take it into the Canadian Arctic. Between that date and 1906 Comer made five voyages, becoming famous for his anthropological interest in the native Inuit with whom he wintered during his voyages, as well as his warm relations with them. On his way to and from his home in Connecticut, Comer would customarily stay at the Luce residence on County Street. On the eve of his 1903 voyage Comer noted in his journal, "Stayed at the home of Mr Thomas Luce where I have always been made at home." But Luce was getting on in years, and Comer noted the next day, "The three voyages before this there have been a party of old men come down with us but this time they feel as though they were too old to come, including our Mr Thomas Luce."[19]

In 1900, Luce expanded his operations further by opening a whaling and trading station in Hudson Bay, manned by whaling master George W. Cleveland. Cleveland spent several winters there trading for furs with the Inuit natives, but unfortunately had little regard for the Luce business interests. In the summer of 1904, he signed on to a Scottish ketch, "leaving almost 400 rotting skins at the abandoned station on Wager Bay for Comer to pick up."[20]

By the 1860s, one notes the presence of Portuguese enterprises grouped along the waterfront, much as would be manifested somewhat later in San Francisco. In 1867, boardinghouses were being operated along South Water Street by Joseph R. Dias (Azores), Antone C. Martin (Cape Verde), Antone Thomas (Madeira), and Francis Perry (Portugal).[21] Along the same stretch of South Water Street, Francis Antone (Portugal) ran a shoe store, while several Azoreans, Joseph De Cunha, Michael Marshall, and Antone Veara, were boot and shoemakers. Also just off South Water were Azorean shoemakers Antone Joseph and John Sylvia. John Francis

Schooner *Era* frozen in at Cape Fullerton, 1900.
Courtesy of the New Bedford Whaling Museum.

(Azores) manufactured and sold cigars at 157 South Water, while Antone Joseph, Antone Sylvia, and Vera & Rose operated clothing stores on the same street. Azorean Manuel Enos (Enas in the directory) was a manufacturer of seamen's chests, while Thomas Joseph and Antone Thomas were dealers in seamen's clothing; and Joseph Costa (Azores) was a merchant tailor—all along South Water and whose business was intended both for the neighborhood and maritime trade.[22]

In 1881, there were five Portuguese-run boardinghouses in town, with four of them located between 89 and 113 South Water Street. Along the same street could be found three saloons operated by Portuguese.[23] The 1892 New Bedford directory lists four boardinghouses, one clothing store, two tailors, and three saloons—all operated by Portuguese near the waterfront.[24]

A "Mere Shadow"

Of the various Portuguese and Brazilians involved in the slave trade out of New York at the time, one appears to have taken on the public face for this group of slave traders. John Albert Machado (João Alberto Machado), born in the Azores, a merchant and lawyer, settled in New York in the late 1840s and there became a U.S. citizen in 1849.[25]

There were many speculators in the slave trade at this time. But as a colleague of Machado explained, "most of them [what he calls "outside Speculators"] were induced to Embark in this business by the large profits reported to be made by others. Neirly all of them so far as I Know them have lost a great deal of money and are not likely to try it again."[26] Machado's connections with Africa go back at least to 1848, at which time he was on a vessel off the coast of Ghana (then known as the British Gold Coast).[27]

In the ensuing years Machado mixed legitimate business with significant activities in the slave trade. Some of the vessels he either partially owned or outfitted were ostensibly whalers. He was a half-owner of the ship *Brutus* (captained by Edward S. Devoll), which sailed from New Bedford in 1860 under a cloud of suspicion and eventually delivered a large number of Africans into slavery in Cuba during at least two voyages, before being burned in 1864.[28]

By the late 1850s Machado had established a close business and personal relationship with Mary Jane Watson, a wealthy widow better known simply as Mrs. Watson. Among the vessels the two sent out from New York for the slave trade was the ship *Thomas Watson* in 1860, supposedly on a whaling voyage. While other vessels in the "whaling" trade had some prior history in the industry, the *Thomas Watson* was a fast Baltimore clipper, ideal for the illicit run across the Atlantic but far from resembling the more awkward-appearing typical American whaler. To avoid suspicion Machado transferred ownership of the ship to Mrs. Watson, who ordered it to New London to outfit and take on a crew.[29]

Edward Manning was one of the many naïve young American green hands enticed to sign onto this "whaling" voyage. Later he recalled the moment in New London when Mrs. Watson appeared: "Not long after we had got the ship in order a tall and fine-looking woman came on board, and engaged the captain in conversation on the poop-deck. This lady was said to be one of the principal owners of the vessel; and, if so, she must certainly have felt some anxiety just then, as after-events will prove."[30] Anxiety must have also been felt by some of the crew members by that time as well, for suspicious local customs officials refused to give the vessel clearance papers. And so it was forced to return to

New York, closely trailed by a U.S. revenue cutter, which followed the *Thomas Watson* until it anchored off the Battery in New York. There the vessel received the necessary papers and slipped out of the harbor, with both Machado and Watson aboard, before it could be visited by the U.S. Marshal.[31] If there had been any doubts up to that point they were dispelled when the crew began to break out the hold, finding "huge quantities of rice, hard-tack, salt beef, pork, etc., in quantities large enough to feed a regiment for a long time. We also found a great amount of light pine flooring [to build the temporary deck to hold the slaves]."[32] Even one of the trypots was full of rice. Someone certainly had been convinced to turn a blind eye to such obvious signs of the trade.

After some 800 slaves were landed in Cuba, the pine flooring was thrown overboard and the crew paid off at Campeche, Yucatan, Mexico. On its return the *Thomas Watson* foundered off the North Carolina coast and was taken into the port of Wilmington in that state. Upon being released it was sailed to London where Mrs. Watson arranged to have it loaded with articles for the Confederacy. As it again approached the U.S. coast, the *Thomas Watson* was pursued by Union forces and run ashore near Charleston, South Carolina, then boarded and burned.[33]

Machado was arraigned in October, 1861, on a charge of fitting out the whaling bark *Mary Francis*, of Warren, R. I., for the slave trade, but released under bail two months later.[34] Doing business under the name of his accomplice, Mrs. Watson, had shielded him, but essentially condemned her; for with three warrants against her, she fled to Cadiz, Spain, where she allegedly continued their operations, outfitting vessels for the trade that had arrived with legitimate cargoes and then sending them on their way to Africa and thence to Cuba, until her death the following June. In the succeeding years Machado would occasionally appear in the news and the courts—accused of being "the ringleader of the Slave-traders" in New York City—but, like many others involved at the time in the slave trade, he was able to escape prosecution. As the *New York Times* wryly noted at the time, "There was abundance of testimony inculpating one Mrs. WATSON, the reputed paramour of the prisoner; but as for MACHADO himself, he appeared in the transaction like the echo of his name—a *mere shadow*."[35]

In September, 1862, Machado was rearrested as he was about to board a vessel for Havana and was taken to Fort Lafayette in New York harbor. There he faced new charges related to the *Mary Francis*, as well as those of outfitting vessels to run the Union blockade of Confederate ports, among them the *Thomas Watson*. U.S. Marshal Robert Murray of New York did not have the least doubt about Machado's central role in the New York slave trade. According to him, "Machado is, and has been for a number of years past, the Master Spirit in the Slave Trade—not only in this Port, but throughout the different ports in the U.S. and Europe." Murray was also as certain about the strength of the case against him, "as his conviction—which is positively certain—will be the finishing blow to the inhuman traffic, in every Port in the United States."[36]

But Machado had friends. His lawyers, the influential firm of Benedict, Burr, and Benedict, claimed he was a political prisoner. And when President Lincoln was informed of Machado's incarceration he apparently requested William H. Seward, his Secretary of War, to release him.[37] A week after being freed, on December 12, according to Marshal Murray, Machado was on a steamer to Havana to take charge of 1600 slaves deposited from two of his vessels, the *Amelia* and the *Clarissa* (neither of them evidently whalers). Upon his return in February 1863, Machado was immediately taken into custody once again, on a new indictment filed the previous month, but then immediately released on bail.[38]

With the coming of the Civil War, prosecutions of slave traders had begun to be carried out more seriously. In November of 1861, Samuel P. Skinner, owner of the whaling bark *Margaret Scott* out of New Bedford, was prosecuted by Massachusetts District Attorney Richard Henry Dana and sentenced to five years in prison. The most serious punishment at this time was meted out to Capt. Nathaniel Gordon of the merchant-slaver *Erie*, who was convicted as a pirate for his role in the slave trade and hanged in February of 1862. In 1863, the owner, agent, and captain of the whaler-slaver *Tahmaroo* of New Bedford were convicted on the testimony of a Portuguese crewman, also named Machado, and a Pacific islander, Joseph Mauri.[39]

As for John A. Machado, he once again evaded punishment with a *Nol Pros* plea entered on November 4, 1865.[40] In spite of an occasional small brush with the law and continued renown as an "ex-slave trader," he continued in business as a merchant, having married into a well-to-do upstate New York family. In the end, like so many of his contemporaries in the slave trade, John A. Machado left, indeed, a "mere shadow" of his illicit activities, protected in part by the ambivalence on the part of law enforcement in the North prior to the Civil War.

The Whaling Fleet Recovers

By mid-1865 the Civil War was over, and the seas were once again safe from the privateers who had terrorized the whaling fleet. In truth, quite a few whaling vessels had been sent out during the War, in spite of the threat of loss. In 1860, the year before hostilities began, a total of 104 vessels sailed from Buzzards Bay. During 1863, in the midst of the conflagration, there were still fifty-six sailings. And as the war wound down in 1865, a total of ninty-two whaling vessels departed the Bay. At this point one notes the continued expansion of Portuguese participation in the industry, based on statistics from these voyages. Just as the percentage had essentially doubled from 1845 to 1855 (6.7% to 13%), it did so again ten years later. Of eighty-eight vessels sailing from New Bedford and registered in several neighboring ports as well (Dartmouth, Fairhaven, Marian, and Westport), 688 of the 2658 men on board were Portuguese—a total of 25.88%. At this point the Azoreans were still dominant, comprising 62% of the total, while Cape Verdeans followed with 22% (six others were from Lisbon, three from Madeira, and twenty-three whose origin was not identified).[41]

Post-1865 saw significant advances in the technology of whaling. Hand lancing now began to go out of practice in favor of the usually safer bomb lances (a time-delay explosive-filled cylinder fired from a gun held on the shoulder) and later the Pierce darting guns that fired a harpoon which, on contact, drove an explosive lance deep into the whale; both allowed the whaleboat to maintain a safer distance from its target. On the other hand, the occasional malfunction of an armed device might seriously injure the user, as happened to Thomas W. Roys on board of the brig *William F. Safford* of Sag Harbor in November of 1856. Early on Roys saw the possibilities in using a rocket launcher to catch rorqual whales, such as the blue, which because of their size and speed were avoided by those practicing traditional hand methods of harpooning and lancing.

While experimenting with an English version of the gun in the Bay of Biscay, Roys lost his left hand when the instrument exploded. The vessel then steered for Porto, Portugal; but due to foul weather it took seventeen days to arrive. There the doctors were forced to amputate Roys's lower arm. But a determined Roys was off again whaling and perfecting his weapons by the middle of the following year.[42] Although his rocket launcher was never widely adopted, Roys nevertheless was a pioneer in what became known as modern whaling, practiced from steam-driven auxiliary "catcher" vessels sporting harpoon guns with explosive heads.

Frank Bullen, who served as a foremast hand on board the bark *Cachalot* of New Bedford at mid-century, recalled the reluctance of its officers to employ the bomb lance. Facing a defiant sperm whale which, after being harpooned, began to attack the two boats attached to it and, alternately, to drag them behind, the mates found themselves unable to lance it. "So Mr. Count got out his bomb-gun, shouting at the same time to Mr. Cruce (Cruz) to do the same. They both hated these weapons, nor ever used them if they could help it; but what was to be done?" Bullen himself discovered one reason why many were hesitant to use a bomb gun when he was knocked unconscious after having fired one.[43]

Similar disdain was affected toward the newer Pierce darting gun (which did not have the same "kick" as a shoulder gun): "But the weapon was so cumbersome and awkward, and the harpooners stood in such awe of it, that in the majority of cases the whale was either missed altogether or the harpoon got such slight hold that the gun did not go off, the result being generally disastrous."[44]

One place where these weapons were more extensively employed in the nineteenth century was in bay and lagoon whaling. Many of the vessels that hunted bowheads in the western Arctic found it profitable to spend their winter months catching gray whales in Baja California. On the open seas, the gray whale, known to whalemen as the "devilfish" for its aggressive behavior when attacked, was usually avoided. The gray whale had a tendency to sink; and its oil was considerably less desirable. But in its breeding ground in the Baja lagoons it was fairly easy prey, as evidenced by the record of whaleman Charles Melville Scammon and others in the late 1850s.

Manuel Silveira Andrade, the young seaman aboard the bark *Pacific* of New Bedford recalled how, in the fall of 1860, Captain Jacob A.

Howland, on his way down from a summer's whaling in the Siberian Arctic, stocked up in Hawaii on some darting guns in preparation "for devil fishing." In early December, the *Pacific* anchored in San Quintín Bay, Baja California, to await the arrival of the migrating grays, as they came down on their long journey from the Arctic. There the men would lower their boats each morning, anchor outside the point, and await the passing grays.[45]

After several days of this activity, the mates got together and decided that they would like their Sundays off. So early that Sunday morning, when Captain Howland ordered the boats to be lowered, no one complied. When he asked if they were going to lower, the mate replied (as recalled in the words of Andrade), "Well sir, we aint going [to] lower the boats here in Sundays, I think." An angry captain asked why. "Because, sir; there is six days in a week for to work and we may leave the seventh for to rest." The captain then said, "But you lower out at the sea, don't you?" The mate then explained to the captain the distinction, as he and the other mates saw it: "Yes sir," he retorted, "but there is a different whaling altogether; here, we have and will seen whales enough in all the weekdays, to be sure; and at the sea, we perhaps would not seen one in the whole week, and then when we see it, we might as well try it; and here, as we have no time at all to wash a shirt, neither to wash ourselves hardly, we most get time for to do this, any way."[46]

This, of course, was mutiny, and in it the foremast hands were in total sympathy with the officers. The second and third mate, in turn, refused the captain's direct order to lower—upon which the captain stated that if they would not lower that day then they could not lower the next, and he stormed down to his cabin. The next morning, amid general apprehension, the captain reappeared and turning to his officers said, "Well, that is over; lower away your boats."[47] And so, evidently, during the next several months of bay whaling, Sundays were a day of rest. It should be mentioned here that whaling on the Sabbath, even at sea, had always been somewhat of a contentious issue. But in general, no matter what the policy of the vessel might have been upon sailing, any restriction on Sunday whaling generally was not followed.

Among the whalemen Sunday was not the only problematic day. There is an old superstition among sailors that it is bad luck to sail on a Friday, the *dies infaustus* in the words of a nineteenth-century compiler of nautical terms.[48] In part to dispel this belief the *Republican Standard* of New Bedford reported in 1854:

> The whaleship Hillman, which arrived here on Friday, seems to disapprove [*sic*] the old superstition that Friday is an unlucky day, most materially. We learn from the first officer of the vessel, that the Hillman sailed from this port on Friday, that she took the first blackfish on Friday, her first whale on Friday, made her first port out on Friday, and finally arrived here on Friday, with 3500 barrels of whale-oil, 180 barrels sperm oil and 61,000 lbs. of bone on board and sent home. Is Friday unlucky? The owners of the Hillman, we presume, are inclined to the opinion that it is not.[49]

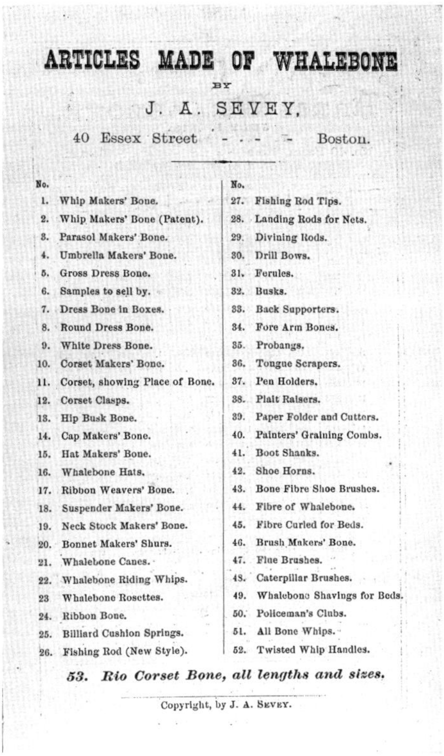

ARTICLES MADE OF WHALEBONE

BY

J. A. SEVEY,

40 Essex Street - - - - Boston.

No.		No.	
1.	Whip Makers' Bone.	27.	Fishing Rod Tips.
2.	Whip Makers' Bone (Patent).	28.	Landing Rods for Nets.
3.	Parasol Makers' Bone.	29.	Divining Rods.
4.	Umbrella Makers' Bone.	30.	Drill Bows.
5.	Gross Dress Bone.	31.	Ferules.
6.	Samples to sell by.	32.	Busks.
7.	Dress Bone in Boxes.	33.	Back Supporters.
8.	Round Dress Bone.	34.	Fore Arm Bones.
9.	White Dress Bone.	35.	Probangs.
10.	Corset Makers' Bone.	36.	Tongue Scrapers.
11.	Corset, showing Place of Bone.	37.	Pen Holders.
12.	Corset Clasps.	38.	Plait Raisers.
13.	Hip Busk Bone.	39.	Paper Folder and Cutters.
14.	Cap Makers' Bone.	40.	Painters' Graining Combs.
15.	Hat Makers' Bone.	41.	Boot Shanks.
16.	Whalebone Hats.	42.	Shoe Horns.
17.	Ribbon Weavers' Bone.	43.	Bone Fibre Shoe Brushes.
18.	Suspender Makers' Bone.	44.	Fibre of Whalebone.
19.	Neck Stock Makers' Bone.	45.	Fibre Curled for Beds.
20.	Bonnet Makers' Shurs.	46.	Brush Makers' Bone.
21.	Whalebone Canes.	47.	Flue Brushes.
22.	Whalebone Riding Whips.	48.	Caterpillar Brushes.
23	Whalebone Rosettes.	49.	Whalebone Shavings for Beds.
24.	Ribbon Bone.	50.	Policeman's Clubs.
25.	Billiard Cushion Springs.	51.	All Bone Whips.
26.	Fishing Rod (New Style).	52.	Twisted Whip Handles.

53. Rio Corset Bone, all lengths and sizes.

Copyright, by J. A. SEVEY.

Courtesy of the New Bedford Whaling Museum.

Evolution of the Industry

With access to whale oil diminished during the Civil War period, the public began to turn to other means for lighting and lubrication, principally through the nascent petroleum industry. Nevertheless, sperm oil, in particular, would remain in demand well into the twentieth century as a fine lubricant. Demand for whale products now turned more and more to the baleen produced by the mysticeti whales, principally the right whale and the bowhead of the western Arctic. Whalebone, as it was called, was employed in women's stays, in parasols and umbrellas, fishing rods, and whips, among other uses.

And it was still possible for forecastle hands to receive a decent remuneration for their time at sea. When the ship *Arnolda* of New Bedford, under Capt. William T. Hawes, returned in April of 1866 from a two-and-a-half-year voyage to the North Pacific, having obtained 2636 barrels of whale oil and 33,000 pounds of whalebone, its crew was relatively well compensated. Seamen Joseph Sylvia and Manuel Veara (alias Henry Carr) netted $733.54 and $871.14, respectively.[50]

Old whaling ports now began to fade: Nantucket sent out its last whaler in 1869; Mystic, ca. 1871; Sag Harbor in 1874; while others had already closed their industry: Bristol, Rhode Island, in 1858; Newport, Rhode Island, 1860; Cold Spring Harbor, 1862. Meanwhile, another, San Francisco, was just beginning to grow. Nevertheless, traditional whaling, ever diminishing in intensity, was destined to last into the third decade of the next century. And the Portuguese of the Atlantic islands were ever increasing their participation—and not only as whalemen, but also as agents and owners.

During this period, notices began to appear encouraging the use of "recruiting" (provisioning) stations operated by Portuguese in various Atlantic ports. In 1858, Joachim P. Coelho and Company advertised its location on the island of Corvo in the Azores. Francisco de Paula Jr. was prepared to supply vessels at Praia da Vitória, Terceira, Azores, as was Antonio B. Fortes at Praia, Santiago, Cape Verde. Awaiting vessels at Barbados in the West Indies, was DaCosta and Co., and at Equimina Bay in Angola was Antonio DeBastos Pina. And from Brazil, Manoel Moreira da Silva begged "to inform owners and captains of whaleships that he has located himself at St. Cruz, (St. Catherines) offers his services to captains of whaleships to provide refreshments and supplies while visiting this port."[51]

The Portuguese of Hawaii

After 1830, Hawaii began to fit out an occasional whaleship of its own; but most activity took place during the 1850s and 60s. In 1858 nineteen Hawaiian whaling vessels sailed; from 1856 to 1870, the fleet averaged twelve voyages per year.[52] In December 1867, Azoreanas Manuel Pico and Jason Perry, together with Capt. Joseph Spencer, purchased the schooner *Alberní*, which became the Hawaiian whaler *William H. Allen*.[53] The brigantine, of 157 tons, had at least one Portuguese master, Captain A. Vera, who sailed in October, 1869 and returned twelve months later from the coast of Peru with one hundred barrels of sperm oil and two hundred of whale-oil. Another voyage is recorded with a Portuguese captain under the Hawaiian flag. In January 1871, the brig *Comet* sailed from Honolulu under the command of a Capt. Silva, returning in April with eighty barrels of sperm oil. The vessel sailed again a day later but was an unfortunate victim of the western Arctic disaster of that year.[54]

Manuel Pico, whose surname became altered to Paiko, was evidently an ex-whaler and native of the island in the Azores for which he was named. He became a prominent citizen in the islands, marrying a native woman in 1853. Later he served as superintendent of roads on Maui.[55] Jason Perry [Jacinto Pereira] was born on the island of Faial in 1826 and arrived in Hawaii as a young man, most likely on a whaler. He owned a dry goods store and, in 1876, became the first Portuguese consul to the Kingdom of Hawaii. The *Hawaiian Gazette* noted upon his appointment—two years before the commencement of mass immigration from Portugal—when already there were some 400 Portuguese, mostly Azoreans and Cape Verdeans, living in the islands and working mainly on cattle ranches and dairies: "The number of Portugese [*sic*] in the country is already quite large, and it is highly appropriate that they should have a representative of their government near at hand. The selection will meet with general approval. Mr. Perry is a resident of many years standing on these islands, and he has by his thrift and industry amassed quite a fortune, and secured the esteem of all."[56]

By the 1870s, the importance of whaling to the Hawaiian economy was diminishing, while at the same time the cultivation of sugarcane was in ascendance. Due to the shortage of labor on the plantations, experiments were made in importing foreign labor. Traditionally this was supplied by men from China. Reluctant for the society to be inundated with Asians and, at the same time, desiring to broaden the immigration pattern by bringing in entire families, the government considered many possibilities. The first to be acted upon was to bring in contract laborers and their families from the Azores and Madeira. Dr. William Hillebrand, a German citizen who had lived in Hawaii, and was temporarily on the island of Madeira, wrote to the government of Hawaii to encourage this enterprise. The tone of the letter is both enthusiastic and subtle, for he was well aware that the life to which these immigrants were committing themselves for three years would be defined by its privations: "In my opinion your islands could not possibly get a more desirable class of Immigrants than the populations of the Madeira and Azore Islands. Sober, honest, industrious and peaceable, they combine all the qualities of a good settler and with all this, they are inured to your climate." With that he added, "Their education and ideas of comfort and social requirements are just low enough to make them contented with the lot of an isolated settler and its attendant privations, while on the other hand their mental capabilities and habits of work will ensure them a much higher status in the next generation, as the means of improvement grow up around them."[57]

In 1878, the *Priscilla*, the first of many ships carrying entire families from the Azores and Madeira, arrived in Honolulu with sixty-three men, sixteen women, and thirty-five children. Between this date and the end of the century, a total of 12,780 Portuguese would arrive and settle on the sugar plantations—5364 men, 2486 women, and 4930 children.[58] Their odyssey, their long struggle for assimilation, and their contributions to Hawaiian society have been related in a number of publications.[59]

"Big Manuel"

Manuel Enos, the Cold Spring Harbor whaleman, had returned to his family in May 1859 and to a new profession as storekeeper. But soon the sea would call once more; and in September 1860, Enos embarked as mate for a four-year cruise on the bark *Java* of New Bedford. It is easy to conflate this passage of time in a brief narrative; but a pastor in Nantucket during the height of the whaling period described some of the

"Cap'n Enos," sketch from Beane, *From Forecastle to Cabin.* Courtesy of the New Bedford Whaling Museum.

passions evoked by these long absences: "The departing father imprints a kiss upon the cheek of his child. Four years will pass away ere he will again take that child in his arms. Leaving his wife sobbing in anguish, he closes the door of his house behind him. Four years must elapse ere he can cross that threshold again."[60]

And so it was with Captain Enos. In June of 1864, after an absence of forty-five months, he returned home to his wife and four-year-old daughter. Two months later, Manuel Enos sailed again on the *Java*, this time as master. Heading first for the Azores, Captain Enos was a most proud man as he brought the vessel into his hometown of Lajes on the island of Pico. Joshua Beane served under Enos and later wrote an account of the voyage. As Beane relates, Enos had visited the island as third mate on another vessel some fifteen years before; and now, "standing more than usually erect in the stern sheets," he "left the ship to surprise his people."[61]

Having departed in August of 1864, in the midst of the War, Enos suffered no problems from the American hostilities; rather, reaching the

North Pacific, he became involved in an international incident while whaling in the Sea of Okhotsk in Russia's Far East. In reading cooper Charles Sherman's journal from those days or Joshua Beane's recounting of meeting their Russian counterparts, all relations were extremely harmonious. The former notes how at one point the captain of a Russian schooner came on board the *Java* and "got a bag of Beans & a bag of Onions."[62] Beane writes of even warmer relations between whalemen representing the two nations when men from the *Java*'s boats and those from a Russian whaler joined on shore in a relatively sumptuous late-night feast of brandy, vodka, salmon, salt beef and pork, and hardtack.[63]

There may have been occasional visits by American whalers in Russian waters during the 1830s and 40s, including one by Captain George Covell of the *Mount Vernon* of New Bedford in 1846; but credit for opening up the Sea of Okhotsk to American whaling generally is given to Thomas Roys, captain of the bark *Superior* of Sag Harbor, who whaled successfully there and in the Sea of Anadyr before passing through the Bering Strait and entering the Arctic Ocean in 1848. Here, according to a Russian complaint in 1863, Americans acted with impunity, not only in the territorial waters but also on land, where "they behave as if they had conquered the country: they burn and chop its forests, hunt its game as well as its whales."[64]

Beane relates one tragic incident that took place in the spring of 1866, when the *Java*'s whaleboats were dispatched in various directions from the bark's anchorage in Southwest Bay in search of whales. There, far from the ship, they fought for two days to survive the severe ice conditions, just missing "having our boats crushed to atoms between masses of ice whirled here and there by the ever-changing tide," and "pulling our boats bodily out of the water on to some large cake of ice in the immense field of dazzling white, which extended farther than the eye could reach." When they finally made the shore they found "a log house, built a few years before by some of our enterprising brother whalemen." Unfortunately someone was careless with the fire; and that night with cries of "Come this way!", all succeeded in exiting the hut but one, Beane's after oarsman, Boy Dick, who was last seen "standing upright within the hut, his arms outstretched as if feeling his way," before collapsing into the flames. The next morning, after scraping out a shallow grave and burying his remains, the men cut his name into a stump: "Josa Sylvia." Then, "without a word, the boys filled the boat kegs, carried the kettle to the boat, and prepared to search for the ship." The next day the *Java* flew its colors all day at half-mast in recognition of the loss of the Portuguese seaman.[65]

It was just this sort of activity by foreign whalemen on its sovereign shores that continued to irk the Tsarist government for at least a decade. Finally, in 1866, the government signed an agreement with three maritime captains from Russian Finland to start Russian whaling there. While initially Enos seems to have found them cooperative (in exchange for allowing his third mate at the time, Antone Perea (Antonio Pereira), to serve with them for the season, Enos was rewarded with a bomb lance gun), eventually he found himself confronting the authorities who questioned his right to whale wherever he wished in Russian territory.[66]

By 1867, attitudes had evidently hardened, for in December of that year the American consul in Honolulu forwarded a deposition by Enos to Washington that described the changing circumstances:

> Manuel Enos, after being duly sworn, deposes, and says, that he is master of the American bark Java; that on the 27th day of July last past, while cruising for whales in Shantar bay, standing towards Silas Richards' bluff, boiling, we raised a smoke towards Shantar gut, which I supposed to be another whaler trying out; soon after, however, we discovered it to be a Russian steamer coming towards us, apparently under full steam, hoisting his flag and throwing open his ports. I ran up my ensign and hauled aback. An officer came on board and ordered me immediately on board the steamer. On arriving on board the Russian vessel, the commander wanted to know what I was doing there. I told him whaling, of course. Without more ado I was ordered to leave the bays within 24 hours. I told him I had boats in Mercury bay, and could not leave until I had them on board. His answer was; 'Boats or no boats, within 24 hours you must leave, or you will be taken to Nikolaievsk or blown out of water, as I shall think proper.' I told the commodore that I had 'whaled it' in these bays for the last 17 years, and had never heard of any one being driven out, nor even a whisper that such might be the case at some future time. My boats very opportunely returned the same night, and I left Shantar bay.

> I was afterwards told by some of the crew of the American bark Endeavor that, knowing nothing of the trouble, they came there a few days after, and had their boats fired at by the same vessel. They immediately pulled for their vessel, and were not troubled further than to receive the same orders that I had.[67]

Manuel Enos was not content with giving up his perceived rights; and the very next day, in company with the captain of the schooner *Caroline E. Foote* of San Francisco, he complained to the Russian authorities that he had whaled previously unmolested in the bays of Okhotsk Sea, and that they had no right to prevent him from whaling wherever he liked, to which Secretary of State William H. Seward felt obliged to communicate to the Russian government that "the captain of the 'Java,' spoke unwarrantably when by implication he denied that the Russian authorities have the right to prevent foreign vessels from fishing for whales within 3 marine miles of their own shore."[68]

This was only a minor problem, however, for Enos and the *Java*. After three seasons in the Okhotsk and a circumnavigation of the globe that lasted from August 1864 until April 1869, the *Java* returned to New Bedford with 112 barrels of sperm oil, 667 of whale-oil, and 5140 lbs. of whalebone—having previously sent home from Honolulu 1075 barrels of whale-oil and 13,500 lbs. of bone. The voyage netted $96,200 (approximately $1.5 million in today's dollars) and earned Enos a $1,000 bonus.[69] Enos could now return to Cold Spring Harbor a relatively wealthy man.

Arriving home in May of 1869, Enos discovered again a four-year-old daughter, Melna having died and been replaced by Ellen. Married now for fifteen years and present at home less than two, enriched from his maritime activities and forty-two years old, Enos decided to settle down. Soon he was operating in the coastal trade with his own 240-ton schooner, the *Flyaway*. Presumably Enos was well acquainted at this time with one of the few other Portuguese to have settled in Cold Spring Harbor, Joseph Sylvia, also in the coastal trade with his schooner *Fernie Bodine*.[70]

Four years later, in 1873, Enos was operating a tannery and kid-leather factory; and in the following year he constructed the large residence that still stands on Main Street, across from the Cold Spring Harbor Whaling Museum.[71] For the next four years he remained in Cold

Spring Harbor, leaving little record of his activities. But, as Robert Farwell has suggested, "Behind the façade of success, however, lay Captain Enos' permanent failure to cope with life ashore, and real inability to transfer his skills as a whaleman to the supervision of other businesses."[72]

In 1877, Enos decided to go to sea again as mate on the bark *George and Susan* of New Bedford. Having returned in 1881, he shipped out immediately as mate of the New Bedford bark *John and Winthrop*.[73] The motives that led Enos to turn around and sail again may only be hypothesized. What is known is that he never returned to Cold Spring Harbor. In an article that appeared in the *Brooklyn Standard-Union* some years later, entitled "Where is Enos?," the writer summed up the facts as they generally have been known. Enos left the *John and Winthrop* at Talcahuano, Chile, and purportedly took command of the former bark *Matilda Sears* of Dartmouth, Massachusetts, now sailing out of Talcahuano as the *Mathieu y Brañas*:

> and here all traces of Captain Enos disappear. It is known that the Matilda Sears was sold and became the Machias Bramas [*sic*], but what became of Enos nobody knows, or if he is dead or alive, for no word has ever come to his friends regarding his fate.... Now and then a company of shipmates of the missing captain will meet and talk of old times. They will recall the time when Enos' ship was stranded up in the Arctic, and they were just about to abandon her when she slipped off with the rising tide and they went on their way rejoicing. They also recall the cask of delicious wine that the hardy whaleman was wont to bring on his return voyage from his father's vineyard at Fayal, in the Azores, but the old salts wait in vain for an answer to the question, "Where is Enos?"[74]

The abandonment of family and hearth implicit in the above ill fits with the widespread image of "Big Manuel," as he was known and admired around the village. Years later, in a 1953 letter to a curious young relative, Enos's granddaughter synthesized the family's sensitivity, referring to the public "conjectures."[75] Even the most exacting historians of Cold Spring Harbor and its whaling industry have insisted that Enos, having left the *John and Winthrop* in Talcahuano, Chile, and taken command of the *Mathieu y Brañas*, had been lost at sea with vessel and crew.

In fact, the *Mathieu y Brañas*, purchased by Chilean interests in 1882, continued to form part of the small Chilean whaling fleet operating out of Talcahuano for more than two decades.[76] Had Manuel Enos died under unknown circumstances, people asked, far from home and family? Or was it possible that the Portuguese presence in the Chilean whaling industry, including Francisco Vieira, the third owner of the whaling enterprise of Mathieu y Brañas, could have attracted him to end his days in this distant South American nation? The truth is that, indeed, Manuel Ignacio Enos de Macedo—as he became known in Talcahuano—met and eventually married a young local woman, María Petronila Araneda, with whom he had six children and thus ended his days as the scion of a Chilean Enos family.[77]

Evidently Enos continued to whale out of Talcahuano on vessels owned by Mathieu y Brañas well into the next decade. His last command is supposed to have been on the whaling bark *Ana Aurora*. Today there is a street in the town named in the honor of "Capitán Enos," who passed away on March 15, 1915.[78] Both the name of Manuel Enos and the seafaring tradition continued with Big Manuel's son, Manuel I. Enos Araneda and a grandson, Manuel Enos Sosa, who, beginning in 1939, spent almost sixty years at sea, many of them as master of merchant vessels. In the fall of 1965, Manuel Enos, as he was known, brought the new Chilean freighter *Aconcagua II* into New York harbor, directly from the Japanese shipyard where it had been built, "and received the traditional maiden voyage welcome, complete with fireboat salute, tugboat escort and helicopter aerial cover."[79]

Other Portuguese Masters

The number of voyages commanded by the Portuguese dipped somewhat during the 1860s just as did the industry as a whole. During this decade five vessels departed New Bedford under their command, and two each from Fairhaven, Nantucket, Newburyport, and Provincetown. It was during the remaining decades of the century, however, that participation grew rapidly. During the 1870s, Azorean and Cape Verdean masters took twenty vessels out of New Bedford, eight from Provincetown, four from Boston, and one from Fairhaven. Again during the 1880s, there was a significant increase, to fifty-seven—forty-seven of which originated in

New Bedford, six from Provincetown, two from San Francisco, and one each from Edgartown and Stonington. During the 1890s, there were forty-seven voyages, of which thirty sailed out of New Bedford, thirteen from Provincetown, and one each from Boston and San Francisco.[80]

Indicative of the increasing presence of Portuguese speakers at the highest level of command is the use of their native language while acting as log keepers aboard ship, normally a routine function of the first mate, or in maintaining a personal journal. The journal of John A. Vieira, a boatsteerer aboard the ship *Gazelle* of New Bedford during its voyage of 1866-70, is an early example. The vessel departed New Bedford on August 13, 1866, with a crew of thirty-one, composed, as Vieira notes, of fourteen Americans, fourteen Portuguese, two Dutch, and one Brazilian. As was already quite typical, the captain was American; the first, second, and third mates Portuguese; and the green hands from places such as Philadelphia and New York.[81]

By 1875, two trends are apparent—the industry is shrinking, and the Cape Verdean presence is increasing rapidly. That year there were fifty-eight voyages out of Buzzards Bay. A compilation of fifty-four of these shows that twenty-nine percent of the whalemen were Portuguese, with more than half of these (fifty-four percent) from Cape Verde. Factoring in the additional Portuguese customarily signed on at the islands, this figure corroborates Elmo Hohman's estimate that by 1880 the crews of the New Bedford whaling fleet consisted of one-third each of Americans, Portuguese, and others.[82]

Whaling captains were now relying increasingly on Atlantic and Pacific islanders to complete their crews. When the bark *Wave* sailed short-handed from New Bedford for a cruise in the Atlantic in August 1879, with just twenty-one crew members, it was already carrying five Azoreans, four Cape Verdeans, a West Indian, three Pacific islanders, and a man from St. Helena.[83] And it would soon be adding more islanders along the way.

At times a whaling crew might be predominantly Portuguese, as in the case of the bark *Lydia*, which sailed out of New Bedford in November 1880 under Portuguese American Captain Edward M. Frazier, with thirteen Azoreans, five Cape Verdeans, and three other Portuguese out of a total crew of twenty-seven, plus Captain Frazier.[84] And when the bark *Sarah* left New Bedford in June 1876 for a two-year voyage under the command of Joseph D. Silva, among a crew of twenty-two—excluding the captain—it carried thirteen Azoreans (from six of the nine islands) and six Cape Verdeans.[85]

Besides the consecrated superstitions such as that related to Fridays, others might spring spontaneously from the minds of crewmen in this hazardous occupation. In 1878, the *Whalemen's Shipping List* reported a communication from the bark *Lucretia* of New Bedford, to the effect that "the third mate of the Lucretia, on striking a whale, got stove, killing the boatsteerer and tub oarsman; they have lost three men this year, all Portuguese, and the remaining Portuguese of the crew think they are fated and propose to desert."[86] Since there is no extant log, we do not know whether the Portuguese ever carried out their threat.

Naturally, name changes continued during all this period. Some were simply harmonic, such as Mendonça > Mandly, or perhaps initially humorous (Acquilla Rodrigues > Henry Clay). Not uncommon was the Hispanicization of Portuguese surnames (Lopes > Lopez, Gonçalves > Gonzales). As previously noted, yet another was the result of the paternal relationship between the ship's master and the cabin boy, often on his first maritime voyage and perhaps only in his pre-teen years. Such was the case of Joseph Sayre, of Riverhead, New York, born José Silveira Bettencourt in 1816 on the island of Faial. Details of his youth are hazy, but he apparently signed on as cabin boy on a vessel commanded by Isaac Sayre, perhaps the ship *Delta* of Greenport, Long Island, in 1832; for it was on Long Island that he settled and from which he whaled.[87]

Thomas E. Fordham, who had some especially profitable voyages to the Pacific Ocean, was already born with the family surname. One surmises that his father, Enos Fordham, had been a whaleman before him. Thomas was born on the island of Faial, ca. 1828. Upon his arrival in the U.S. he settled in Edgartown. However, his voyages as master however were all made from New Bedford. They include the ships *Milo* (Nov 1859-May 1863) and *Florida* (Jul 1864-Sep 1868). The catch of the former totaled 263 barrels of sperm oil, 3346 barrels of whale-oil, and 28,500 pounds of whalebone. As was common with voyages of this length, much of the cargo was transshipped, in this case from Hawaii.[88] The return from the *Florida* was even more impressive: 241 barrels of sperm oil, 1688 barrels of whale-oil, and 48,680 pounds of whalebone. One also notes as well that the *Milo* returned to port in the middle of the Civil War, while the *Florida* sailed as the Confederates were threatening the capital and Sherman was marching through Georgia.

Fordham's other commands were: bark *Tamerlane* (New Bedford, Jul 1869-Jun 1873); schooner *Lottie E. Cook* (New Bedford, May 1884-Aug 1885); and (as replacement master) brig *Eunice H. Adams* (Edgartown, Oct 1887-Sep 1890). The four-year cruise on the *Tamerlane* was not particularly remunerative and had its difficulties, probably related to the weak catch. Typically, Fordham spent the short summer season in the western Arctic and during the rest of the year cruised in the South Pacific, occasionally making ports such as New Zealand's Bay of Islands and Sydney, Australia. There, in January of 1872, nine men, including two Portuguese, deserted. Worse yet, that November, after the *Tamerlane* returned from the Anadyr Sea and anchored in San Francisco, a total of twenty-three, out of a crew of thirty-one, deserted.[89]

The voyage of the *Lottie E. Cook* was even less fortunate. During his fifteen-month voyage, Fordham followed a typical pattern for schooners in the Atlantic. He was first on the Western Ground; then in July at the Azores; in October, at the Cornell Ground and Abrolhos Banks; November on the coast of Africa; Abrolhos again; at Barbados in February; Dominica in March; the Western Grounds again in April; Charleston Grounds in June; and home in August. For all of Fordham's efforts the schooner returned with a mere eighty barrels of sperm oil.[90]

Entering a vessel as replacement master brought its own problems. While in the West Indies in February of 1890, Fordham was contracted to bring the brig *Eunice H. Adams* back to Edgartown. Upon coming aboard, he discovered "all hands dissatisfied with every thing Brig leaking bad and provisions bad and long layes so ends this day."[91] By the next month he had the brig back in its home port. Evidently still sailing in 1892, Fordham died at Bahia Tarta, Angola (known to the whalemen as Salt Pond) early that year.[92]

Severino D. Pierce, who had sailed as master of the ship *Magnolia* of New Bedford in 1858-62, later served as replacement master on the ship *Europa* of New Bedford, during its voyage of 1862-67. He must have joined somewhat later, for he does not appear on the outgoing crew list.[93] His next voyage was as mate and log keeper of the schooner *Washington Freeman* of Fairhaven (Nov 1868-Sep 1870). Upon his return in the fall of 1870 he took command of the schooner *Thriver* of Boston.

The schooner sailed on November 21. The brief log ends in April of the following year. There is a note of the sale of one-half gallon of an unspecified liquid, presumably whiskey, dated 25 Sep 1871, charged to Captain Pierce.[94] Twice the *Thriver* was forced to seek repairs at Bermuda, in October of 1871 and September of the following year. Finally it was lost on the island of Haiti in February 1873. It has been reported that Pierce was killed by a whale on this voyage, but the documentation is scanty.[95]

One of the interesting common experiences of many of these men who reached the apex of their industry was the young age at which they left their native isles. Joseph D. Silva, born in Faial in 1823, left home at the age of nine to serve as cabin boy on the whaleship *Brandt* of New Bedford under Captain Warren Howland.[96] He commanded a number of vessels over a thirty-year period, beginning with the bark *Milwood* of New Bedford (Jul 1854-Jul 1857) and ending as replacement master on the schooner *Hattie E. Smith* of Edgartown during its voyage of Oct 1885-Aug 1887.

Another young arrival was John E. Luce, who came to the U.S. in 1834 at the age of twelve aboard the ship *Chili* of New Bedford, in which he had been serving as cabin boy for Captain Lot Luce. As was not uncommon at the time, once ashore, the young man went to live with the captain and his family on Martha's Vineyard, there adopting the surname of his benefactor. Luce whaled for the next several decades. He served as mate under Capt. John P. Praro Sr. on the bark *Crowninshield* of Fairhaven, Massachusetts, during its only voyage, 1869-73. In a letter he wrote from Hobart, Tasmania, to a Captain Hathaway, in the fall of 1871, Luce furnished particulars regarding the catch, as well as his concerns about the profitability of the voyage, and his low opinion of the master. "Since leaving the bay [Bay of Islands, New Zealand] in April Last," he noted:

> we have taken 450 sperm 100 whale I am in hopes to Pick up 300 More so that I shall Make booth ends Meet but I am afraid That we shall all looze in this … for oil is low & expenses is Going to be very high I don't Sea My way clear yet I shall Go as far as the bay & sea how Things is there all keeps Me here is My Property this is the hardest Time I ever saw a whaling.… We officers Get along first rate. We have not lost but one whale This voyage by our boats.… We have Got a very abusive Man

> for a Master & we should have all Left long a go if we had been like some Men but we all Want to Make a voyage.[97]

Later Luce took out the New Bedford schooner *Fannie Byrnes*, of which he was part owner, for a twenty-month voyage (Dec 1877-Aug 1879).[98]

Henry Mandly Sr. (Henrique Mendonça, 1848-1944) was born on the island of Graciosa in the Azores. In 1862, just short of fifteen, he signed on as cabin boy aboard the ship *Bartholomew Gosnold*, on its way to the Indian Ocean.[99] His first command came fourteen years later on the schooner *Quickstep* out of Provincetown (Nov 1876-Aug 1877, Nov 1877-Sep 1878, and 1878-?). He later sailed the schooner *Mary E. Simmons* out of New Bedford on twelve voyages between 1879 and 1904.

Henry's younger brother, Antone J. Mandly (1863-1929), shipped as cabin boy at the age of eleven on the packet bark *Kate Williams* and later spent a career as master of whaling schooners out of the same two ports from 1883 until 1925, when he took out the *John R. Manta* on New Bedford's last whaling voyage. His first command was as replacement master on the schooner *Franklin* out of New Bedford (Oct 1883-Aug 1885). He then took the vessel out as master on its next voyage (Oct 1885-Jul 1887). While his hunt for oil on this voyage was not very successful, Captain Mandly was able to claim credit, along with Captain George S. Johnson of the schooner *Antarctic* out of Provincetown, for a lump of ambergris valued at the time at $14,000.[100]

Joseph Thompson (1830-1906) was born on the island of Faial and at the age of eleven signed on as cabin boy aboard the bark *Garland* of New Bedford.[101] As happened at times to other young Portuguese, Captain Albert Scranton took a liking to the youngster and, once back home, brought him to live with him and his wife, who began to tutor him. In time, Captain Scranton died, and young Thompson married the widow, five years his senior. The couple were very religious and belonged to the Second Advent Church in Fairhaven. Aboard ship Captain Thompson prided himself on never drinking, besides reading only the Bible. But he was enough of a pragmatist not to hesitate to lower on the Lord's Day.[102]

Thompson had a long and distinguished whaling career, commanding the brig *Rosa Baker* out of Boston (May 1874-May 1875 and Jun 1875-Sep 1876) and then the bark *Minerva II* of New Bedford, first in Apr 1877-Sep 1880, in which he returned with 2410 barrels of sperm oil. His second voyage in the *Minerva II* (Feb 1881-Jun 1883) was less fortunate, as the vessel was lost on the coast of Mozambique.[103] Thompson's next command, the bark *Sea Queen*, of which he was part owner, suffered a similar misfortune when it was condemned at Mahe, Seychelles, in the Indian Ocean in 1888, four years after leaving the port of New Bedford. Fortunately Captain Thompson was able to recover 1370 barrels of sperm oil and 335 barrels of whale-oil.

Henry Clay's commands included the brig *Star Castle* of Fairhaven, which was lost in 1869 after a two-year sail, but managed to send home 1020 barrels of sperm; also the bark *Cicero* (May 1870-Oct 1873); the bark *Seine* (Nov 1876-Apr 1877), and the schooner *Golden City* (Jun 1877-Sep 1878)—all of New Bedford.

John P. Praro Sr. (Pico, 1831-1876) was another Portuguese whaling master who sailed out of New Bedford/Fairhaven. There are indications that as a young man he sailed off to California from Nantucket during the Gold Rush.[104] He was keeper aboard the ship *Potomac* of Nantucket from April 1855 to March 1856. In 1866, Praro was residing in Nantucket; and in June of that year he signed on as mate of the ship *Bartholomew Gosnold* of New Bedford (Jun 1866-Jul 1870).[105] Later he commanded the barks *Crowninshield* (Fairhaven, May 1869-Aug 1873) and *Lydia* (New Bedford, Jun 1874-Oct 1877). The *Crowninshield* came back into the Atlantic from her four-year voyage to the Pacific in poor condition. She was condemned and sold in Bermuda and her oil sent home.[106] Later, while serving on the *Lydia*, Praro died of fever on December 15, 1876.[107]

The Bentons

By the decade of the 80s there was an even greater increase in the number of Azorean and Cape Verdean whaling captains, mostly sailing out of New Bedford. Anthony P. and Joseph P. Benton—presumably brothers—were both from the island of Maio in Cape Verde. Anthony began a series of commands on New Bedford vessels commencing with the well-known schooner *Pedro Varela* (Nov 1876-Sep 1878) and progressing through the barks *Peru* (Nov 1878-Sep 1881), *Ohio* (Nov 1881-Apr 1885), and *Wave* (Jun 1885-Sep 1886); (at this point he had spent some six months ashore during the past ten years); then the *Tamerlane* (Jan-

Oct 1888), the *Clara L. Sparks* (Jul 1891-Aug 1895), and the *President II* (Nov 1896-Mar 1899).

The voyage of the bark *Peru* in 1878 was an early example of the significant participation of Azoreans and Cape Verdeans in the industry. The fact that the master was Cape Verdean probably intensified this tendency. Whereas the mate was an American named Durfee, most all the remaining crew members were from the Portuguese islands. Specifically, eleven were Azoreans, and eight were from Cape Verde, along with one Irishman, a Brazilian, an African, and a Portuguese resident of New Bedford.[108]

The makeup of the crew on Benton's next voyage, on the bark *Ohio* (Nov 1881-Apr 1885), was more varied; in fact the keeper declared a few months after sailing that, "I think that we Have gut Forward the meanest Set of men that Ever sailed from New Bedford, as Ever will again Thay is some of them in Hot water all the time. We Have English Dutch Irish Germans and Pourtugeas French. And a number of outher different Nations."[109]

Maintaining discipline on board ship was an art. Every hand was necessary; and there was no calaboose to send a man off to. Serious transgressions were often treated with a deft hand, while minor ones might bring out the ire of a frustrated captain. In January of 1884, after the *Ohio* had been out over two years, there was a serious altercation between a forecastle hand and the second mate. The officer instructed Ignacio Gomes to go to the wheel barn and learn how to steer; but he twice refused, whereupon the officer began to push him aft. Gomes then stabbed the officer in the back and was in turn struck twice by a belaying pin. Captain Anthony Benton then put the seaman in the rigging and in irons in the hold. Two weeks later Gomes was let out of the hold to scrub the trypots during the day and returned to the hold in irons at night. Finally, after another week had passed, the captain "gave him talking and send him foreward."[110]

One might compare this rather humane treatment of a serious act of transgression with that of the master of the ship *Chelsea* of New London in the mid-1830s after a Portuguese by the name of Antonio drew a knife on a shipmate. As shipmate William Davis later reminisced, "But, however much we condemned Antonio's act, when we saw and heard the rope's-end cut the writhing flesh, and heard the man's moaning supplication to 'Santa Marie' and 'Jesu Christo,' we forgot it. Very serious thought occupied my mind as I looked on the cruel scene; and for the first time in the voyage the life became horrible and hateful to me."[111]

Hudson Bay

While activities in the western Arctic are the best known, it was on the eastern flank of Canada that Americans first experienced the rigors of Arctic whaling. They were preceded of course by the Basques, who roamed the coast of Labrador early in the sixteenth century, and by the British and Dutch, who began to send their whaleships into Davis Strait early in the following century. By the 1730s, the Americans were also whaling there. And by 1851, vessels had begun wintering over. In the 1860s, as whales became scarce in the area, the Americans pushed into Hudson Bay. During the Civil War this had the advantage of protection against Southern privateers, but it required surviving extreme conditions. There the whalers enacted the same routine that would be followed later in the western Arctic. In the late fall, as the vessels were becoming frozen in, the natives began to arrive and settle for the season. There they would trade deer meat and furs for tobacco and cloth, and they would take sailors along on their hunts. In the more organized of these winterings a theatre might be built on shore, in which dances and masked balls would be held, just as they would be later on Herschel Island in the western Canadian Arctic.[112] After the New Bedford brig *Isabella* went into winter quarters on Marble Island in Hudson Bay in the fall of 1880, together with the *Abbie Bradford* and the *Abbott Lawrence*, the crews formed a company called the Hudson Bay Minstrel Troupe, which gave weekly performances during the months of darkness.[113]

As the ice began to melt in late spring, but before the vessel was able to break free, the whaleboats would be hauled across the ice to open water (for so-called "floe whaling"), where they then began searching for whales. But the floating ice and severe conditions presented a particularly hazardous moment for whaling. The ship *Hannibal* of New London spent the winter of 1860-61 in Davis Strait. One morning in June, while the vessel was still frozen in, a boat's crew left the ship in search of whales. When they were about fifteen miles distant they captured a whale and were towing it back to the ship when a gale arose, and they were obliged to cut from it and make for land, which lay some eight miles distant. As the *Republican Standard* of New Bedford later reported, "The crew went ashore among the natives, but one of them, a Portuguese, whose name is not recollected, refused to stop, saying that he was going to walk to the ship across the ice. But it proved not to be solid, and the last seen of him he was drifting North on the loose ice, and no assistance could be rendered him. He must have perished miserably."[114]

The intent of wintering over in the Arctic was to give the whalers early access to the cetaceans as they migrated back to the region the following spring. Meanwhile the men must be kept active and healthy to the extent possible. In the fall of 1885 Anthony P. Benton sailed the small bark *Wave* of New Bedford into Hudson Bay and took the customary precautions to prepare for the winter harshness as they settled in at Marble Island. A house was built over part of the deck, and snow was banked up against the vessel to insulate it against the cold. By January, the temperature had fallen to forty degrees below zero. And the following month one of the men died of scurvy. But the winter of 1885-86 was apparently especially harsh, with the whales in no hurry to return, for by mid-May, with the boats readied to be hauled out to open water for floe whaling, there were no whales to be espied. And so, by the end of the month, attempts began to free the bark from its winter prison. To make matters worse, eight crew members of the *Wave* were now down with scurvy. Working together with the crew of the bark *George and Mary*, also from New Bedford, they attempted to saw through the ice. On May 28 they resorted to using their bomb guns to try to break it up. It was not until June 17 that the *Wave* finally set sail.[115]

On July 3, the bark lowered its boats and struck a whale, which they brought alongside and began to cut in, just when the wind came up, and they were forced to cut loose. Three days later, they encountered the whale again and finished processing what would be their only catch. A month later, with the ice still a danger and impeding the movement of the two barks, it was time to sail for home. The *Wave* returned with 150 barrels of whale-oil from its one whale. The *George and Mary*, under the command of Erastus Church, had seen one small whale in the last days in Hudson Bay, but had failed to capture it, and returned clean.[116] Such were the vagaries of whaling.

The *Tamerlane*, under the direction of Anthony P. Benton, left New Bedford in January of 1888 headed for the North Pacific, evidently with the intention of removing its home port to that of San Francisco at the end of its voyage. While in the South Atlantic the vessel's steering gear became disabled, and it put into the Falklands for a new rudder. There, although it is not clear why, the first, second, and third mates abandoned the vessel. One might suspect that this was a rather rare case of racism against a competent captain of African heritage.[117]

With new officers, the *Tamerlane* then proceeded around Cape Horn and into the Pacific. That October, at St. Paul Island, in the Pribilofs, Benton noted that seven men were ill with scurvy and the vessel was unfortunately out of medicine. And thus they "could not whale it if we saw the chance," he noted, as he was now keeping the log.[118] So by the twenty-seventh of the month the *Tamerlane* ended its voyage in San Francisco Bay, now by far the most active whaling port in the nation.

Joseph P. Benton, twenty years old, first appears on the bark *Ohio* of New Bedford (Nov 1881-Apr 1885), captained by Anthony Benton.[119] Joseph's first command was just ten years later on the schooner *Clara L. Sparks* of New Bedford (Jul 1891-Aug 1895).[120] Given the acute racism of the period, it is not at all surprising that episodes of prejudice would appear when the officers of a whaling vessel were black and some of the seamen white. Such occurred on Benton's voyage on the *Clara L. Sparks*. Under normal circumstances, the captain's privacy on the quarterdeck was always respected, and a crewman would not appear there without being invited. Benton alludes to this in his narrative of a racially tinged incident between two men in the forecastle (Charles Williams, a native of St. Helena, and Antone Evora, a Cape Verdean):

> At 5 P.M. as I stood on the weather side of my quarter deck talking with the mate, the port [watch] was on deck before 4 P.M. this man Charles Williams belongs to the Starboard watch, and Antona Evore belongs to Port watch. Antona Evore went down into the forecastle and had a little trouble with this seaman Charles Williams whatever the trouble was the complaint was not brought to me in the manners, on the first place he walked right up on my weather side and said Capt. Bento that makes 3 times that I would be sitting down on my chest and this Antona Evore would come and fart right in my face. Charlie called him a black Nigger, and Antona cursed his Mother, but when I asked both men one cursed Just the same as the other. As I did not pay

> any attention to it after a while I concluded to shift the thing over and see who was to blame, so I called Charles Williams and asked him how many times I ordered him forward if he has any trouble among them to bring the complaint to me in a manner and I will look out for it, in the first place I do not want any man to call another one a bad name on board my vessel for I do not want any espesily Nigger because I am a coulered man myself and all of my offercers are coulered but one.[121]

The year 1899 found Joseph as mate of the New Bedford bark *A. R. Tucker* (Jun 1899-Sep 1901).[122] It was not long after the beginning of this voyage in 1899 that Captain Martin Van Buren Millard became ill. In the vessel's log, Benton noted over a period of time the captain's health and the meager treatment that he could offer him. On July 10, 1899, he remarked that he "gave him a doce of spirit-Of Nitre and 20 drops of Lowdnam [laudenam]."[123] A week later, with the captain on his death bed, Benton encouraged him to write a last note to his family, but to no avail: "he Sais bento thise is my last-So I Sais Captain why not you write you farmily few lines he sais yes I like to write my wife but I cand now."[124]

With the death of Millard, mate Benton took over command of the vessel. But two years later his luck would also run out. On June 15, 1901, Captain Benton made his last entry in the log: "at 4 whalles of[f] 2 points of weather bow." The log then continues in Portuguese, with the notation that three boats were sent after the whales, both the captain and mate fastening. But then the lines from the two boats entangled, and the order was given to cut the lines. "But it didn't turn out well," noted laconically the new log keeper and replacement captain, Joseph Avilla, for "the Captain was taken away and we had to lose the five whales that we had killed because it was very late and we arrived aboard around eight o'clock at night and thus ended the day with such bad luck." And so passed the ignominious end of the career of Joseph P. Benton.[125]

What happened to the earnings of Portuguese crewmen who died during a voyage? That is a question that the Acting Vice-Consul of Portugal in New Bedford, John F. Tucker, posed to the Consul General in New York in 1867. Tucker explained that the earnings for these men were turned over to the public administrator, who in turn would hand them over to the Commonwealth of Massachusetts. Tucker then asked if those funds could be turned over to him so that he might direct them to the proper heirs. Some of those earnings from a voyage of three or four years, "are often very large," he reiterated.[126] It is not known whether this was acted upon.

More Portuguese Mariners

Jacintho S. Leal, a native of Faial, where he was born ca. 1857, was master of the *Pedro Varela* during one of its earlier voyages out of New Bedford (May 1887-Aug 1889).[127] The schooner sailed with twelve men, including ten Azoreans, evidently bound to the islands to make out the remainder of its crew.[128]

Clarence J. Silvia was born on the island of St. Eustatius in the Netherlands Antilles, much frequented by the New England whalers. While neither strictly Portuguese nor Portuguese American, he is emblematic of the larger Portuguese presence in the Americas. Silvia's father was a whaleman from the island of Faial. His great uncles were brothers John D. and Joseph D. Silva and his uncle was Henry Clay. As a young man he settled in New Bedford and in his twenty-sixth year became captain of the brig *Eunice H. Adams* out of Edgartown (Apr 1893-Sep 1894). Silvia became ill with tropical fever in March of 1894 and was relieved by John T. Gonsalves. Returning home he died later that year at the age of twenty-seven from a recurrence of the fever.[129]

Those mariners who arrived early in the U.S. were the most likely to move up to the rank of officer and master for a simple reason—either with the master while serving aboard as cabin boy, or for some period of schooling in New England—they learned to read and write in the English language. For many had not even the training to read their own native language. To be promoted to mate, and eventually captain, one would almost always have to learn the art of navigation and the keeping of a log. This lack of an adequate education was a major factor in keeping more Azoreans and Cape Verdeans out of the captain's cabin. In extraordinary circumstances, however, it was possible to become master of a whaling vessel while semi-illiterate. What it took, essentially, was superb seamanship and skill in handling men. More commonly some of the captains had limited but passable written skills in both Portuguese and English.

Azorean-born Joseph T. Enos was an example of a superb mariner and whaleman who had achieved command in spite of his limited lit-

Bark Greyhound at the wharf, New Bedford, ca. 1920.
Courtesy of the New Bedford Whaling Museum.

eracy.[130] Enos made three successful voyages out of New Bedford on the bark *Greyhound* (Jul 1885-Aug 1887, Oct 1887-Jul 1892, and Oct 1892-Nov 1897). He was assisted in navigation by his wife, Penelope, a native of the island of St. Helena and evidently quite literate.[131]

Enos's first voyage aboard the *Greyhound* was especially eventful. While at the Azores mate Joseph D. Silva noted in the bark's log the smuggling in which it was engaging: "Ship lay off and on at Flores Selling tabacco."[132] Word must have gotten back to the captain, who was on shore at Faial with his wife and child, for a few days later he sent a letter—presumably transcribed by his wife—to the mate admonishing him to inventory the number of cases of tobacco aboard. If he found more than the number that had been declared on the manifest, to "take it way with out the gards seeing, and either hide it so as not to be found, or else trow it over board some way or other, without been seen by the gards if you can, and to not pay much attention to those rascals because all they want is to get the ship in to troble, there fore we must do the best we can to deceive them," which they evidently succeeded in doing, as they were soon off for the coast of Africa in search of prey.[133]

Two serious incidents marred this voyage of the *Greyhound*. The first was the loss of five crew members when a bomb gun burst while they were in pursuit of a whale not long after leaving the Azores. In June of the following year, Captain Enos was himself the victim. As he was practicing with a new whaling gun the cartridge got stuck, and when he tried to force it in with the ramrod it exploded, taking off four fingers of his right hand.[134] Harry Allen Chippendale, who sailed under him, claimed that he had lost three fingers, and that in his good-natured way Enos would call attention to it: "He was quite a card, but his eccentricities were mostly on the humorous side. He wasn't the hard-boiled type. If he wanted anything done he would invariably raise up his hand with the thumb and little finger extended."[135] The voyage itself was a success, yielding 1000 barrels of sperm oil and 390 of whale-oil.

Enos again sailed the *Greyhound* out of New Bedford on October 20, 1892, this time on a five-year voyage. Problems arose only a few months later, on February 25, 1893, while the vessel was anchored at St. Helena. Twenty-one members of the crew refused duty, complaining of bad treatment and inadequate food. Perhaps the reality of a five-year voyage in the Atlantic had sunk in. The American Consul came aboard to hear their complaints, and soon all were in the local jail. Consul James B. Coffin found the men "fat and healthy, and do not show the least Sign of being insufficiently fed but rather the Contrary."[136] Capt. Enos and his wife, Penelope, were evidently in no hurry to proceed on the voyage, but rather spent the time ashore with her family. And so it was not until the twenty-second of March that the majority of the recalcitrants were carried on board of the *Greyhound* by the local police and most of them immediately put in irons, since they continued to protest their conditions. The boatsteerers and four of the foremast hands were left on shore to be discharged, as they appeared to be the ringleaders of the mutiny.[137]

Enos later served as master of the New Bedford bark *President II*, which sailed in August of 1899 and was later wrecked on the coast of southern Angola. The vessel had been cruising near Porto Alexandre on May 7, 1904. Suddenly, about midnight, the watch sighted land and the vessel quickly struck. The sea was heavy, with breakers "mast high," according to the account of the second mate.[138] The crew first lowered the captain's boat on the larboard side, but a wave smashed it to bits. Then the starboard boat was sent down and seventeen or eighteen men crammed into it. About halfway to shore a huge wave overturned the boat and broke it apart. Luckily, all the occupants made it to land. As the crew was next attempting to lower the waist boat the davits broke off; and the boat, with one man in it, dropped and was swept intact to shore.

There were now ten left on board the vessel, including Mrs. Enos. The vessel immediately began to break up, and all huddled on the after-deck. When it sank all ten floated off on a piece of the deck barely large enough to hold them, their weight also submerging it somewhat. On that improvised raft, they floated for six hours, until at dawn the men on shore brought out the surviving boat and took them in. Soon their raft broke up and disappeared. Most happily, all had survived the ordeal. Now they looked around and found themselves surrounded by sand—no greenery, no water, no habitation. And they began sorting through remnants of the wreckage that had washed ashore: a few canned goods, a cask of bread, and others of molasses and vinegar, which they mixed together with a partial cask of brackish water to make the latter drinkable. Half-clothed, they welcomed remnants of the slop chest that floated in. Barefoot, except for Mrs. Enos, who had on a pair of light slippers, they started off for Porto Alexandre, shells cutting their feet along the beach and the heat burning them. Three days later they reached town and for a week were taken care of by friends of the captain. Then it was off again on foot to the larger town of Moçamedes (now Namibe). From there they proceeded to Luanda, Lisbon, and finally to the U.S. As the second mate told the reporter upon arriving home, "It was the toughest experience that I have ever passed through, and there wasn't one of us who ever expected to see New Bedford again."[139]

Penelope Enos was not the only wife to accompany her Portuguese husband on a whaling voyage. In 1898 Kitty Edwards sailed with her husband, Joseph T., aboard the *Greyhound* for a four-year voyage. And Anne Joseph accompanied her husband, William F., on the bark *California* in 1902. Maria Corvello went along with her husband, Antonio C., on three voyages of the same *Greyhound* in 1913, 1917, and 1918.[140]

As happened with the *Wanderer* in 1924, catastrophe could occur much closer to home. Three Cape Verdean crew members, Joseph G. Reig (Reis?) and Henry Gonsalves, both from Brava, and Manuel Deandra, from neighboring Fogo, were the fortunate survivors of the bark *Sarah* in October of 1878. The *Sarah*, a small vessel of 128 tons departed New Bedford on Saturday, October 12, 1878, with a crew of twenty-three, under Capt. Henry M. Peakes of Edgartown. In spite of threatening weather on the day of sailing, Capt. Peakes opted to depart. At noon, scarce hours later, the *Sarah* ran full into a hurricane. At four p.m. it hove to and at seven capsized off Block Island. At the moment when the vessel gave a lurch to starboard and rolled over, the three survivors were in the forecastle awaiting the call to watch. With them were four other crew members. Water quickly rushed in, leaving only some eighteen inches of breathing space and cutting off any chance of escape. The men clung to the rail tier to save themselves, but one by one they gave up until only the three survivors were left alive.

The following afternoon Deandra dove down through ten feet of water and made his way up the companionway to the deck. There he found himself alone, except for his fellow countryman and boatsteerer Joseph A. Pena whose body he found lashed to the fore rigging. Monday morning Reig followed. Gonsalves remained alone and, unable to swim, had given up hope. Finally, on Tuesday afternoon, three days after the vessel capsized and all but the men in the forecastle were swept into the raging sea, they were discovered by a pilot boat. When a hole was cut into the bluff of the bow the rescuers discovered Gonsalves clinging to a floating chest. All three soon recovered.[141]

Edward M. Frazier was born in New Bedford in 1852, the son of an Azorean named Joseph Frazier who conducted an outfitting business at 123 Water Street. Frazier went to sea as a teenager aboard the bark *Reindeer* of New Bedford. The bark was one of those many that were lost in the Arctic disaster of 1871. Later he commanded two voyages out of New Bedford: bark *Lydia* (Nov 1880-Aug 1884) and bark *E. B. Phillips* which sailed in October, 1884, and was condemned the following year at Praia on the island of Santiago, Cape Verde.[142] The *Lydia* carried a heavily Portuguese crew. Of the twenty-seven crew members, thirteen were Azoreans and five were from Cape Verde, along with one Portuguese whose origin is not listed. Frazier and his wife later conducted a grocery business on Acushnet Avenue, New Bedford. In the census of 1900 he is, in fact, listed as a grocer.[143]

Anthony J. Frates (Freitas, 1844-1903) was a native of the Azores.[144] He had three commands, the first being from New Bedford and the other two from Provincetown. In October 1881, he sailed as master of the

Brig Frances A. Barstow. Charles S. Raleigh, 1877.
Courtesy of the New Bedford Whaling Museum.

schooner *Golden City*, returning in September of the following year with a disappointing catch of 23 barrels of whale-oil and 100 of sperm oil. His next commands were on the schooner *Joseph A. Manta* (Provincetown, Nov 1898-Aug 1900 and Dec 1900-Oct 1903). The latter voyage ended in disaster. After having sent home 1900 barrels of sperm oil from Horta, and instead of heading south through the channel between Faial and Pico, Frates chose to sail around to the north of Pico in order to wave adieu to his wife at the village where she was staying. There the vessel was caught in a hurricane and dashed against the rocks and was lost with all hands.[145]

Forbidden Intimacy

Among men who spent the great majority of their working years aboard all-male vessels, an occasional physical attraction was all but inevitable. The practice of any homosexual act in the crowded milieu of the forecastle, however, would have been virtually impossible not to be detected. On the other hand, among those aft there was somewhat more opportunity. Such an incident was recorded in the tersest of terms in the log of the brig *Frances A. Barstow* of San Francisco while under the command of Joseph T. Edwards. Captain Edwards was bringing the vessel on a three-year whaling voyage (Jun 1894-May 1897) back to its old port in New Bedford. As it was cruising off the coast of Brazil in 1895, "After dinner. Ship was hauled aback,–and boat was lowered,– The boat Steerer Philip Forte was put into it with his chest, and put ashore – in consequence – of the crime logged on Friday Jan 4."[146] On the final page of the log, at the end of the voyage, is an entry that should have been included with the original. Perhaps a dutiful mate, while wishing to spare the details to any who might perchance peruse the logbook during the voyage, nevertheless felt duty bound to include the details. It reads, "Jan 4 Friday One of the Boatsteerers, Philip Forte, – was caught in the act of sodomy with one of the crew. – The case was reported – by one of the Fore mast hands. – Jim Keifer. – who witnessed the act on the night of – Monday. – Dec 31 – 1894. – In consequence of this, the Captain moved him out of the steerage, and put him in the Booby Hatch."[147] Forte was thirty years old and from Cape Verde.[148]

Perhaps the most vulnerable individual on board, in this respect, was the young cabin boy. Another Manuel Enos, a native of the Azores, was the repeated victim of the advances of veteran whaling master Nathaniel W. Sowle on board the ship *Montreal*, which sailed out of New Bedford in November of 1857. Enos had shipped as cabin boy, and was soon fending off the captain's advances. He managed to desert in California and then signed on to another vessel; but he unfortunately crossed paths with Captain Sowle at Lahaina and was forced to return to the *Montreal*. There, on a cruise to the North Pacific, he was sodomized. On the return to Hawaii in 1858, Enos again ran away and entered the whaleship *Dromo* of Warren, Rhode Island, on its way to Nagasaki, Japan. But there seemed to be no escape from Captain Sowle, who found him there in April of 1860 and took him on one last cruise before Enos could finally flee in Hawaii and file suit in the courts of the Kingdom. There, Captain Sowle faced two libellants. Besides Manuel Enos, another Azorean, by the name of Manuel Vieira, also sued. Vieira, from the island of Pico, had signed on in 1859 at Honolulu for a year's cruise. He too was then repeatedly sodomized by the captain. The verdicts of the Admiralty Court of Hawaii—in spite of questions of jurisdiction brought forth by the American Consul and which reached the Supreme Court of the Hawaiian Islands—were for the victims, and Sowle was

The Old Mill, Nantucket, ca. 1880s.
Courtesy of the Nantucket Historical Association.

charged $1500 plus costs in the case of Vieira and $2500 plus costs for Enos.[149] It is worthy of note that these were suits for damages and not criminal trials. They evidently had some effect on Captain Sowle's career, however, for he never commanded another whaling vessel.

A Nantucket Whaling Master

One of the last of the whalemen to sail out of Nantucket was John Murray (1823-99), a native of the island of Graciosa in the Azores.[150] Murray's parents were Francisco A. Mendonça and Florinda Bettencourt, so apparently the new surname was picked up somewhere along the way. Murray served as second mate on the New Bedford whaling ships *Isaac Howland* (Dec 1854-Jun 1859), and *Lapwing* (Jun 1860-1863). Later, as master and part owner, he sailed the schooner *Abbie Bradford* from Nantucket on two Atlantic voyages (Apr 1867-Aug 1868 and May 1869-Oct 1870). This latter voyage was marred by a scuffle between two Portuguese crew members. Joseph Manuel ended up striking the other man with a handspike, and he soon died. Manuel was subsequently sen-

John Francis Sylvia and wife, Nantucket. Courtesy of the Nantucket Historical Association.

tenced to three years in state prison. On his return from this last voyage Murray brought his young son back from Graciosa. Ten years later the two were operating a grocery store that remained in the family for many years. Son John Jr. became an influential member of the island's Portuguese community and, according to one newspaper account, "one of the best-known citizens of Nantucket."[151]

One of the most important historic attractions on the island is the Old Mill, built in 1746—the oldest functioning mill in the country. In 1854, it was purchased by George Enas, a mariner born on the island of Flores in 1815. In 1865 Enas sold the mill to John Murray, who in turn sold it a year and a half later to John Francis Sylvia, another mariner from the Azores. Sylvia operated the mill for many years. He died in 1896, and the following year it was purchased for its current owner, the Nantucket Historical Association.[152]

Provincetown Whalemen

Portuguese and Cape Verdeans were particularly active out of Provincetown, which historically supported essentially a schooner trade restricted to the Atlantic. Joseph B. Foster, born on the island of São Jorge, Azores, in 1838, took advantage of these "plum-pudding" voyages to sail the schooner *N. J. Knight* out of Provincetown each season from February through late summer from 1873 through 1881. After that he continued almost annually on various schooners sailing out of New Bedford through 1895.

Antone E. Mello was master of the Provincetown schooner *Alcyone* during two voyages (Aug 1888-Jun 1889 and Aug 1889-Aug 1890). Contrary to the experience of young Manuel Silveira Andrade—who in 1859 was not allowed to step ashore from the bark *Pacific*'s boat to visit his native town of Faial—when a sick ship keeper begged to be put ashore on his native Brava, Captain Mello relented. As the mate Joseph P. Thompson related, "At 1 P.M. Sighted the Island of Brava Bearing South about 30 miles distance at 7 Pm hove too on the lee of the Island in the Calm the sick man ship keeper Benjamen Rose Beg the Captain to land him on shore.... The Captain did not want to land the man but he beg so hard that he is sick and so Clost to his home and sick did not want to [go] any farther with the Vessel so the Capt took pity on him and landed him on his native land."[153]

The Sealers

Vessels sailing out of New London area ports continued their dual activities as whaler-sealers during this period. Joseph J. Fuller most certainly belonged to that old breed of sailor who spent little time at home. In his twenty-five years as master of New London area whaler-sealers, from 1870-95, he was almost always at sea or anchored at one of the desolate islands of the South Indian Ocean. It was during this period that both American whalers and sealers began to depend heavily on Portuguese islanders, especially Cape Verdeans, to make out a good portion of their crews. When Fuller left New London on the 194-ton schooner *Pilot's Bride* in April of 1880 he carried only eight men, finally rounding out the crew at the Azores and Cape Verde by adding twenty more.[154]

As Briton Cooper Busch has noted, "Towards the end of the trade, practically the only men who would put up with conditions to be found at Heard Island and such dismal places, aside from crimped sailors, were Cape Verde Islanders, eager to escape the periodic famines and epidemics which visited their islands off the west coast of Africa. They were willing to work and work hard, for little pay in the hope of eventual discharge in the United States."[155]

When Fuller landed on his way out at the island of Flores he discovered that the man to do business with was an "oldish" priest, who would "do all of my business—the getting of my potatoes and the getting of the men.... There was some Portuguese that could speak good English and I asked them about the priest. They said that he had more to say about the island than the governor, and I think what they said was true."[156] As frequently happened, an offer of American tobacco soothed the spirits not only of the priest, but of the custom house officer as well. Later, after calling at Brava to fill out the crew, the *Pilot's Bride* headed for Desolation (Kerguelen) and a fate that would see the vessel lost on the island and a wait until September of 1882 for the crew to be rescued by the New London schooner *Francis Allyn.*

In the journal that he kept during the long ordeal, Captain Fuller recalled a time when one of the two boatsteerers, a native of Groton, Connecticut, by the name of George Manice (Moniz) asked the captain for permission to move to another part of the island and there build a house with three Portuguese crew members. Manice's paternal grandfather had been an immigrant from Portugal; and so young Manice evidently felt some affinity for these men.[157] About a month later Manice appeared once again at the main camp with his baggage. "Manice, what is the matter," asked Fuller. "Got homesick? I thought you like living down there."

"Oh, I like it well enough but I [would] rather be with the crowd. It is too lonesome for me. Them three Portuguese do nothing but jabber Portuguese from morning to night."

"Well, Manice, why did not you learn to speak Portuguese? Then you could jabber with them."

"I cannot get that damn twist on my tongue, Captain," he replied.[158]

In August of 1887 the *Francis Allyn,* under Captain Joseph J. Fuller, left New London with a skeleton crew and then shipped a total of seventeen men (plus two stowaways) at Brava.[159] Fuller visited Desolation but also took time to drop off a number of men on isolated, uninhabited, and nearly inaccessible Gough Island (also known historically as Diego Alvarez) in the South Atlantic.[160] Almost all were Cape Verdeans. Listed later by the log keeper they were: "Peater Babtist, killed by a whale, fidel Montague, Manual Bueque, Josey De Barro, Antone Martiles, Antone Babtist, Josey Gomes, Died on the Island frozen and Exhausted, Jack Paw Manual, Francis De Sant, and Manuel de Sousa."[161] There they remained, under the command of the third mate, Charles H. Moor, until April of the following year with little luck. When the schooner came by to check on them, it "stood in towards the Island and met the 3rd mate coming off he had only 55 seals this last clamed that if he had had more men he could have taken several hundred more an what oil and few skins we have will not more than pay expenses it is thought best to leave the 3rd mate here with 3 more men 8 in all and we are to stay out another year."[162] And so there the men remained for another season.

New London, in fact, became more and more dependent on Cape Verdeans to fill its elephanting/sealing voyages. In its instructions to Captain Simeon Church, who was about to take the schooner *Charles Colgate* on such a voyage in 1877, the owners directed, "You will proceed direct for the Cape de Verdes Isls where you will ship Portugues enough to make your number (30) all told ship them before the Vice Consul and give them the (1/265)th lay each." In order not to lose anyone at the islands they further counseled that he "not let Jose de Pena Boatsteerer go on shore at Cape de Verdes, also look out of your other Portigues."[163]

The Smugglers

As noted in Joseph Fuller's visit to the island of Flores, American tobacco was indeed still highly sought at this time.[164] In September of 1877, the log keeper of the ship *Arnolda* of New Bedford noted that, while the vessel was cruising "in the passage between Pico and St. George a boat came up and took 70 boxes of tobacco."[165] As it had been for over a century, however, the enforcement of laws against the smuggling of tobacco was unpredictable, a fact that Captain Joshua G. Lapham of the bark *Mary Frazier* of Edgartown was to discover. While he was cruising off

Faial in October of 1887, the Portuguese gunboat *Açor* seized his vessel and towed it into the port of Horta. Lapham was detained, charged with running tobacco at Flores and giving a false manifest at Pico. He was levied a heavy fine and his vessel impounded. Lapham appealed, stating, in defense of the transaction at Flores, that he was in international waters twelve miles off the island and did not know the nationality of the traffickers. Seven months later the most serious charges were dropped by the Lisbon Tribunals, freeing him to sail. It was now early May of 1888, and the *Mary Frazier* departed with fourteen men for the island of Flores to fill out her crew. There Lapham anchored in the bay off the village of Santa Cruz (unlike Faial there was no harbor in which to seek shelter) and while ashore shipped nine more men, sending them off to the bark. The bay at Santa Cruz is only somewhat protective. The next day, with the captain and mate onshore, a southeast gale started to blow; and Lapham, fearing for his vessel, offered $50—an immense sum in the islands at the time—to anyone who would put him aboard. But there were no takers. Then,

> suddenly the wind backed from east to northeast and the vessel tailed inshore, with her stern right on toward land. The second officer, an experienced man, finding the weather growing worse, paid out all chain on the port and starboard anchors, made everything snug and ordered the ship abandoned.
>
> The crew started ashore through the terrible sea in two boats, and with the sea which was running it seemed as if they would never reach land. The first boat contained 11 men, and when they struck the surf the boat was capsized but fortunately all reached the shore in safety. The second boat's crew was not so successful. There were eight men aboard and the boat went all to pieces in the breakers. Five men, including the third officer, John Adams, a Kanaka, and four Portuguese sailors, shipped at the islands, were drowned.[166]

The next day there was no sight of the *Mary Frazier*. Having dragged her anchor she had been drawn seaward. Later she was discovered at sea by the steamer *Peconic* in a waterlogged condition and burned as a derelict.[167]

Lost at Sea

One of the more harrowing experiences of whaling was the so-called "Nantucket sleigh ride." After one of the whaleboats had fastened, the whale would tend either to sound or immediately begin to run. In either case line was let out in order to give the whale a chance to tire itself out. But at some point if the captive continued to run, then the men must sit back and face being towed many miles away from their vessel. Such was the case of a boat's crew of Portuguese from the bark *Kathleen* of New Bedford. The story was told later by the one non-Portuguese among the six, Robert Ferguson, boatsteerer for the second mate, whom he refers to only as "Mr. Roderick."[168] On the morning of Thursday, March 1, 1883, while about sixty miles from St. Helena in the South Atlantic, Ferguson fastened to a sperm whale that soon carried them some distance away at a tremendous speed. The whale finally tired and it was soon dispatched with a bomb. Looking around, the men saw no other boats nor the bark. Hoisting a lantern on an oar they took stock of their supplies, which consisted of a bit of hard-tack and some water, all of which they readily consumed. Soon all settled down for the night.[169]

The next morning they found themselves still alone. And thus they remained exposed to the heat of the day and the cool of the night until Sunday evening, when Ferguson, and perhaps others, began to think about the infamous story of the *Essex* and the cannibalism of its lost crew. Evidently foreseeing the possibility, Ferguson had already hidden the knives and hatchets while the men were asleep. By Monday the wind was blowing, and the sea was turning rough. At that point mate Roderick asked Ferguson to take over, since the latter was an accomplished navigator.

Although presumably the whale could have provided some sustenance to the men, it was decided to cut it loose and head for Africa, some 1200 miles away. On Tuesday, the men spotted smoke on the horizon, but it was soon gone; and that evening some lights were seen but again soon faded. By Thursday, now a week with neither food nor water, Ferguson observed that "the men say nothing but look wistful and sad. Their lips are dry and cracked."[170] The sight of a flock of albatross on Friday caused consternation among the superstitious seamen.

It was now Saturday, March 10—nine days without food or water. Thoughts of death permeated the atmosphere. Mate Roderick kept thinking of his wife at home in New Bedford. On Monday, now eleven days out, with a sense of hopelessness among the men. Ferguson awoke with the mate calling him. "He said, 'Bob, I thought you were dead.' He had been calling me for some time. The tears were running down his cheeks. It hurt me to see this big strong man in tears and made me feel bad. I had a long talk with him this afternoon and he felt better. At night I had him kneel down and pray alongside of me. He had his beads around his neck and counted them over. It is his belief and he is sincere. When he lay down to go to sleep, he said good-bye as tender as a woman and as if it was his last night."[171]

At last, on Tuesday morning, they were spotted by an English bark that quickly took them on board and treated them so that all recovered from their great ordeal. The next day the captain of the bark took the Scotsman aside and inquired as to how a second mate could neither read nor write nor have any skill in navigation. Ferguson had to explain the peculiarities of the whaling trade, "that it was his whaling qualifications that let us have a mate who could not navigate," and that their ship, "though small, had four mates and that only two needed to be navigators."[172]

While it was quite possible for a mate to be illiterate and incapable of navigation, it was a long step from there to captain. Joseph Roderick, age forty-nine, from the island of Brava, first mate of the bark *Charles W. Morgan* on its voyage of 1908-1910, is in all likelihood not the the same person as Mr. Roderick of the *Kathleen*; but his limitations were the same. While the vessel was sailing in the Indian Ocean in June of 1909, Captain Arthur O. Gibbons was taken ill, and the vessel made port at Durban, South Africa. There, on June 18, mate Roderick was made master of the vessel; but first the entire crew refused to weigh anchor. Finally, on June 22, the American consul persuaded them to sail under Roderick after having agreed with them to provide "another Navigator."[173]

It was a brief voyage, however, as most likely had been arranged at the time of sailing, for on August 2, the *Morgan* was once again back at Durban, with the command to be assumed by veteran whaleman Charles H. Church. One can imagine the humiliation of Captain Roderick on being so quickly replaced. On the eve of the vessel's return, the logkeeper noted that the captain had been drunk all night; and the next day Roderick was discharged to go home.[174]

In general, in an age of significant prejudice against people of color, the idea of white men sailing under a black captain was nevertheless accepted, as seen with masters such as Severino D. Pierce and the Bentons. It would have been another matter, however, to set sail with an illiterate captain quite deficient in navigational skills, as the above suggests.

San Francisco and the Western Arctic Whalers

As whaling activity shifted increasingly to the North Pacific and western Arctic, vessels—even if owned by New Bedford interests—began to favor San Francisco. Thus for the first time, in 1882, this far-flung port on the West Coast sent out more whaling vessels than did New Bedford, and would continue to do so for the next two decades. Ten years later, there were just ten sailings from New Bedford, and the two vessels that headed for the Pacific that year, the *Gay Head II* and the *Navarch*, would soon make San Francisco their home port. New Bedford vessels, like their counterparts in Provincetown, now tended to limit themselves to the Atlantic, with occasional forays into Hudson Bay.

	1880	1885	1890	1895	1900	1905	1910
New Bedford /Buzzards Bay	36	18	9	6	6	5	14
San Francisco	8	39	42	15	15	14	6

Number of Voyages of Whaling Vessels from New Bedford/Buzzards Bay and San Francisco in the Years 1880 to 1910[175]

It should be noted that, contrary to the tendency of the whaling fleet in New England, where by the last decades of the nineteenth century, Azoreans and Cape Verdeans had worked their way up to master, there was no such equivalent in the West Coast fleet. Captains, and even mates, tended to be New England Yankees, while the Portuguese, either Azoreans or Cape Verdeans, did not usually rise above boatsteerer or junior officer. On the other hand, a large number of foreigners of various nationalities often filled the forecastle.

It was not uncommon for whaling vessels sailing out of San Francisco during this period to have on board a half-dozen or more Azorean or

Cape Verdean crew members. A number of these vessels, although no longer on the East Coast, continued to maintain their original registry. Thus, when the bark *Napoleon* of Edgartown left San Francisco in December of 1884, it carried fourteen Cape Verdeans, all residents of San Francisco. Unfortunately the vessel was lost in the Arctic with twenty-two men the following year.[176] The New Bedford registered bark *Abraham Barker*, at the time also sailing out of San Francisco, left that port in November of 1887 with fourteen Portuguese crewmen, all residents of San Francisco as well. The same number of Portuguese, again all residents of San Francisco and equally divided between Azoreans and Cape Verdeans, sailed from there in 1893 aboard the steam bark *Jesse H. Freeman*.[177]

John Rogers (da Rosa, Faial, 1837-1907) was one of the few natives of Portugal to command a whaling vessel out of San Francisco. Rogers arrived in the U.S. at the age of fifteen aboard the whaling bark *United States* of Stonington.[178] The majority of the vessels he sailed on out of New Bedford during the early years of his career frequented the western Arctic, and so it was natural that his two commands would be from San Francisco. The first was on the bark *Mary and Susan* (Dec 1884-Nov 1885), during which he was listed as mate and acting master.[179] Rogers returned from the North Pacific with 775 barrels of whale-oil and 17,950 pounds of whalebone. A few years later he sailed again to the North Pacific as master aboard the bark *Tamerlane* (Mar 1889-Nov 1889), now of San Francisco.

The *Tamerlane* was a veteran whaleship, having been built in Maine in 1824. It reached its tragic end in 1892 in the surf on the island of Hawaii. The voyage had begun in January of 1891, under the command of Walter F. Howland, with a crew of thirty-eight officers and men. This was Captain Howland's first command in the Pacific. The night of February 1, 1892, was dark and stormy, and the waves exceptionally high. The captain evidently thought he was much farther from land, when sometime after midnight the bark suddenly struck on the rocks south of Hilo. Howland soon appeared with a valise, and he and the mate and several others were lowered into one of the two available boats—and then they disappeared into the raging sea, never to reappear. All in all, eighteen men lost their lives that night. Fortuitously, toward morning someone discovered a plank wedged between the wreck and a rock; and those who had not been washed away were able to make their way

Vessels at Point Barrow waiting to break out of the ice: Steam brigantine *Jeanette*, steam schooner *Jeanie*, revenue cutter *Bear*, steam barkentine *Newport*, and steam bark *Fearless*, 1890s. Courtesy of the New Bedford Whaling Museum.

ashore and up a high cliff to safety. The next month the vivid tale of loss and miraculous survival was related to the *San Francisco Call* by the fifth mate, an Azorean by the name of John San Francisco. San Francisco later settled for a time at Point Hope, Alaska, and worked as a shore whaler.[180]

Whaling, from the point of view of the capitalist, was a risky business, with wildly varying success rates for any particular voyage. The Civil War and its predators had been a disaster for the industry. Scarcely a few years later, with a great number of American whalers now plying their trade in the western Arctic, the risks had not abated. In order to navigate in the Arctic Ocean vessels had to follow the leads that opened up between the fast ice along the shore and the pack ice as it began to break up. The picture that presented itself to the approaching vessels at this moment was nothing like what the observer of a thawing bay or lake in the "lower forty-eight" might envision. As the log keeper of the bark *Mount Wollaston* of San Francisco observed while the vessel was making its way through the Beaufort Sea near Point Barrow in 1877: "The 'Helen Mar' has been up to the head of this bight and reports the ice 80 feet high and forming a perfect wall with vertical sides...."[181]

A week later he gave an even more vivid description of the Arctic ice:

> The ground Ice which has the apearance of an Imense Mass of Marble extending North and South some 12 miles and of perfectly vertical sides, from 20 to 80 feet high, and of variagated strata, which by some convultion of Nature has been cleft In twain, the entire length of the coast the outer portion having drifted away to sea and left this Imense portion standing out in bold releif forbiding nearer approach to what Is Now an Inaccessable coast In some portions It Is much discolored with earthy matter which gives It a somber cast while In others It Is of Imaculate whiteness enhanced by the glitterny rays of the Sun resembling the frosting of an enormous fruit cake or just what It acTualy is consolidated Snow and Sea water. The heaviest Ice I ever saw In the Arctic Ocean.[182]

Any contrary movement of the pack ice could be disastrous, as happened in 1871 when thirty-three whaleships were imprisoned in the ice and then crushed. Fortunately, during this catastrophe over 1,000 men, women, and children were saved without a single loss of life.

Another unhappy possibility was being caught in the pack ice as it moved north away from land. Such a disaster occurred in October of 1879 when the ship *Mount Wollaston* of San Francisco, under Capt. Ebenezer Nye, and the bark *Vigilant* of New Bedford, Capt. Charles R. Smithers, were carried away in a storm. On board the *Vigilant*, when it sailed from New Bedford, were three Azoreans and four Cape Verdeans.[183] Lost on the *Mount Wollaston* were two Cape Verdeans, fourth mate Theotorio Sabino and boatsteerer Joseph S. Oliver.[184] In the summer of 1881, the U.S. revenue cutter *Thomas Corwin* learned from some natives on the Chukchi Peninsula that the previous fall a ship had drifted close to shore, and upon boarding it they had discovered several corpses. From their description it was supposed that it was the *Vigilant*.[185]

There are many tragic tales of whaling in the western Arctic. The 206-ton bark *Ohio* (often referred to at the time as the *Little Ohio*, to distinguish it from the 363-ton *Ohio II*), when it left San Francisco in December of 1887, carried at least ten Portuguese on board, including three boatsteerers and the third and fourth mates, Manuel Lopes and Joe Enos. After sailing to Guam and spending an unsuccessful summer in the western Arctic, catching only one whale, the bark ran ashore at Point Hope, Alaska. Of the twenty-nine men on board only thirteen reached land. The mate soon died from the cold, as well as one of the boatsteerers, Big Pete, a Cape Verdean, who had only had time to put on an overcoat over his naked body. But five more men, including Lopes and Enos, were destined to die when, frustrated by the ships that continued to pass by without noticing them, they entered a boat and attempted to reach a whaler passing in the distance—and were never seen again.[186]

Longtime Barrow resident Charles Brower related a harrowing story that took place in August of 1897 when he was on board the whaling bark *Navarch.* When the bark unfortunately collided with the pack ice and began to be carried away, Capt. Joseph Whiteside ordered the vessel to be abandoned. Upon attaining a certain distance from the *Navarch*, Whiteside, his wife, and a few of the men who were hauling a whaleboat with all the provisions, halted and—without informing the others—returned to the vessel. Once Brower and the rest of the men discovered what had happened they too wished to return to the *Navarch*, but now fissures had opened in the pack, making a retreat impossible. So they continued on, without provisions, on a trek that ultimately took the lives of half of the thirty-one men who accompanied him. One by one they gave up, fell, and died. As Brower noted later in his diary:

> On the fourth and fifth day we lost six more men they just played out as they droped behind, not being able to keep up some would cry out not to be left behind this was heart rending, I had never deserted a man in my life before, there was nothing we could do, some of us were nearly in thier condition,, It was only a matter of a day or so, If we did not reach the edge [of the floe], we were all bound to go, Just the same it was hard to listen to thier voices dying away in the distance beging not to be left, always they called on me, reconizeing me as the leader, saying Charlie, do not leave me behind. These were mostly Portuguese, or from some of the South American countries,
>
> One of the Portuguese we left was a young man, strong and able when we left the ship, he played out by not wanting to leave his good clothes behind, When we left he had on three suits of clothing all new, these when they got wet were a weight to carry, he finaly had to discard some of them not before he was all in, a day later we left him.[187]

On the eighth day the remaining men opted for a cake of ice that was moving away from the ice field on which they found themselves. On this "raft" there was at least the remote possibility of reaching land. The next day they were escorted by a large pod of whales. On the eleventh day a schooner passed without noticing them, upon which the third mate Enos N. Dias, from Cape Verde, lost his mind, and they were forced to subdue him to keep him from leaping into the frigid waters. Finally, after twelve days, they were seen by the whaling bark *Thrasher*—all sixteen of the remaining men floating on a cake of ice in the open sea. Although rescued, not all of the men survived intact. Four subsequently lost their feet; and Enos Dias died on the vessel that was taking him back to San Francisco. "The others," Brower reflected, "just disappeared, as sailors do everywhere."[188]

Steam Whalers

Although the British had been whaling with steam vessels since the 1850s, the Americans were much slower at adopting the technology. In 1866 the veteran New London bark *Pioneer*, with a new steam engine installed, sailed into Davis Strait. But it would take over a decade before the experiment would be repeated. Sensing the limitations of the sailing vessel in Arctic waters, William Lewis, a New Bedford whaling agent and former whaleman, coordinated the construction of the steam bark *Mary and Helen I*, which sailed from New Bedford to the western Arctic under the command of Martin Van Buren Millard in September 1879. The *Mary and Helen I* arrived in San Francisco in October of the following year with 2350 barrels of whale-oil and 45,000 lbs. of whalebone, one of the largest and most valuable catches in the history of western Arctic whaling. Soon a series of steam whalers were built, first on the East Coast: the *Belvedere*, *Mary and Helen II*, *Lucretia*, *North Star*, and the *Balaena*; and then in California: the *Bowhead I*, *Alta California*, *Orca*, and the *Narwhal*. These latter anticipated the factory ships of the next century in their substitution of steam tryworks and iron oil tanks for the traditional blubber-fired tryworks and wooden casks.[189]

The West Coast vessels belonged to the newly formed Pacific Steam Whaling Company of San Francisco, which came to dominate western Arctic whaling. That same year the Arctic Oil Works was established in the same city with modern facilities to process both oil and whalebone, eliminating expensive transshipment to New England. Modernization of the technology and methods was made necessary in part by the fact that already by 1883 whalemen in the western Arctic had killed 15,000 bowhead whales, which amounted to eighty percent of those that would be taken in the region up to the end of the fishery.[190] Vessels customarily left San Francisco during the winter months and spent some time right whaling, often on the Japan Grounds, before heading to the Bering Strait in the spring, where they hoped to encounter the bowheads. Finally, as the ice would begin to break up, by early August the whaleships would reach Point Barrow. But time was very short, and they would be forced to leave shortly in order to miss the freeze-up and not be crushed by the oncoming pack ice.

Noting that the ideal moment for the capture of the bowhead was when they passed by Point Barrow in June on their way to their feeding grounds to the east, in 1884 the Pacific Steam Whaling Company dispatched eight men from San Francisco to Point Barrow with the charge to establish a shore-whaling station. The station operated without great success, and four years later the company decided to send a team of Portuguese to take over. A German who would drop by to visit them commented that "it was like living in a foreign country at the station where the talk was all Portugese, he could understand nothing."[191] According to Charles Brower, some of the Portuguese remained only a short time at the station in Barrow before selling their whaling gear to him and heading for Point Hope, evidently desirous of finding a ship to return them south. But here their luck ran out. As Brower recalled, "Weeks later, those chaps tried to reach a ship off Point Hope and that was the last heard of them. There were eleven in the little boat."[192] Brower does not identify these infelicitous individuals, nor can it be supposed that they were all Portuguese.

The next year the company closed the station; but some of the men employed there at the time opted to remain in the area. One of them was Antone Bett (António Bettencourt), a native of the island of São Jorge. Engaged in both shore whaling and trading, he moved occasionally between Point Barrow, Point Hope, and Cape Smyth. In 1890, he was operating the Cape Smyth Whaling Company; and in 1891 he organized four teams of Inupiat shore whalers at Point Barrow. Bett's presence was an attraction to other of his countrymen. The steam bark

Inupiats going out on the ice for whales in April, ca. 1890s.
Courtesy of the New Bedford Whaling Museum.

Narwhal left San Francisco in March of 1892 and four months later weighed anchor at Point Barrow. The next day boatsteerer Joseph Peters disappeared from the vessel. When Captain Horace P. Smith discovered that the man's trunk was empty, one reads "[I] made up my mind that I wood travil a piece after Mr 'Joe' I went to Cape Smyth stopped at the Hut of Antone made inquires which convinced me that this Country man of Joes knew something of him and his where abouts and that in time I should get my man. Then I returned to my ship." The following day, Smith and two of his officers searched Antone's place and found deserters concealed in a loft, whom they took into custody, but not Joseph Peters. A day later word arrived that Charles Brower had apprehended the fugitive boatsteerer at Cape Smyth. When Captain Smith arrived there he "was told that the deserters object in leaving the ship was to join Antone in his whaling station."[193]

"Soon after the new year 1894," reminisced Arctic adventurer Christian Klengenberg, "there mushed into Point Hope an old Portuguese sailor by name of Antone Bates (Bett), with an Eskimo woman who had helped him along all the way down from Point Barrow, which is about seven hundred miles around to the north and east from Point Hope. Antone was a restless survivor from a whaleship which had been crushed in the ice."[194] Antone instructed Klengenberg in the art of shore whaling and hired him at a 1/5 lay to supervise one of the teams. In charge of another was a friend of Klengenberg, Joe Ferero (Ferreira). After each party had killed a whale, the natives lost interest, for they now had their meat and blubber. But, with whalebone selling at a premium at the trading posts, Joe and another man decided to stay on the ice with their wives in hopes of getting at least one more whale. The four managed to kill another whale and remove the bone, but discovered, as the lead opened, that they were adrift and heading north. Rather than abandon $3,000 worth of whalebone (some $75,000 in today's dollars) the men decided to remain while their wives, Mrs. Joe and Glemnia, went off to the village in their umiak. Fortunately it ended well—the women came back with help, and all returned to Point Hope.[195]

The Inupiats had many beliefs related to the hunt for whales, seals, and walruses. Several of these reflected attitudes toward women. A young woman who was menstruating was not allowed to approach the shore ice during whaling season. Worse yet was a woman urinating on the ice—a virtual death sentence. Like many indigenous Americans the Inupiats operated under a code of vengeance. Ignorance of local beliefs by strangers thus led at times to disastrous results, as in the case of a whaler named Portuguese Joe. In the spring of 1890, Joe was working as a shore whaler at Point Barrow when he began to live with a native woman with whom he was working, who had abandoned her Inupiat husband. It was near the end of the season, and Joe began to advise her to return to her husband. She insisted, however, on waiting until she was paid by the whaling company. One day, upon leaving their umiak and heading across the ice toward home, they decided to sit down and rest. At that moment her husband, who lived in another village, appeared. After a very pleasant conversation, Joe arose and said goodbye to the couple. When he was a short distance away the husband lifted his rifle and shot Joe in the back, killing him instantly.[196]

It was not until toward the end of the 1880s that the idea of wintering in the western Arctic became common. The coastal zone of the North Slope of Alaska (and northern Canada to the east) is devoid of trees, except in a few river valleys. It has consisted, historically, of tundra whose earth lies perpetually frozen a few inches below the surface.

Only in the brief summer months does the ground brighten with the variegated colors of the Arctic flora. To the south lies the Brooks Range that forms, at a height of over 9,000 feet, a formidable natural barrier. With an average high temperature of 46° F in Barrow in the month of July and -8° F in January, and capable of dropping to –50° F, it is a most inhospitable climate.

In August 1890, a new era in western Arctic whaling commenced when three San Francisco vessels sailed east into the Canadian Arctic and entered Pauline's Cove on Herschel Island, four hundred miles east of Point Barrow, and were soon frozen in. They were the famous tiny steam brigantine *Mary D. Hume*, the steam barkentine *Grampus*, and the schooner *Nicoline*. Their gamble was a huge success. After wintering over at Herschel the *Grampus* returned to San Francisco in the fall of 1891 with 30,000 pounds of baleen and 1200 barrels of whale-oil. The little *Hume*, too small even to carry tryworks, remained another year in the Arctic, and with a catch of thirty-eight bowheads, arrived in San Francisco with 70,000 pounds of whalebone, valued at the time at $400,000 (some $9,000,000 in today's dollars). As for the Portuguese presence, there was only Azorean boatsteerer John Sands on the *Grampus*.[197] But there would be many more to follow.[198]

Soon these pioneering vessels were followed by many others, and a pattern was set that would be followed by the Arctic whaling fleet for the next decade. As the sea ice began to form, the vessels were anchored in the Cove and protected from the encroaching pack ice that had crushed less-protected ships in the past. As in the eastern Arctic, snow was piled up against the sides of the vessels to insulate them from the cold, and a structure was built on deck to allow some work to proceed. And so the long wait until the following summer proceeded. But the whaleships could not bring everything with them; and quickly a trading pattern developed with both Inupiats and Indians from the interior, who would provide the fresh caribou necessary to prevent scurvy.

And then they waited for the darkness and the cold—and discovered that the worst enemy was probably the boredom and the strange sensation of spending weeks in which, in the semi-darkness, night and day were almost indistinguishable. (Although the sun itself did not make an appearance it was not too far below the horizon, and thus some light was apparent in the sky.)

At Herschel up to 500 men wintered in relative comfort. A large building was constructed in which they could get together. To keep the men occupied over the long months bands and theatrical companies were formed. The presence of wives and children of many of the captains helped to moderate masculine behavior. One of the Portuguese boatsteerers on the *Jesse H. Freeman*, for instance, spent time with little Dorothy, the daughter of Capt. William P. S. Porter and his wife Sophie. As Mrs. Porter noted in February 1896, "Dorothy enjoyed her sleigh today with 'Roxy' to pull her and her good friend Antoine to guide them, for over two hours."[199]

In October 1894, Sophie Porter described the birthday party of three-year-old Bertie Sherman, son of Captain and Mrs. Albert C. Sherman of the steam bark *Beluga* of San Francisco. There was a five-course dinner and then card playing and dancing to the Herschel Island Orchestra. Later, on Thanksgiving, the crew members from various vessels organized a theater company dubbed the "Herschel Island Snowflakes," and gave a performance aboard the *Beluga*.[200]

On January 17, 1894, the reappearance of the sun brought a moment of joy for all of the whalemen, for it had not been seen since the previous November. "The sun showed its lower limb above the hills to day it looked fine and for a few minutes she was gazed at by every eye at Herschel Island," commented Captain Horace Smith of the steam bark *Narwhal*.[201] While whalemen certainly would have looked forward to the summer sun, they also adapted to the long winter cold, as seen from the lament of Captain John McInnis of the steam bark *William Baylies* in January of 1895: "The Southard Thermometer Stands 14 above and I wish it would get Down where it belongs as there seems to be quite a number of Cases of Sickness since it has moderated."[202]

In fact, temperatures usually remained well below zero during the winter months, leading to the common ailment of frozen extremities. In such cases, it fell to the captain of the vessel to perform whatever emergency surgery might be in order. Hopefully, he would have the requisite experience. As the *Narwhal*'s master noted in his journal, "I have been studying up the essentials of sugery; amputations, ligature of Arteries. untill I have got some parts which I think I may need so that I can repeat them."[203]

After returning from a hunting expedition, one of the Portuguese crewmen aboard the *Jesse H. Freeman* was fortunate enough to escape

"Winter Quarters at Herschel Island," depicting typical winter activities, including baseball, soccer, and skiing. John Bertonccini, ca. 1894.
Courtesy of the New Bedford Whaling Museum.

his captain's scalpel. Sophie Porter noted in her journal that "on the 18th [December, 1895] the weather became colder… dropping lower each day and most of the time a strong S. W. wind blowing Most of the dog teams that had left for the country, had to come back to shelter & on the 24th it being 46° below zero several of the men who came in were frozen, one was Antonio Roderic our boatsteerer, but we were glad to find he had only slightly frozen his feet, which are already getting well."[204]

During this long period with little necessity for productive work and much idle time, ships' captains encouraged their crews to participate in sporting activities. Tobogganning and skiing were natural sports for the climate. In November of 1894, Capt. McInnis noted that the temperature was 25 degrees below zero and the men had spent the day "Slideing on Sleds & Skees…."[205] Soccer, and even baseball, were common pastimes. The crew of the *Narwhal* were out playing the former sport in the Arctic gloom of mid December, 1892.[206] On February 22, 1895, with the temperature at 32 degrees below zero, Sophie Porter, aboard the *Jesse H. Freeman*, noted, "Flags were all waving in honor of Washington's Birthday. In the afternoon a game of Base Ball was played between the foremast hands and officers of different ships. The fore-

Steam barks Bowhead II, Beluga, and possibly the Thrasher and Orca in Oakland Creek (Estuary), 1890s; Photo B11.9021PL. Courtesy San Francisco Maritime NHP.

mast hands won 7 to 6." Even the ships' masters at times got into the spirit of the sport: "The Captains played their first game of 'Ball' and the ladies were among the spectators," she wrote in February 1895.[207]

For some of the wives, at least, the western Arctic was a great and positive adventure. As Captain McInnis noted, "Who Would have thought a few years ago of Ladies Wintering in the Arctict Ocean and seem to take Pleasure in it but so it is."[208] Sophie Porter was certainly one, as can be seen from her reflections published in the *San Francisco Chronicle*:

> Winter quarters in San Francisco Bay is very different to winter quarters at Herschel Island. So it is difficult to recognize in the weatherbeaten and greasy whaleships now in the mud of Oakland creek [Oakland Estuary], the comfortable and picturesque homes which for two and a half years have been the center of civilization in the Arctic Ocean. Seven months spent in cruising in the Behring Sea, along the bleak and rugged northeast coast of Siberia, getting caught and jammed in the ice, released to be caught and squeezed again till the vessel's timbers creaked and groaned under the pressure, then on through the straits, following the Alaskan coast to the northeast, at times driven by the pack on the shoals along the shore or creeping for shelter behind some huge pile of

> ground ice, then working off shore into a lead in hopes of finding open water ahead, finally reaching the whaling grounds, between the mouth of the McKenzie River, Cape Bathurst and Bank's Land. So it was with a feeling of relief we dropped anchor about the 1st of October in the well sheltered harbor at Herschel Island, and began preparations for our winter life.[209]

While the officers and other veterans of the western Arctic climate had become relatively inured to its hardships, the green hands who had signed on—or perhaps had come on board not entirely of their own volition—often spent their time devising ways to flee. This was exacerbated during the winter of 1895-1896 when gold fever, as it had almost half a century before in California, reached the isolated whaling community at Herschel. Abandoning all common sense, many men deserted for the reputed goldfields to the south. In January 1896, thirteen men, from several of the fourteen vessels frozen in there, took eighteen dogs and provisions and headed toward the Yukon River, some 250 miles away. Soon a curfew was imposed, requiring all men to be on board their respective vessels at 10:00 p.m. A month later the deserters were still not making great headway. Boatsteerer Antone Roderic, returning to the *Jesse H. Freeman* from a trip inland with a large load of deer meat, reported that he had met up with them; but fortunately they did not bother him, since they had already acquired by force from the natives three dog teams and much meat.[210]

With the idea of getting the men in condition for the rigors of the approaching whaling season, the captains at Herschel in 1894 organized a series of competitions among all the crew members that took place on the Fourth of July. There was a tug of war, won by the crew of the *Newport*, who earned a box of tobacco for their success. Other competitions included the long and high jumps, running races, a sack race, and some amusing activities such as a blindfolded wheelbarrow race, a greased pole climb, and a three-legged race that was won by Captain Benjamin F. Tilton and one of his officers. But there were practical contests as well, tailored to shipboard duties: harpoon darting for the crewmen and hand lance darting for captains and officers. The natives themselves, having dedicated the season to providing for the needs of the whaling community, were not forgotten as they participated in a 100-yard race, with the winner receiving ten pounds of prized tobacco. Within a few days the various vessels lifted anchor and set out awhaling.[211]

One of the last blows to western Arctic whaling came with the disaster of 1897 when the *Navarch*, *Orca*, and the *Jesse H. Freeman* were crushed in the ice near Point Barrow; and five other vessels, the *Belvedere*, *Rosario*, *Newport*, *Fearless*, and *Jeanie*, were frozen in with inadequate supplies to last for the long winter months. The U.S. government unfortunately had abandoned the refuge station at Point Barrow; but the displaced whalemen were fortunate to be able to take shelter at the whaling station operated there by Charles Brower.

An Era for Crimps

Just as the leviathan itself became harder to find, so it was with prospective crew members. As one veteran of the Pacific whaling industry observed, "Men shipped in Honolulu were generally sailors and, although they were probably broke, they knew what they were doing. When the outfitting of whalers shifted to San Francisco, there was a radical change in the type of men who were available for foremast hands on either whalers or merchantmen."[212] Anyone familiar with the industry knew that a voyage was not only hazardous to one's health and limbs, but also a gamble for any profit—least of all for the foremast hands. As an example, the bark *Triton* sailed out of San Francisco on December 15, 1891, for a cruise in the North Pacific, with a crew of thirty-eight under Captain Elihu G. Gifford. When it returned exactly eleven months later it had on board 160 barrels of sperm oil and 600 pounds of whalebone. Profits for the captain and mate were not recorded, but of the crew, those with smaller lays—such as the other mates, boatsteerers, blacksmith and carpenter—received $5.00 for their services, while the foremast hands were paid off with $1.00.[213]

Thus in the port of San Francisco the abuses increased. For an industry that in its infancy employed predominantly local youths from New England, then later reached out to the cities of the Northeast, to the Portuguese and other islanders of the Atlantic, and to those of the Pacific, by the late nineteenth century it had become more and more difficult to entice a sufficient number to make up a complete crew. Increasingly it became necessary to resort to a system of recruiting that was at times

Capt. Manuel E. Costa and wife Philomena. Courtesy of the New Bedford Whaling Museum.

brutal. The managing agent was ultimately responsible for obtaining both a captain and a crew, and as for preparing the vessel for the voyage; but there were other layers to this enterprise.

Often acting as crimps were the owners of sailor's boardinghouses, clothing stores, and saloons. It took a mighty astute sailor to avoid the clutches of these men. Already, as a whaling vessel passed through the Golden Gate on its way to anchor in San Francisco Bay, they—or someone in their employ—had come on board, cajoling the foremast hands with the advantages of a certain little place to stay, while providing them with the liquor and tobacco—or perhaps even a certain beauty with dark eyes—that they craved after a long whaling season of months or years.

Philomena Nunes Costa, the wife of Manuel E. Costa, from her vantage point aboard her husband's vessels around the end of the nineteenth century, was accustomed to observing the life of the whalemen. She left a short narrative poem in which she described the act of hunting and processing the whale, and also mused on the life of these maritime men. "After the ship is full," she noted, "Everyone will be happy to go home to see their parents and relatives." But as soon as they arrive in the city, their money will be gone. "On what I ask? It is spent on women and much more." And so, "There they go again, strolling with the whales."[214]

After a few nights in the boardinghouse, with its elevated prices, the whaleman would find himself quite in debt to the proprietor. The only remedy was to ship out again, with everything neatly arranged by the crimp. From the advance that each crew member received upon signing the ship's articles was deducted the cost of his boardinghouse stay, as well as the cost of his outfit—again at elevated prices—and the crimp's commission. This was, of course, all money that the sailor would never see, because it would only be paid by check to the crimp on the day of sailing.

In itself this system was often not sufficient to provide a full crew. Other enticements were tried, such as newspaper ads that promised "a voyage to the South Seas, experience unnecessary." The last resort was force. It is supposed that it was in San Francisco that the word Shanghai became a verb, indicating the kidnapping of a person with the help of alcohol, drugs, or simply force. As Henry A. Clock related in "The Narrative of a Shangaied Whaleman," the crimp's "favorite hours for work are between dark and daylight, and if a ship is to sail with the morning tide he is busy indeed.... They generally roam in pairs, these man-hunters, and when the saloons show no prospective captives they ensconce themselves in some dark alleyway and await the appearance of their prey."[215]

Lisbon House Hotel, *União Portuguesa*, San Francisco, 1898. Courtesy of the J. A. Freitas Library, San Leandro, Calif.

It might be thought that some naïve Portuguese islander would fall occasional victim to such insidious practice, but rather we find the Portuguese anything but ingenuous. The center of commerce of the Portuguese community of San Francisco at the time lay at the edge of the Bay on the periphery of the infamous Barbary Coast, at Jackson and Drumm streets. There in 1885 a Portuguese sailor could put himself up in the Lisbon House Hotel, run by John Cardozo and José Baptista; the Vasco da Gama House at 116 Jackson Street, Frank O. Mello, proprietor; the Lisbon Hotel, 75 Jackson Street, José Christiano da Silveira, proprietor; the Portuguese Exchange Hotel, Joseph L. Enos, 117 Jackson Street; the Luzitania Hotel, 119-121 Jackson Street; or in the boardinghouses of Bello and Costa, 111 Jackson Street; Anna Gomes, 208 Jackson Street; Freeman (Firmino) J. Cunha or Manuel Frates (Freitas). He could get a haircut at the shop of John J. Silva and a new set of clothes at the haberdashery of Antonio Rodrigues and Manuel S. Raposo or the Jackson Street Clothing House of Antonio Vieira. Or he could even quench his thirst at the Azores Saloon, whose proprietor was Raulino Vitorino da Silveira, from the village of Fajã Grande on the island of Flores.[216]

In 1890, with the few funds at his disposal, our mariner could have a drink or two at the bar of the Balaena Restaurant and Guest House on Jackson, which was "furnished with the best quality of wines, liquors, beer, cigars, etc.," and whose Azorean owner promised to find work for its guests, including seamen. Or he could wander up to the Avenue Music Hall of Manuel Correa Machado, offering "good music," the best liquors and a billiard table "for the entertainment of our customers." And if he went a few blocks up to Kearny Street and entered the Golden Gate Saloon he would be well received by Azorean barman Manuel Silva. Finally, venturing deeper into the area of the Barbary Coast, he could try the House of All Nations, where another Azorean, Luiz Gomes, hired dancers from "all the civilized nations."[217]

What did all these entrepreneurs have in common? Most were ex-whalers and served also as crimps. For example, in 1887, when Firmino Cunha placed five sailors on board a whaler he received $580 in advances from the well-known shipping master James Laflin. That same year John Cardozo received $250 and Joseph Enos $350, also in advances.[218] In 1890, Luiz Nunes placed nineteen crew members and received $1509. For the placement of sixteen crewmen Manuel Freitas (sometimes listed as Frates) received $1450. And Raulino Silveira received $990 for twelve individuals. Clothiers Luiz Nunes and Manuel Freitas were among the most important of crimps supplying sailors to Laflin.[219] The latter, Manuel Teixeira Freitas, was born on the island of São Jorge in 1853 and—with only three years of formal education—arrived in San Francisco in 1870. He began as a dishwasher on the waterfront, then became a cook and restaurant owner. Freitas then purchased a hotel, established the Portuguese American Bank on Front Street, and later served as Portuguese Vice-Consul and as a director of the Bank of Italy (which later became the Bank of America). Freitas worked as a commission merchant, buying produce from farmers to sell to urban comsumers and was also involved in the dairy industry. His home ranch in Marin County to the north of the Golden Gate was sold to developers in the 1950s and now forms the town of Terra Linda, reached by the Manuel T. Freitas Parkway.[220]

Other Portuguese participating in this activity included Antonio Vieira, Rodrigues and Raposa, Luiz Gomes, and two women—one perhaps the latter's wife—Mrs. Anna Gomes, and the other, Mary Duarte.[221] Indeed, one of the whalemen himself, the Cape Verdean, Delfino Gomes, functioned as a crimp in the interval between his voyages to the Arctic as mate.[222] Most of the above did business with other Portuguese. By the end of the first decade of the new century, Gomes, whose name had segued into Louis Gomez, became one of the more prolific providers of crewmen for the San Francisco whaling fleet. While he often was responsible for placing his compatriots aboard ship, very frequently he was signing up green hands of all nationalities.[223] Gomez certainly was assisted in his contact with his maritime clients through the operation of his liquor store on the waterfront.[224]

"Portuguese Joe" Gomes was operating a sailor's boardinghouse on Commercial Street in 1888. A rival, John T. Callender, who ran a boardinghouse for men of color, discovered two of his men missing one morning. Mariano Torres, a Portuguese, and Manuel Jo, a Mexican, were each in debt to Callender for $30. The boardinghouse keeper soon found that the men had been shipped in a drunken state aboard the whaling bark *Cape Horn Pigeon*, at that time sailing out of San Francisco under Captain Thomas Scullun. Evidently Gomes had shipped these men and two others aboard the vessel after getting them intoxicated. Callender then vowed vengeance on Gomes for the loss of his $60, and not long after was arrested "on a charge of threats against life." As the *Daily Alta California* observed at the time, there appeared to be no honor among thieves.[225]

More Beachcombers

The description that ex-whaling captain Philo Holcomb—who later settled as a trader on Yap, in the western Carolines—gave to his sister in 1871, succinctly portrays the romantic attraction of westerners to these exotic spots in the central Pacific. It was a land, Holcomb wrote, "where the natives go nearly naked and appear to enjoy life as well as the fashionables of New York, where the earth produces all that is required to sustain life without labor, and where although it is sometimes warm there is never any cold to freeze a person, where the fruit is always in season and is free to all."[226]

As a yachtsman observed about the Pacific in 1875, "Go to what island you will where the natives are at all friendly, in one of the first canoes that visits your vessel there will be a white man…."[227] Joshua Beane, sailing on the bark *Java* under Captain Manuel Enos in the mid-1860s, remarked on the Portuguese presence in the western Pacific when the bark called at the Bonin Islands. He noted that there "were living a curious crowd of some fifty or sixty people, white, black and brown. There were Americans, Chimoros, Portuguese and Kanakas. Mr. Brava, a Cape de Verd Islander from the island of Brava, was said to be the richest man among them, owning a beach and many acres of fertile land, which he had occupied for more than twenty years."[228]

One reason that Portuguese and others were found on many of the Pacific islands at this time was because of the practice called "blackbirding." By trickery or by force tens of thousands of South Sea islanders were taken from Melanesia, Micronesia, and Polynesia and carried to such places as Australia and Peru. Although often men, women, and children were taken, there was a general preference for able-bodied men. And so the decimation of the male population on many of the islands offered an additional encouragement to whalemen to settle in these remote regions. Such was the case in the four-island Tokelau group: Fakaofo (Bowditch Island in whaling days), Nukunonu, Atafu, and Olohega, situated to the north of American Samoa.[229] And so it transpired that a Cape Verdean trader by the name of Antonio Pereira (known locally as Atone) arrived in Fakaofo in May of 1863, three months after half the population had been removed by slavers. There he, and later his son Peleila (José Pereira), became major figures in the island's economic life.[230]

Similarly, on Nukunonu, the only established trader was a Portuguese named Hehe Perez (Pires). Like Atone, he had arrived just after the depopulation. Perez and his native wife gave rise to what became known as the *kaiga potiki* or "Portuguese family" of the island.[231] Tokelau is currently a colony of New Zealand, but in two recent referendums it barely failed to gain its autonomy. At the time of the first vote in February of 2006, Ian Parker of the *New Yorker* magazine traveled to these tiny coral atolls comprising in total 10.8 km^2. There he met Luciano Perez, a descendant of the original settler, "a man in his late sixties, [who] has a kind of chuckling, restless sociability that I didn't otherwise see on the islands."[232] A retired school principal, he and his wife run a small hotel on an island that has virtually no tourists. In a society that is deeply communitarian, observed Parker, the couple stands out as being "Tokelau's village capitalists."[233]

Joshua Slocum gained fame at the turn of the twentieth century for being the first person to circumnavigate the globe singlehandedly, sailing from Nova Scotia in July of 1895 in his tiny sloop *Spray*, and appearing again three years later in Newport, Rhode Island. In April of 1896, he reached the island of Más a Tierra, the largest of the Juan Fernández Islands, lying some 400 miles west of Chile and now known as Robinson Crusoe Island. It is of course the setting for Daniel DeFoe's famous novel, based on the real-life adventure of Alexander Selkirk, who had spent four years isolated on the island after abandoning the privateer on which he was serving. There Slocum was met by a small boat carrying the "king." Not to be confused with the governor of the island, which belongs to Chile, the man "was king only by courtesy. Having lived longer on the island than any other man in the world,—thirty years,—he was so dubbed." The "king" was, in fact, an old Azorean boatsteerer who was settled there comfortably with his Brazilian wife. Slocum was especially impressed by the whaleman's home: "The house of the king of Juan Fernandez, Manuel Carroza (Cardoza) by name, besides resembling the ark, wore a polished brass knocker on its only door, which was painted green. In front of this gorgeous entrance was a flag-mast all ataunto [fully rigged], and near it a smart whale-boat painted red and blue, the delight of the king's old age."[234]

To name some of the Marshall Islands, which lie to the east of the Carolines and some 2100 miles southwest of the Hawaiian Islands, is to bring up memories of World War II and its nuclear aftermath: Kwajelein, Enewetok, and Bikini—a legacy with which the Marshallese are still encumbered. But 150 years ago it was a very different world. A Portuguese who was most successful and who left a lasting legacy in the Marshall Islands was Antone Jose De Brum (António José de Brum). De Brum was born in 1837 on the island of Pico in the Azores and served as boatsteerer on the whaling bark *General Scott* of Fairhaven from 1858 to 1862. October, 1864, finds him on the atoll of Ebon in the Marshalls. De Brum soon formed a partnership with Adolph Capelle a German who had arrived at Ebon five years before at the age of twenty-one, and who was charged with exploring commercial opportunities in the region. There on Ebon, the two established a trading post working first with a Scotsman, by the name of Edward J. Milne, and later an American, Charles Ingalls. Soon Adolph Capelle and Company, as it was known, became engaged in the extraction of coconut oil, and later copra, which they sold to European firms.[235]

Descendant Alfred Capelle, when interviewed, was the Marshall Islands' Ambassador to the United Nations. He related a family tale of how, when it became apparent that the firm needed a more reliable source, it acquired an entire atoll, Likiep. De Brum had married the niece of the chief who had control over the atoll and so, in 1877, was able to negotiate with him to purchase it for himself. Soon after, De Brum turned over title to the firm. Residents were allowed to remain if they agreed to work for the firm. Typically these islands had a few coconut trees around the native settlements, but beyond that much was brush. This was soon cleared to be replaced by a coconut plantation, with the land divided among the three men. After Ingalls became ill and returned to the U.S., ownership fell to the two remaining partners.[236]

Antone's first son, Joachim, was born in 1869. Joachim was also a mariner, serving for a time as captain on inter-island trading vessels. He later developed a boat-building company on Likiep that turned out fast sailing vessels for inter-island use. He also operated a store and was active in the family copra business. Spending his life on the tiny island of Likiep, Joachim was, in his own way, a man of the world. He has been described as a "true renaissance man" of the western Pacific, a "self-taught businessman, builder, artist, engineer, shipbuilder, and amateur scientist," who left the only extensive private library in Micronesia, the Joachim de Brum Library. He was also an amateur photographer whose important collection of over 2,000 glass plate negatives documented life in the Marshalls during his lifetime. Among his many correspondents were relatives in the Azores.[237]

Likiep today, in spite of some historic litigation, continues to belong to the descendants of Capelle and De Brum. And copra still forms the basis of the island's economy. The two families intermarried, and today their descendants are among the most prominent citizens of the Marshall Islands. Besides Ambassador Capelle, the Hon. Banny de Brum was for many years the Republic's ambassador to the U.S.

Many of these Portuguese became traders in the region. In 1868, John Silva, probably a Cape Verdean, and known as "John the Portugee," began trading on the island of Pohnpei in the Carolines, and was still living on the island in 1887.[238] Manuel Silva was less fortunate. Setting himself up as a resident trader in 1881 on Ulul Atoll, also in the Carolines, he and his business partner, an Irishman named Powers, who lived on neighboring Pulowat Atoll, were killed within a few months of each other at the instigation of the people of the latter island.[239]

A man of less savory reputation was Christian Barbus [Cristiano Barbosa], a Cape Verdean who lived at Pohnpei from 1883 to 1891. During the Spanish administration of the island he spied for them and was later killed by a leading chief, whom he had betrayed to the Spaniards.[240]

During the last decades of the nineteenth century, the Pacific and especially the western Arctic became critical to American whaling. In the 1860s, Hawaii became a major refitting station for American whalers, intent on avoiding the temptations of the Golden State. And later Hawaii developed its own short-lived native whaling industry. Also, what had begun as a minor settlement of Azoreans and Cape Verdeans in the Hawaiian Islands as a result of their service on American whalers, evolved into a major influx, as thousands of Portuguese were brought in to work the sugarcane plantations.

As the century ended, San Francisco had become, at least temporarily, the whaling capital of the United States. Here, the Portuguese would perpetrate some of the most egregious cases of exploitation, ironically, against Americans and northern European immigrants.

While the halcyon days of American whaling were certainly over by the end of the century, there was still life in the industry. In 1899, nine vessels sailed out of New Bedford, three out of Provincetown, and one from New London—all but one headed for the Atlantic grounds (the schooner *Francis Allyn* went into Hudson Bay). Nineteen whaling vessels left San Francisco that year, all bound for the North Pacific. As for the Portuguese, they continued to predominate as boatsteerers and junior officers, while three whalers sailed out of New Bedford under the command of Azorean masters: Joseph T. Enos, Manuel F. Gomes, and Joseph Thompson. In truth it was only in the final era of the industry that the great growth in the percentage of Portuguese whaling captains would be seen.

Notes

[1] *New York Times*, 26 Aug 1859; Ellis, *History of New Bedford*, 311-14.

[2] Unidentified newspaper, 1861, Scrapbook T2:173, ODHS.

[3] Bark *Cleone*, Luce, New Bedford, Mass., 11 Jul 1865, Log 280, Ships' Logs Coll., MSS 220, NHA.

[4] Joseph F. Francis, Memoir, dated Vila das Lages [*sic*] – Ilha do Pico, 1915, Joseph and Frank Vera Papers, MSS B97-13, ODHS. Whaling master Joseph F. Francis was born on the island of Pico in the Azores in March of 1847 and first arrived in the U.S. in 1865. He later served as master on three voyages of the whaling bark *E. B. Phillips*. In 1900, he was residing in Provincetown and continuing to whale (U.S. Census, Barnstable Co., Mass., 1900). Later he returned to his native island. Amaral lists him as Joseph C. (*They Ploughed the Seas*, 66-67). He is not to be confused with another Joseph F. Francis, born on the island of São Miguel, Azores, in 1852 who came to the U.S. in 1873 and settled in New Bedford. This latter Francis became a merchant and shipowner in the city (Amaral, 67-68.)

[5] Rev. Moses How, Private journal, 5 February 1853, NBL.

[6] Lovins, et al., *Winning the Oil Endgame*, 5.

[7] Davis, Gallman, and Gleiter, *In Pursuit of Leviathan*, 15.

[8] Taber, "Condensed Data," 71-96; Crapo, "The Story of Cotton,." Actually, capital from New Bedford whaling interests had been flowing into the nearby Fall River mills since the 1830s.

[9] *Ship Registers of New Bedford*, 1:814 and 3:737.

[10] U.S. Census, Bristol Co., Mass., 1860 and 1870; *Ship Registers of New Bedford*, 1:814 and 887; *New Bedford Directory* (1867-68), 226.

[11] *New Bedford Directory* (1867-68), 226; *Ship Registers of New Bedford*, passim.

[12] Ellis, *History of New Bedford*, 465 and 521.

[13] U.S. Census, Bristol County, Mass., 1870; *New Bedford Directory* (1849), 151; *Directory of New Bedford* (1871), 207; *Directory of New Bedford* (1881-1882), 272; *Ship Registers of New Bedford*, passim.

[14] Joseph and Frank Vera Papers, MSS B 97-13, ODHS.

[15] *Ship Registers of New Bedford*, passim.

[16] Thomas Luce, U.S. Passport Application, 12 Jan 1895, RG 59, NAW; "Thomas Luce & Company," Wikipedia, http://en.wikipedia.org/wiki/Thomas_Luce_&_Company (accessed 2 Oct 2008).

[17] Ship *Roman II*, Barker, New Bedford, Mass., 1844, Seamen's Register, ODHS.

[18] *Ship Registers of New Bedford*, passim.

[19] Ross, *Arctic Diary*, 41.

[20] Ibid., 57n.

[21] *New Bedford Directory* (1867-68), 218; U.S. Census, Bristol County, Mass., 1870. In the directory Martin's first name is misspelled as "Antoine."

[22] Ibid., passim. Costa is listed as "De Costa" in the directory.

[23] *Directory of New Bedford* (1881-1882), 307 and 337.

[24] *New Bedford and Fairhaven Directory* (1892), 512-13, 556, and 560.

[25] Cf. Power of attorney of José Maria Ferreira, 19 Jun 1863; Common pleas Court, New York City, vol. 92, record 49, 7 Apr 1849, NANY.

[26] H. Cohen to J. A. Machado, Fort Lafayette, 21 Nov 1862, Jacob Morris Papers, courtesy of NYHS.

[27] Governor Wincett, Cape Coast Castle, British Gold Coast (Ghana) to Capt. Machado, 20 Jun 1848, Jacob Morris Papers, courtesy of NYHS. No other extant documentation has been located referring to Machado as a maritime officer.

[28] Reilly, "Slavers in Disguise," 189; Agard Coll., MSS 23, Ser. A, f. 43, ODHS; Stuart M. Frank, "False Whalers—American Slavers Disguised as Whalers," MSS, ODHS. Devoll left the vessel in the Azores.

[29] It is not known whether there was any connection between Mrs. Watson and Captain Thomas Watson, a Scotsman who was a naturalized American citizen, nor whether the name of Machado's vessel is linked to him. In 1858 Watson's schooner, the *Lydia Gibbs*, was impounded on the coast of Africa under suspicion of being involved in the slave trade. Capt. Watson soon died there of the fever. U.S. Census, New York City, N.Y., 1860; Consul Bunch to the Earl of Malmesbury [Extract], British Consulate, 1 Jun 1858, "African Slave Trave," H. R. Ex. Doc. 7, 36th Congress, 2nd Session, 88; *New York Times*, 4 Nov 1858.

[30] Manning, *Six Months on a Slaver*, 11-12.

[31] Reilly, "Slavers in Disguise," 67.

[32] Manning, *Six Months on a Slaver*, 21.

[33] Howard, *American Slavers*, 252; deposition of John W. Baker, 16 Oct 1862, Turner-Baker Papers, RG 94.2.5, NAW.

[34] Known as the *Mary Frances*, its name appeared briefly altered during 1861 and 1862 (v. *American Lloyd's Register* [1862], 190.) Howard, *American Slavers*, 234; Turner to Stanton, 8 Nov 1862, Turner-Baker Papers.

[35] *New York Times*, 21 Sep 1862. The 'ch' is pronounced as 'sh' in Portuguese. Mrs. Watson allegedly died from drink while in Cadiz. (Murray to Seward, 11 Nov 1861; Stanton to Turner, 8 Nov 1862.)

[36] Murray to Turner, 13 and 23 Oct, 13 Dec 1862; Turner to Stanton; deposition of Henry M. Smith, 30 Oct 1862. Turner-Baker Papers; *New York Times*, 21 Sep 1862.

[37] Cf. "I have seen your friend on 21st last eve. She stated that Mr Cochran had spoken to the President & all will be arranged." And "Mr W. had promised that you should be released last week & again we thought that under the order from the Secry of War you would be out anyhow." H. Cohen to J. A. Machado, 21 and 26 Nov 1862, Jacob Morris Papers, courtesy of NYHS.

[38] Turner to Murray, 10 Dec 1862; Murray to Turner, 24 Dec 1862. Turner-Baker Papers; The United States vs. John A. Machado: Indictment, 01/09/1863, U.S. Circuit Court for the Southern District of New York, RG 21, NANY; *New York Times*, 22 Feb 1863.

[39] Howard, *American Slavers*, 233-34; Reilly, "Slavers in Disguise," 97.

[40] "The United States vs. John A. Machado, 9 Jan 1863," U.S. Circuit Court for the Southern District of New York, Records of District Courts of the United States, 1685-1991, NANY.

[41] Seamen's Register, ODHS (data courtesy of NBWP).

[42] Schmitt, de Jong, and Winter, *Thomas Welcome Roys*, 71-72.

[43] Bullen, *Cruise of the Cachalot*, 100-01, 195. As the bark *Greyhound* of New Bedford was cruising on the Western Grounds in June of 1900 the keeper noted that, "the second mate shot the whale and the gun exploded and he got his fingers blowed off." (Bark *Greyhound*, Enos, New Bedford, 9 Jun 1900, Log Wh G845 1898L, Nicholson Whaling Coll., PPL.)

[44] Bullen, *Cruise of the Cachalot*, 126.

[45] Andrade, "True Remarks," 23 Oct and 18 Dec 1860. See also pp. 105-06.

[46] Ibid., 23 Dec 1860.

[47] Ibid., 24 Dec 1860.

[48] Smyth, *Chapman the Sailor's Lexicon*, 324.

[49] *Republican Standard (New Bedford)*, 23 Mar 1854. Incredibly, over half a century later it could still be difficult to get a crew to sail on a Friday. For that apparent reason three whaling vessels, the schooners *Ellen A. Swift* and *John R. Manta*, and the bark *Bertha*, were held over until Saturday before setting sail out of the port of New Bedford in April of 1916. ("Whalemen Refuse to Sail on Friday," *New Bedford Times*, 28 April 1916).

[50] Ship *Arnolda*, Hawes, New Bedford, J. B. Wood & Co. Records, MSS 75, Ser. A, S-s 1, v. 1, ODHS.

[51] *WSL*, 13 Jul 1858, 19 Jan 1864, 21 Dec 1869, 19 Mar 1872, and 15 Aug 1882.

[52] Kuykendall, *Hawaiian Kingdom*, 1:139.

[53] *Pacific Commercial Advertiser (Honolulu)*, 28 Dec 1867.

[54] The latter's name is variously recorded as "I. de Silva" (Records of the Honorable Harbor Master, Registry of Entries and Clearances, Jan 1868-Jul 1880, MSS, Ser. 104, vol. 6, HSA), and simply "da Silva" (Hegarty, *Returns of Whaling Vessels*, 50). He is most likely either Joseph D. Silva or possibly his younger brother, John D. Silva.

[55] Knowlton, *Portuguese in Hawaii*, 46-47.

[56] *Hawaiian Gazette (Honolulu)*, 10 May 1876.

[57] Hillebrand to the Minister of the Interior, 6 Jun 1877. Quoted in Kuykendall, *Hawaiian Kingdom*, 3:123.

[58] Knowlton, *Portuguese in Hawaii*, 11.

[59] Cf. Knowlton, *Portuguese in Hawaii*; Dias, *Presença Portuguesa no Havai*; Santos, *Portugueses no Hawaii*; Felix and Senecal, *Portuguese in Hawaii*; Felix, Nunes, and Senecal, *The 'Ukulele.'*

[60] Rev. John S. C. Abbott, "Trials and Dangers of the Whale Fishery." *New Bedford Weekly Mercury*, 31 Mar 1843.

[61] Beane, *From Forecastle to Cabin*, 51-52.

[62] Charles H. Sherman and William Martin, Journal, bark *Java*, Enos, New Bedford, Mass., 15 Aug 1867, Log 787, KWM.

[63] Beane, *From Forecastle to Cabin*, 280.

[64] Parry, "Yankee Whalers in Siberia," 42.

[65] Beane, *From Forecastle to Cabin*, 189-192; bark *Java*, 31 May and 1 Jun 1866, Log 787-88, KWM.

[66] *National Era (Washington, D. C.)*, 10 Dec 1857; Beane, *From Forecastle to Cabin*, 208.

[67] Parry, "Yankee Whalers in Siberia," 46.

[68] Senate Ex. Doc. No. 106, 50th Cong., 2nd Sess., 255.

[69] Valentine, "'Big Manuel,'" 57.

[70] "List of Vessels Belonging to Cold Spring Harbor" (1874), in ship *Alice*, Penney, Cold Spring Harbor, 1854-58, Log, Hewlett-Jones Coll., CSHM.

[71] U.S. Census of 1870, Suffolk County, New York; "List of Vessels Belonging to Cold Spring Harbor" n.d.; Custom House Records, 1848-1874, Hewlett-Jones Coll., CSHM; Valentine and Valentine, *An Island's People*, 6; Building-Structure Inventory Form, Division for Historic Preservations, New York State Parks and Recreation, Town of Huntington, Summer, 1979.

[72] Farwell, "Manuel Enos," 109.

[73] Bark *George and Susan*, Colson, New Bedford, Mass., 1877-81; bark *John and Winthrop*, Shiverick, New Bedford, Mass., 1881-85, Seamen's Register, ODHS.

[74] *Brooklyn Standard-Union*, MSS, n.d., CSHM.

[75] Mrs. Alfred L. West to Estelle, 16 March 1953, Valentine Collection, CSHM. (In her letter Mrs. West identifies Capt. Enos as being her grandfather.)

[76] There is some confusion over the new name of the *Matilda Sears*. The fact that the whaling firm itself was known as Mathieu y Brañas probably contributed to this. It has been stated, for instance, that the American bark *D(r). Franklin* became the Chilean whaler *Mathieu y Brañas*, ca. 1864 (Starbuck, *History*, 581), and that the *Matilda Sears* had been rechristened the *Gabriel Toro* (unidentified newspaper, 8 Jan 1905, NBW Scrapbook, 2:410). However, a Chilean source seems more reliable, identifying the *Mathieu y Brañas* as the old *Matilda Sears* (Sandoval Hernández, "Talcahuano y los últimos balleneros," 234), as do the informal sources quoted here. In any event, it is clear that both the *Gabriel Toro* and the *Mathieu y Brañas* were still whaling out of the port of Talcahuano as late as 1904 (Sandoval Hernández, "Talcahuano y los últimos balleneros," 234; *New Bedford Evening Standard*, 6 Dec 1904).

[77] There must have been some communication with his old town, for Enos and María Petronila did not marry until March of 1904, several months following the death of Susan Brush Enos in Cold Spring Harbor. Rafael Enos Aguirre, "Mis notas al Capitán Enos" (private document). Since 1867 Francisco Vieira, a Portuguese, had been managing partner of the firm of Mathieu y Brañas, the other eponymous partners being French and Galician. ("Efemérides nacionales y mudiales 'julio'" http://www.inpesca.cl/EFEMERIDES/JULIO.htm, consulted 30 Sep 2009). Also, ca. 1880, Juan Macaya and João da Silva formed a partnership that later became known as the Empresa Ballenera Macaya, supposedly the oldest whaling company in the world still operating in the 1970s. (Sepulveda Ortiz, "La epopeya," 545-46.)

[78] Rafael Enos Aguirre to Donald Warrin, email, 29 Sep 2007 and 18 Mar 2009; Rafael Enos Aguirre, "Orígenes e historia de la familia Enos" (private document).

[79] Rafael Enos Aguirre, "Orígenes"; "El Mar fue mi universidad," Manuel Enos Sosa, interview by Macarena Rojas Gómez, in the digital magazine "La Columna," 19 Sep 2004 (no longer online); *New York Times*, 16 Nov 1965.

[80] None of these statistics include those taking over as replacement master during the voyage.

[81] Ship *Gazelle*, Gifford, New Bedford, Mass., 1866-70, Log 361, KWM; Log 96A, ODHS; *WSL*, 21 Aug 1866. It was on this voyage that the *Gazelle* carried off the Irish revolutionary John Boyle O'Reilly after he had escaped from his imprisonment in western Australia (cf. Roche, *Life of John Boyle O'Reilly* and Evans, *Fanatic Heart*).

[82] Seamen's Register, ODHS (data courtesy of NBWP); Hohman, *American Whaleman*, 301.

[83] Bark *Wave*, Briggs, New Bedford, Mass., 1879-82, Seamen's Register, ODHS.

[84] Bark *Lydia*, Frazier, New Bedford, Mass., 1880-84, Seamen's Register, ODHS.

[85] Bark *Sarah*, New Bedford, 19 Jun 1876, Seamen's Register, ODHS.

[86] *WSL*, 22 Oct 1878.

[87] Unidentified newspaper, Sanford Scrapbook, 146, SCHS; *Brooklyn Eagle*, 24 Sep 1900.

[88] *WSL*, 11 Jun 1861.

[89] Bark *Tamerlane*, Fordham, New Bedford, Mass., 16 Jan-11 Feb and 11 Nov-17 Dec 1872, Log 1224, ODHS.

[90] Schooner *Lottie E. Cook*, Fordham, New Bedford, Mass., 1884-85, Log Wh L884 1884L, Nicholson Whaling Coll., PPL.

[91] Brig *Eunice H. Adams*, Edgartown, Mass., 20 Feb 1890, Log Wh E88 1887L, Nicholson Whaling Coll., PPL.

[92] Catherine M. Mayhew gives the date of death as February 17 (110 Whaling Captains Collection, Genealogical Files, MVH). Amaral states April 17 (*They Ploughed the Seas*, 60).

[93] Ship *Europa*, Milton, New Bedford, Mass., 1862-67, Seamen's Register, ODHS; Lund, *Whaling Masters*, 254, 471. Between this and his next voyage, Pierce managed to fit in a command of the Cape Verde packet schooner *Blue Bell* of New Bedford, which returned from Brava in September of 1868. (Halter, *Between Race and Ethnicity*, 179.)

[94] On 18 Sep 1871 João Roza was charged for five gallons of whiskey, and during the next seven days six other individuals, including Pierce, were noted as purchasing from one "painte" to one "galon." (Schooner *Thriver*, Pierce, Boston, 1870-71, Log 100, ODHS).

[95] *Thriver* log; Thompson, *American Reports*, 441-42. The only reference to his death is in the Federal Writers' Project, *Whaling Masters*, 222.

[96] Amaral, *They Ploughed the Seas*, 120.

[97] John E. Luce to Captain Hathaway, 23 Nov 1871, Collection of the author.

[98] Amaral, *They Ploughed the Seas*, 85; Joseph Sequeira Vera, "Capt. John E. Luce (1822-1909)," MSS, Collection of the author.

[99] Amaral, *They Ploughed the Seas*, 89-90.

[100] Ibid., 92.

[101] Amaral, *They Ploughed the Seas*, 145. The family name was Abreu; obviously there was some American influence in the choice of his adopted name.

[102] Ibid., 145-46; *New Bedford Evening Standard*, 5 Aug 1905.

[103] Amaral, *They Ploughed the Seas*, 147.

[104] See p. 139. According to João Gomes Vieira his original name was José Pereira Leal (Gomes Vieira, *Man and the Sea*, 255).

[105] Ship *Potomac*, Ackley, Nantucket, 1853-57, Log 189, Ships' Logs Coll., MSS 220, NHA. *WSL*, 26 Jun 1866; ship *Bartholomew Gosnold*, Nichols, New Bedford, Mass., 1866-70, Whalemen's Shipping Paper, 18 Jun 1866, mic 179, NBL.

[106] Dennis Wood Abstracts, 4 vols. Bound MSS copied from NBL, v. 4 pt. 2, p. 385; for more on the *Crowninshield* under Capt. Praro see p. 194.

[107] Federal Writers' Project, *Whaling Masters*, 305.

[108] Bark *Peru*, Benton, New Bedford, Mass., 1878-81, Seamen's Register, ODHS.

[109] Bark *Ohio*, Benton, New Bedford, Mass., 26 Mar 1882, Log Wh 037 1881L, Nicholson Whaling Coll., PPL.

[110] Bark *Ohio*, Log, 18 Dec 1883-10 Jan 1884.

[111] Davis, *Nimrod of the Sea*, 198.

[112] See pp. 226-27.

[113] Brig *Isabella*, Blossom, New Bedford, Mass., 1880-81, Log 218, ODHS.

[114] "Whaling in Davis' Strait," *Republican Standard (New Bedford)*, 24 Nov 1859.

[115] Bark *Wave*, Benton, New Bedford, Mass., Jun 1885-Sep 1886, Log 221, ODHS.

[116] Ibid.

[117] Bark *Tamerlane*, Benton, New Bedford, Mass., 9 May 1888, Log 39, ODHS; bark *Tamerlane*, Seamen's Register, ODHS.

[118] Bark *Tamerlane*, Log, 6 Oct 1888.

[119] He is listed as "Joseph P. Bento" (Seamen's Register, ODHS).

[120] See p. 52.

[121] Joseph P. Benton, Journal, schooner *Clara L. Sparks*, New Bedford, Mass., 24 Aug 1892, Log 73, ODHS.

[122] Whaling crew list card file, NBL.

[123] Bark *A. R. Tucker*, Millard, New Bedford, Mass., 15 Jul 1899, Log 456A, NBL.

[124] Ibid., 18 Jul 1899.

[125] "Mas não foi bem neum P que o Capitão fose embora e tivermos que perder as 5 baleias que tinhamos mortas P que ser mut° tarde e chegarmos abordo pelas 8 horas da noite e a sim se fecho o dia com tanta má sorte." Ibid., 15 Jun 1901.

[126] John F. Tucker, Acting Vice Consul of Portugal, New Bedford, to Acting Portuguese Consul General, New York, 1 and 23 Apr 1867, Journal, ship *Hibernia II*, Whitfield, New Bedford, Mass., 1865-69, NBL, 64.

[127] Passenger Lists of Vessels Arriving at Boston, Massachusetts, 1820-1891, RG 36, NAW.

[128] Schooner *Pedro Varela*, Leal, New Bedford, Mass., 1887-89, Seamen's Register, ODHS.

[129] Amaral, *They Ploughed the Seas*, 124, 129, and 131. Since Silvia is technically neither Portuguese nor Portuguese American he is not included in any of the related statistics.

[130] When Enos signed the shipping articles for his voyage as second mate aboard the schooner *Roswell King* in 1857 he did so with his mark, indicating an inability to write his name (schooner *Roswell King*, 29 Apr 1857, Whalemen's Shipping Paper, mic 185, NBL).

[131] Chippendale, *Sails and Whales*, 135.

[132] Bark *Greyhound*, Enos, New Bedford, Mass., 18 Aug 1885, Log 84, ODHS.

[133] Joseph T. Enos to Joseph Silva, bark *Greyhound*, Log, 22 Aug 1885.

[134] Bark *Greyhound*, Log, 28 Jun 1886.

[135] Chippendale, *Sails and Whales*, 134.

[136] Statement of James B. Coffin, U.S. Consul, St. Helena, bark *Greyhound*, Enos, New Bedford, 28 Feb 1893, Log Wh G845.1892L, Nicholson Whaling Coll., PPL.

[137] Bark *Greyhound*, Log, 22 Mar 1893.

[138] "Survivors of President," unidentified newspaper, Scrapbook 2:314, ODHS.

[139] Ibid.

[140] Joan Druett, "Women Who Went A-Whaling: An Annotated List," MSS, ODHS. Druett states that the voyage of the *California* took place in 1899, but in that year Joseph served as mate and not master; and the couple was not yet married.

[141] Bark *Sarah*, Peakes, New Bedford, Mass., 11 Oct 1878, Seamen's Register, ODHS; *WSL*, 22 Oct 1878.

[142] Hegarty, *Returns of Whaling Vessels*, 16.

[143] Bark *Lydia*, Frazier, New Bedford, Mass., 1880-84, Seamen's Register, ODHS; U.S. Census, Bristol County, Mass., 1900.

[144] U.S. Census, Bristol County, Mass., 1880.

[145] Amaral, *They Ploughed the Seas*, 70-71.

[146] Brig *Frances A. Barstow*, Edwards, San Francisco, Log, 9 Jan 1895, CSHM, 79. Another record of the same voyage noted succinctly that on that date they "landed boatsteerer on beach" at Lat. 20:00, Long. 38:18 (brig *Frances A. Barstow*, Log 356, KWM).

[147] Brig *Frances A. Barstow*, Log, CSHM, 4 Jan 1895, 191.

[148] Brig *Frances A. Barstow*, 1894-97, Crew List, NASB.

[149] Dennis Wood Abstracts, 3:216 and 516; *Polynesian (Honolulu)*, 15 Dec 1860; WSL, 12 Feb 1861; Enos vs. Sowle and Vieira vs. Sowle, in *Hawaii Reports*, 332-48.

[150] The surname occurs as "Murry" in the federal censuses of 1850 (Essex Co., Mass.) and 1880 (Nantucket Co., Mass.). Most references, however, employ "Murray."

[151] John Murray, Death Index, Office of the Nantucket Town Clerk, Nantucket, Mass.; ship *Isaac Howland*, Hobbs, New Bedford, Mass., 1854-59, Whalemen's Shipping Paper, mic 183, NBL; *WSL*, 19

Jun 1860. From his mark on the shipping paper it appears that Murray was illiterate; *WSL*, 12 Jul and 13 Dec 1870; U.S. Census, Nantucket Co., Mass., 1880; Kartunnen, *Other Islanders*, 115; obituary of John Murray Jr., unidentified newspaper, 21 Feb 1920, Edouard A. Stackpole Coll., MSS 335, f. 525, NHA.

[152] Seager, "Portuguese Islanders," 10-14.

[153]Joseph P. Thompson, Journal, schooner *Alcyone*, Mello, Provincetown, Mass., 8 Oct 1889, Log 630, KWM.

[154] Busch, *Master of Desolation*, 127.

[155] Busch, *War Against the Seals*, 177. While the bark *Union II* was at the island of Fogo in September of 1864 the keeper noted getting men was much easier than supplies: "could not trade eny for the natives have not enough for themselves to eat. got one Pig one Goat two Chickens and three men." (Bark *Union II*, Rogers, Sag Harbor, N.Y., 23 Sep 1864, Log, RML.)

[156] Busch, *Master of Desolation*, 128.

[157] U.S. Census, New London Co., Conn., 1880.

[158] Busch, *Master of Desolation*, 199.

[159] Schooner *Francis Allyn*, Fuller, New London, Conn., 2-5 Sep 1887, Log 170, MSM.

[160] Gough Island lies 230 miles southeast of Tristan da Cunha. It is in the direct path of the Roaring Forties, receiving in excess of 100 inches of rain per annum. It is seven miles long by three and one-half wide, and of volcanic origin, rising nearly 3,000 feet above sea level. Most of the coastline consists of cliffs some 500-1,000 feet high, and there is no sheltered harbor or anchorage.

[161] *Francis Allyn*, Log, n. d.

[162] Ibid., 14 April 1888.

[163] Lawrence and Co. to Captain Simeon Church, 31 July 1877, Lawrence and Company Records, Coll. 25, MSM.

[164] See p. 211-12.

[165] Ship *Arnolda*, Howland, New Bedford, Mass., 29 Sep 1877, Log 591, ODHS.

[166] *Vineyard Gazette (Edgartown, Mass.)*, 26 Jul 1889.

[167] Samuel W. Dabney, U.S. Consul at Faial to William F. Wharton, Assistant Secretary of State, Wash., D.C., Doc 533, 14 Jun 1889, Despatches from United States Consuls in Fayal, 1795-1897, vol. 10, 2 Jan 1889-30 Dec 1893, RG 84, NAW; *Vineyard Gazette (Edgartown, Mass.)*, 12 Jul 1889.

[168] The *Kathleen* did not sail in 1880 with a mate named Roderick. But mates were often picked up in the Portuguese islands during a voyage. He is presumably Frank Roderick, an Azorean who served as third mate on the *Kathleen*'s prior voyage of 1875-79 (Bark *Kathleen*, 1875-79 and 1880-84, Seamen's Register, ODHS).

[169] Ferguson, *Harpooner*, 164-65.

[170] Ibid., 169.

[171] Ibid., 169-70.

[172] Ibid., 170.

[173] Bark *Charles W. Morgan*, Gibbons, New Bedford, 1908-10, Seamen's Register, ODHS; bark *Charles W. Morgan*, Jun-Aug 1909, Log 974, NBL.

[174] *Charles W. Morgan*, Log, 1 and 2 Aug 1909.

[175] Hegarty, *Returns of Whaling Vessels*, 9-41.

[176] Ibid., 17.

[177] Bark *Napoleon*, Smith, Edgartown, 1884-85; bark *Abraham Barker*, Tobey, New Bedford, 1887-88; steam bark *Jesse H. Freeman*, Fraser, San Francisco, 1893-93, Crew lists and Whalemen's Shipping Papers, NASB.

[178] Amaral, *They Ploughed the Seas*, 110.

[179] Bark *Mary and Susan*, Rogers, San Francisco, Calif., 30 Dec 1884, Crew List, NASB.

[180] *San Francisco Call*, 19 Feb 1892; bark *Tamerlane*, Howland, San Francisco, Calif., 13 Jan 1892, Crew List, NASB; Jabbertown Resident Whalers, Notes, Don C. Foote Collection, UAF.

[181] Bark *Mt. Wollaston*, Barker, San Francisco, Calif., 29 Jul 1877, Log 935A, ODHS.

[182] Ibid., 5 Aug 1877.

[183] Bark *Vigilant*, Smithers, New Bedford, 21 Jan 1879, Seamen's Register, ODHS.

[184] Bark *Mt. Wollaston*, Nye, San Francisco, Calif., 15 Apr 1879, Crew List, NASB.

[185] Bockstoce, *Whales, Ice, and Men*, 179.

[186]*San Francisco Chronicle*, 15 Sep 1889; steam bark *Navarch*, Whiteside, San Francisco, Calif., 1897-97, Crew List, NASB.

[187] Charles D. Brower, "The Northernmost American [microform]: An Autobiography," (University Microfilms, 1971), 504.

[188] Brower, *Fifty Years Below Zero*, 195-200; "The Northernmost American," 505-08.

[189] Bockstoce, *Whales, Ice, and Men*, 213-15.

[190] Ibid., 220-21.

[191] Brower, "The Northernmost American," 344.

[192] Brower, *Fifty Years Below Zero*, 132.

[193] U.S. Census, Tigara, Point Hope, Alaska, 1900; Michael Healy to the Secretary Treasurer of the U.S., 1 Sep 1890, Michael A. Healy Coll., M/F 92, UAF; Bockstoce, *Whales, Ice, and Men*, 238.; Brower, "The Northernmost American," August and September, 1888; Capt. Horace P. Smith, Journal, steam bark *Narwhal*, San Francisco, Calif., 20-22 Jul 1892, Log 1051, ODHS.

[194] Klengenberg, *Klengenberg of the Arctic*, 81.

[195] Ibid., 104-18. For Ferreira see also pp. 285-86.

[196] Brower, *Fifty Years Below Zero*, 145-46, 149; "The Northernmost American," 187-92.

[197] A Cape Verdean by the name of Joe Freitas sailed as third mate but left the vessel four months later at Port Clarence, Alaska. Crew List and Whalemen's Shipping Paper, steam barkentine *Grampus*, Norwood, San Francisco, 1890-91, NASB.

[198] The following year, among the eleven men sailing on the *Nicoline* out of San Francisco were three Cape Verdeans: a third mate, boatsteerer, and steward (Crew List and Whalemen's Shipping Paper, NASB). While there were no Azoreans on this voyage, a number of them could be found that year on West Coast merchant vessels. When the merchant steamship *Newbern* of Portland, Oregon, left San Francisco in June of 1892, it carried, among its crew of thirty-six, six natives of the Azores.

[199] Sophie Porter, Journal, steam bark *Jesse H. Freeman*, Porter, San Francisco, Calif., 9 Feb 1895, Log 1080, ODHS. Just which of the Portuguese boatsteerers Porter is referring to is uncertain. "Antoine" was either Antonio Frazer, from the Azores; Antone Rosa, from Cape Verde; or Antonio Roderic, who is not on the outgoing crew list. (Steam bark *Jesse Freeman*, Porter, San Francisco, Calif., Crew List, March 1894, NASB).

[200] Porter, Journal, steam bark *Jesse H. Freeman*, 16 Oct and 22 Nov 1894.

[201] Capt Horace P. Smith, Journal, steam bark *Narwhal*, San Francisco, Calif., 17 Jan 1894, Log 1079, ODHS.

[202] Captain John McInnis, Journal, bark *William Baylies*, San Francisco, Calif., 2 Jan 1895, Log Wh W7163 1886L, Nicholson Whaling Coll., PPL.

[203] Smith, Journal, steam bark *Narwhal*, 4 Mar 1894, Log 1079.

[204] Porter, Journal, steam bark *Jesse Freeman*, 29 Dec 1895.

[205] McInnis, Journal, bark *William Baylies*, 18 Nov 1894.

[206] Smith, Journal, steam bark *Narwhal*, 14 Dec 1892, Log 1051.

[207] Porter, Journal, steam bark *Jesse H. Freeman*, 9 Feb 1895.

[208] McInnis, Journal, bark *William Baylies*, 9 Nov 1894.

[209] Sophie E. Porter, "A California Woman's Winter Experiences on a Whaler at Herschel Island," *San Francisco Chronicle*, 22 Nov 1896.

[210] Albert C. Sherman, Journal, steam bark *Beluga*, San Francisco, Calif., 27 Jan 1896, log 980B, ODHS; Porter, Journal, steam bark *Jesse Freeman*, 20-22 Mar 1896. The Portuguese, mostly officers or boatsteerers, were often depended upon to venture forth with the natives and assist in the hunt for game.

[211] Smith, Journal, bark *Narwhal*, 4 Jul 1894, Log 1079.

[212] Williams, *One Whaling Family*, 272.

[213] Bark *Triton*, Gifford, San Francisco, Calif., Season of 1892, Crew List, J. and W. R. Wing Account Books, no. 3050, NBL.

[214] Philomena Nunes Costa, "A Triste Vida do Marinheiro," in "Women and the Sea," MSS B 90-9, ODHS. Translated by the author.

[215] Clock, "Shanghaied Whaleman," 456.

[216] *Langley's San Francisco Directory*, (1885), *Progresso Californiense (San Francisco)*, 21 and 28 May 1885; Whaling crew list card file, NBL.

[217] *União Portuguesa (San Francisco)*, 3 May, 7 Jun, and 24 Nov 1888; *San Francisco Call*, 12 Nov 1894; Asbury, *Barbary Coast*, 288.

[218] Although Bill Pickelhaupt has listed boardinghouse keeper John Cardozo as one of the city's "famous shanghaiers," he apparently is using the term to refer to the role of boarding agent, which was a legal activity. (Pickelhaupt, *Shanghaied in San Francisco*, 163.)

[219] Ibid., 165.

[220] *San Rafael/Terra Linda News*, 15-21 and 22-28 Oct 1986.

[221] Pickelhaupt, *Shanghaied in San Francisco*, 81, 161.

[222] Ibid., 219-30; James Laflin, "List of Officers Composing Whaling Fleet," MSS SH 381.95 L5 1885, and "Recipients of Sailors' Advances, 1886-1890," MSS SAFR 15005, SFM.

[223] Laflin, "List."

[224] *San Francisco Directory*, (1907), 694; *San Francisco Directory*, (1910), 728; U.S. Census, San Francisco County, Calif., 1920 and 1930.

[225] *San Francisco Daily Alta California*, 29 Nov and 7 Dec 1888.

[226] Hezel, "A Yankee Trader in Yap," 6.

[227] Wood, *Yachting Cruise*, 159.

[228] Beane, *From Forecastle to Cabin*, 280.

[229] The latter, known to nineteenth-century mariners as Swain's Island, continues with this name, since it has been under American control since about 1900 and now is considered part of American Samoa.

[230] Huntsman and Hooper, *Tokelau*, 184-275 passim.

[231] Ibid., 226.

[232] Parker, "Birth of a Nation," 72.

[233] Ibid., 73.

[234] Slocum, *Sailing Alone*, 136-37, 141. This was not the first time that a Portuguese settler had been discovered on the island. When the ship *Potomac* of Nantucket tarried there briefly in January of 1846, collecting fruit, at what the ship's carpenter, Thomas Atkinson, described as "the most romantic spot on Earth," he noted in his journal that, although at one time the island had held a Chilean convict colony, there were only three persons living there, "an English boy who left an American whaler," "a Portugue who was sick and put Ashor to Take Care of him self by an American Whale Ship," and one [North] American. (Thomas Atkinson, "A Journal of a Sperm Whaling Voyage to the Pacific Ocean in the Ship Potomac," MSS HM 68457, HLCA.).

[235] Bark *General Scott*, Huntting, Fairhaven, Mass., 1858-62, Seamen's Register, ODHS; McGrath, "Joachim deBrum Papers," 181-84; Hezel, *First Taint of Civilization*, 213-15.

[236] Adolph Capelle, interview by Donald Warrin, New York, N.Y., 7 Apr 2005.

[237] McGrath, "Joachim deBrum Papers"; Spenneman and O'Neill, "Curating Books on Remote Islands," 398-99; Petrosian-Husa, "De Brum Photo Collection," 44-49.

[238] Fr. Francis X. Hezel, "Beachcombers, Traders & Castaways in the Marshall Islands," http://www.micsem.org/pubs/articles/historical/bcomber/pohnpei.htm (accessed 19 Jan 2010).

[239] Westwood, *Island Stories*, 129-30.

[240] Hezel, "Beachcombers, Traders & Castaways."

CHAPTER 5

End of American Pelagic Whaling, 1900-1927

Whaling Enjoys a New Life: Money in the Business Now—New Bedford Industry Has a Little Boom—Big Catches with Large Profits and Whales Plentiful.
(*New York Times*, 14 September 1902)

As the new century dawned, there appeared a brief renewal of enthusiasm in the American whaling industry. In September 1902, the bark *Canton II*, at the time the oldest active whaler—having been built in Baltimore for the merchant service in 1835 and converted to a whaler in 1841—under the command of Nicholas R. Vieira, had returned to New Bedford from a seventeen-month cruise in the Atlantic with over 2,000 barrels of sperm oil, worth about $50,000 (over $1,000,000 in today's dollars). According to a report in the *New York Times*, profits for the voyage were sufficient to pay the entire crew "well for their trouble."[1]

The following month the *Boston Sunday Herald* explained how the much-reduced fleet was having a banner year. There were significantly fewer vessels, and a higher percentage of them were now schooners, such as the *Pedro Varela*, not venturing much past the North Atlantic, but bringing in good catches of sperm oil. And barks like the *Sunbeam* and *Morning Star* were already reporting good catches from Faial.[2]

Of particular interest in the *Herald*'s article are the diverging viewpoints of landlubber and mariner—with perhaps a bit of tongue-in-cheek on the part of the latter. After informing the reporter that the voyage of the *Canton II* had been "uneventful," Captain Vieira—with the assistance of officers and crew, and the vessel's log—described what to the journalist was anything but: Mate Sylvia "got smashed" by a maddened whale; lowering for a whale at 7:00 a.m. and the trypots were still

Bark *Canton II*, Valentine Roza, master, 1907, with Cape Verdean officers and green hands from Boston.
Courtesy of the New Bedford Whaling Museum.

boiling at 11:00 p.m.; a gale near the Azores that sent the bark on her beam ends, to be saved, perhaps, by the oil from two casks poured into the raging waters; whaling on Christmas Day, with the great sharks all around taking their share of the booty; the burial at sea of two of their comrades; and a waterspout that "in a moment had stripped her clean of mainsail, foresail and main staysail with all their ropes—all these tales were told, haltingly and briefly, by Nicholas R. Vieira, captain of the *Canton II*, with apologies for not being able to tell anything interesting because it was only a very ordinary voyage."[3]

But the owners of the *Canton II* had not found it easy to round up a fledgling crew of Portuguese islanders—or most other islanders, for that matter. And so one saw among the whaling crews of the time a motley selection of green hands, collected from New York or Boston, with a sprinkling of immigrants from northern Europe. When the bark sailed from New Bedford in April of 1901 under the command of Captain Vieira, it carried a crew of thirty-three. Other than mate Arthur Gibbons, the officers were either Azorean or Cape Verdean. The boatsteerers were Azorean except for one man from St. Helena. With two exceptions all the seamen were from the Portuguese islands. Among the green hands, however, could be found an Englishman, an Austrian, an

"Last of the Whaling Fleet," barks Charles W. Morgan and Wanderer, New Bedford, 1920. Courtesy of the New Bedford Whaling Museum.

Irishman, and several city dwellers from the Northeast.[4] Certainly, by the end of the voyage, these latter would have seen little of the "romance" in whaling inferred by the journalist.

Actually the New Bedford whaling fleet was only a shadow of its former self. In 1902, it sent out a mere five vessels, while sixteen sailed from San Francisco, including well-known former New Bedford vessels, such as the *Charles W. Morgan* and the *Wanderer*. It would take until 1909 for the latter port once again to exceed the West Coast in sailings. Typical of the New Bedford voyages was that of the schooner *Adelia Chase* under Capt. Ayres J. Senna in 1901-02. The schooner touched at Faial in the Azores, Brava in Cape Verde, and Dominica in the West Indies. This was pretty much the range of its cruising, which included as well the Hatteras Banks off the coast of North Carolina. Ayres J. Senna was born in Brava, Cape Verde, in October 1857, and first came to the U.S. in 1878.[5]

While the numbers of the whaling fleet were severely diminished from just a few decades previous, Portuguese participation continued to grow. In the first decade of the new century, fifty-three whaling vessels left American ports under the command of Azoreans and Cape

Verdeans, thirty-eight from New Bedford, ten from Provincetown, and five from San Francisco.

Adventure in the Canadian Arctic

Manuel Francisco Santos was born on the tiny island of Corvo in the Azores in 1869 and shipped around the age of fourteen as cabin boy on the New Bedford bark *A. R. Tucker*. His ten schooner commands spanned almost two decades of the twentieth century. From Provincetown he made three voyages on the *John R. Manta*. Out of New Bedford he sailed the *Charles H. Hodgdon*, *Francis Allyn*, *Pedro Varela*, *Golden City*, and the *Bertha D. Nickerson*. Typical of many voyages at this time, virtually the entire crew of this latter voyage (Oct 1909-Aug 1911) consisted of Cape Verdeans (plus two West Indians).[6]

Few are the narratives left by Portuguese whalemen. Manuel Santos figures in one of the most interesting, a story in verse by Manuel Tomaz Lopes from the island of Flores.[7] The poem's initial stanzas could be put in the mouth of almost any young Azorean who clambered aboard an American whaler:

> They speak of going awhaling,
> as if it were nothing,
> now I will tell you,
> the story of my life!
>
> I was 16 years old!
> a well-mannered young man.
> I fled from my mother,
> to avoid being made a soldier.
>
> The way that I fled,
> was in a whaling ship;
> I had lots of clothes,
> but I had no money![8]

In the year 1896 the author left his island, as he states, to escape universal conscription, leaving with plenty of clothes but no cash. Unhappy with the vessel and master, Lopes, along with two companions, after touching upon Africa, finally deserted in Brazil. There he relates how he began to sell off his clothes in order to eat. Finally, barefoot, he, along with his two companions, signed on to a vessel that took them to Fernando de Noronha, off the coast of northern Brazil. Desirous of again finding an American whaler, the young men were elated to see one come into the harbor. Discovering that one of the mates was Portuguese, Lopes entered into conversation with him, and soon all three had signed on for fifteen months—the remainder of the unnamed vessel's voyage.

Lopes, writing some years later, was using a bit of poetic license. Both he and his friend João Ventura had signed aboard the bark *Greyhound* at the Azores in September of 1898; and the two deserted at Portsmouth, Dominica, in April of 1899.[9] The mate who saw that they got signed on to his vessel there was fellow Azorean Manuel Francisco Santos, of the schooner *Francis Allyn*, which had departed New Bedford two months after the two young sailors had deserted in the West Indies.

The *Francis Allyn* soon sailed north and, not long after surviving a ferocious gale, ran straight into the Arctic ice pack. With the captain at the helm, Santos giving orders and the sails furled, the vessel was soon trapped. At that moment Lopes remembered the mate saying "Life is short; we are fast in the ice. If it breaks up the ship we will all die."[10] For two weeks they remained caught, doubtful of their survival, before the wind shifted and the ice moved off. Finally the vessel reached its destination in Hudson Bay and all settled in for the winter. Lopes was particularly intrigued with the natives. When ten of them came on board, with their long hair, he "didn't know if they were men or women." And he marveled at their houses of ice and their beds of animal skins.[11]

Lopes's mentor, mate Santos, a veteran of previous Arctic voyages—and soon to have his own command—was not so intimidated by these Arctic Inuit. In fact, like quite a few of his fellow officers and masters he fathered a daughter on this voyage with a native named Minnie Dyer, herself perhaps the offspring of a whaling captain. The daughter, born in 1900, was known as Peperguy.[12]

By May of the following year, as the temperature moderated, the crew loaded up the boats with provisions, to be towed away over the ice by the native sled dogs until they hit open water some twenty-five miles away. Accompanying the mate with his boat were Lopes, João Ventura, and two other Portuguese. At night they slept in the whaleboat, and each had to prepare his own meal. When on open water, some eighty miles from the *Francis Allyn*, mate Santos and the third mate in his boat managed to kill a whale, which was towed to shore and only the baleen saved. After a stormy and hazardous trip back to the vessel, the crew set about sawing it out of the ice. By the end of June, they were able to cut free and began to make their way to New England, arriving, after some difficulties, in New Bedford. "After a week," Lopes relates, "I went to get my wages/but they were thieves/they gave me almost nothing//twenty-three dollars and a half/was all my wages/fifteen months on a ship/with my life at risk."[13] The remainder of this tale of wanderlust, not atypical of Portuguese immigrants in the U.S. at the time—especially those who had passed through such adventures—takes place in Montana, Nevada, California, etc., and is outside the bounds of this narrative.

When the *Francis* Allyn sailed again to Hudson Bay, in the summer of 1901, its master was Manuel Santos. The explorer David Hanbury recounted a visit to the schooner that fall as the crew was preparing to winter in. Hanbury's observations on the crew were a harbinger of troubles to come. "The captain, mates, and boat-steerers (harpooners)," he wrote, "were all old hands at the whaling business, but the rest of the crew were novices. They had shipped in response to an advertisement inserted in a Boston paper, some from a spirit of adventure, others because they were 'broke,' while a few thought that there was big money to be made out of it. They were all youngsters, and the experience would doubtless be of benefit to them, if it only taught them to remain at home and stick to regular work."[14] Hanbury remarked as well on the apparent lack of concern for scurvy, as the vessel carried only five gallons of lime juice for its nine-month stay.

Hanbury returned again in December. He found the temporary deck-house built over the quarter-deck, a place of great activity, the hands being occupied in overhauling all the equipment that would be needed as the thaw began some months on. What particularly impressed him was how these veterans of the hunt for the leviathan lived, ate, and breathed whale:

> We had whales at breakfast, lunch, and supper, and all the hours between. If the subject of "whaling it" commenced, it continued the whole day. The captain would get fast to a whale long before break-

> fast, and would not kill it until that meal was finished. Then the mate would start in with a sperm and fighting whale, which, after killing all the men and smashing all the boats, was at length despatched, and ultimately gave the record number of barrels of oil. Now it was the second mate's turn. One could see that he had been holding himself in readiness. His experience had been with a bow-head which, after getting itself and the boats hopelessly tangled up in icebergs, was finally slain by the intrepidity of one of the men, who died shortly afterwards, as the wonderful men usually did. The third mate, who had been with difficulty restrained from breaking in with his experience, now got a show, but he had hardly got his boat clear of the vessel when some one else was fast to another whale, which yielded the record amount of bone. And so it continued, yarn after yarn, the stock of whales being inexhaustible.[15]

In mid-July 1902, as the *Francis Allyn* came out of winter quarters, a conflagration destroyed all hopes of a successful voyage and set off a race for survival. With almost all of the crew out floe whaling, a foremast hand supposedly knocked over a tallow lamp while in the process of washing clothes, and soon the schooner was engulfed in flames. Fortunately, the boats were saved; and all set off in the hope of locating another whaler, rowing at times and hauling their craft at others as they traversed the Arctic ice. Within a few days they came upon the Scottish steam bark *Active* and then were able to transfer to the schooner *Era* of New Bedford under the command of veteran Arctic whaleman George Comer and were thus carried home that fall.[16]

Sunk by a Whale

The ultimate fear on a solitary whaleship far out to sea was, as in Melville's *Pequod*, that of being stove by an angered or disoriented whale. Such was the fate of the bark *Kathleen* of New Bedford. The 206-ton vessel, under the command of Captain Thomas H. Jenkins, left port on October 22, 1901, bound for an Atlantic cruise. The make-up of the crew was typical for the time. The experienced men were largely islanders, mostly from Cape Verde, but also a couple of men from St. Helena. The second mate, Paul Gomes, age 46, was from Santiago, Cape Verde. Manuel D. Viera, age 44, also from Cape Verde, was third mate. Of the four boatsteerers two were Cape Verdean: Phillippe J. Viera (from Fogo) and Cecelia [*sic*] Manuel Delgardo (from Santo Antão). The cook was from Brava, as were seven others of the crew. Altogether fourteen of the crew hailed from Cape Verde. And finally there were the green hands from Boston, Nova Scotia, Edinburgh, etc., certainly unprepared for their fate. The captain's wife also sailed aboard, signed up, as often happened, as assistant navigator; she had brought her pet parrot with her.[17]

At the Cape Verde islands the *Kathleen* added some additional crewmen. Then it was off to the coast of Brazil, where Captain Jenkins expected to find sperm whales in abundance. But such was not to pass. Actually, it was not unusual at the time for a whaleship to go many weeks without even sighting a whale. Thus it was that in March of 1902, after five months at sea, the *Kathleen* had not a barrel of oil in its hold. Capt. Jenkins decided that he might have better luck to the north; and so it was there, on March 17, that a huge pod of sperm whales was discovered.

In the little volume produced that same year, entitled *Bark Kathleen Sunk by a Whale*, the reader is enticed by the frontispiece that begins, "The most thrilling episode ever known in the history of the American Whale Fisheries has just occurred. It is full of the mystery and thrill and terror of the deep sea."[18] And yet the laconic words of Captain Jenkins, in his narrative, scarcely evoke the drama of that day. Essentially, Jenkins's story is as follows: On March 17, 1902 a large pod of sperm whales was discovered, and as the bark approached, all four boats were lowered. Soon mate J. W. Nichols was fast to a whale. By this time the other three boats were fast as well. Jenkins brought the *Kathleen* alongside the mate's whale, and as the mate was ready to hoist his boat, the captain, now in the topmast crosstrees, saw a large whale not five hundred feet away and swimming in the direction of the vessel. Jenkins then sent Nichols out after it. The mate met the whale head on, but the boatsteerer failed to throw his iron. And instead of going down or turning to windward the whale continued toward the *Kathleen*, gathering speed as it went. At the last moment, it may have attempted to go under the bark; but it was too late, and "he struck the ship forward of the mizzen rigging and about five or six feet under water," shaking it considerably.[19] Jenkins then called the mate away from the dazed whale, for he was concerned about the other three boats, which were out of sight. At that point one of the hands, descending into the forecastle, discovered that the floor was covered with water. Jenkins, realizing the danger, ordered flags set at the mastheads, informing the boats to return immediately to the vessel.

Unfortunately, Viera—who was in sight about a mile and a half away—did not wish to cut loose from his whale. Captain Jenkins entered the cabin and found his wife reading, unaware that the bark was sinking. Soon she and the parrot were on deck. And so, with only one boat in which to store bread and water for the entire crew, and nineteen of the crew plus the captain, and his wife, and her pet parrot, they all entered, sinking the whaleboat almost below the waterline. As they approached Viera's boat the party watched the *Kathleen* roll over and go under. The men and provisions were soon divided between the two boats, and all set off in the dark in search of the others. About nine o'clock that evening the remaining two boats were sighted. Provisions and men were now more evenly apportioned among the boats. Jenkins ordered the other boats to keep him in sight, and he mounted a lantern so that they could make him out. And then all started for the island of Barbados, according to his calculations some 1060 miles to the west. Fortunately, early the next morning the steamer *Borderer* of Glasgow came upon the second mate Paul Gomez's boat; and soon the captain's boat, along with that of the fourth mate, Morris Murray, of St. Helena, were located. Nine days later all were landed at Pernambuco, Brazil; and then it was on to Philadelphia and finally home to New Bedford.

Captain Jenkins ends his brief narrative by invoking the fate of the fourth boat, under the command of third mate Manuel D. Viera: "The other boat containing one of the mates and 9 seamen landed safely at the Barbados after being in the boat 9 days [actually eleven] with but 5 gallons of water and a little ship bread."[20]

As for Viera, having lost contact with the other boats during the night, he decided that it was best to continue on to Barbados on his own. And, using his pocket compass, he pointed for the island. For ten days each man was given a ration of one bottlecapful of water, morning and evening, as well as one biscuit. "That kept up for eleven days," Viera recalled, "during which we had sung all the songs we knew and told all the yarns we could make up. After a while, though, we were so played out that we could hardly handle the sails, and we all wondered how long we could last. On the eleventh day, when our last drop of water was gone, we sighted land, and got up enough pluck to give a little cheer. The island was Dominica, and it wasn't long before eight happy men [plus Viera] were dragged out of the boat."[21]

By late April Viera and his companions had reached Manhattan and were ensconced there in the Sailor's Home, awaiting transportation to New Bedford. Suddenly in great demand from the media, Viera gave a series of interviews. In speaking to a reporter from William Randolph Hearst's *New York American and Journal*, he offered a version of the events that closely resembled the subsequently published narrative of Capt. Jenkins.[22] Probably, after this interview, Viera and his crewmates decided to have some fun, first with a journalist from the venerable *New York Times*, who perhaps showed a bit of condescension toward these Cape Verdean mariners, "all of whom," he noted, "have names that suggest fine cigars."[23] It was now he, "chief bowman" Viera, who,

> put a large whale on the grapples ready to be pulled aboard, when the Captain espied an enormous "bull" some distance from the ship. He immediately ordered the chief bowman [Viera himself, referred to by the reporter as "De Veira"] to start in pursuit. De Veira soon caught up and, standing in the bow of the boat, harpooned the monster. De Veira, who is a big mulatto and speaks excellent English, told his story yesterday afternoon, soon after the docking of the Madiana [a steamship on which the men had just arrived from the West Indies].[24]

Here anyone familiar with traditional whaling at the time would recognize that mates did not normally harpoon the whale, that being the responsibility of the boatsteerer, who then would change places with the officer so that the latter might deal the death blow to the whale with a lance or bomb gun. Nor was the term "chief bowman" part of the whaling lexicon. Viera's narration continued:

> Right after I struck him he sounded. He stayed down quite a while, and he must have made up his mind to do something desperate, for as soon as he came up he made straight for the bark, which was about 200 fathoms to leeward. He kept on at such a speed that we could smell the rope burning as it tore through the cleats. We would have been swamped in another minute, so I cut the rope. The next thing I saw was the whale bang up against

> the Kathleen's starbord side, just abaft of 'midships.' The whale went down and the bark showed a big hole in her side, and splintered planks. I brought my boat alongside and took off the Captain and his wife, the steward, and "Hot Tamale," who, being an educated bird, swore something terrible.[25]

Viera's opportunity to elaborate on the incident was expanded when he was interviewed by a reporter from yet another newspaper. Now more details were elicited. He recalled that, as the bark was about to go down,

> Mrs. Jenkins, with the parrot clasped to her breast, was the first to get into the small boat. The veracious sailormen who came in yesterday solemnly state, that only one sentence of the parrot's remarks were quotable.
>
> "Damn hard luck, that's what I say! Damn hard luck!" he repeated over and over again....
>
> The skipper's wife lifted the parrot up after they had been rowing a little while to take a last look over the billows at the hulk of the Kathleen, which the weight of the cow whale had pulled over on her side. The parrot closed down a gray lid over his left eye, and said simply:
>
> "Damn hard luck, that's what I say!"[26]

As Viera had recounted to the *New York Times* reporter how the men had survived their travail not only with a few drops of water and a bite of biscuit, but also by singing "all the songs we knew," and telling "all the yarns we could make up," it is easy to see how this hero of the moment could continue to elaborate on the adventure. Whatever its truth, Viera's narrative certainly helped bring to life to thousands of readers the dramatic events of the loss of the bark *Kathleen*, "sunk by a whale."

Mutinies and Desertions

Most whaling vessels sailing around this time contained either a predominantly Portuguese crew (the bark *Sunbeam* sailed from New Bedford in 1904 with twenty Cape Verdeans aboard), or at least the officers and boatsteerers were Portuguese. They might also sail shorthanded with the intention of making up a crew at the islands. Such was the case of the schooner *Ellen A. Swift* of Provincetown in 1902. Sailing with nine

Whaling masters aboard the Ellen A. Swift, New Bedford, ca. 1910.
Left to right: Manuel F. Santos, George L. Dunham, Antone J. Mandly, unknown Cuttyhunk pilot, James A. Tilton.
Courtesy of the New Bedford Whaling Museum.

crew members under the command of Arthur O. Gibbons, it signed on fifteen men at São Vicente.[27]

When Antone J. Mandly prepared to take the *Ellen A. Swift* out on its next voyage, in 1905, it was held in port for a time awaiting a complete crew. As the press reported at the time, it was usually Cape Verdeans who would make up a crew, "and in the spring of the year they are not inclined to ship as crews, preferring to work as laborers and as seamen on coasting vessels. Whaling agents have had no end of trouble this year in shipping crews, so great is the scarcity of men, and when the Canton sailed she made up the balance of her complement with seven men who were sent here from Boston. The same is true of the Bertha, while the Wanderer, which sailed yesterday, shipped fourteen who were sent to New Bedford by agents in Boston."[28]

In fact, many a green hand discovered that the first port of call, usually Horta, Faial, in the Azores, was an ideal place to abscond. Two days after the bark *Morning Star* of New Bedford arrived at Faial in October of 1898 eight men deserted, mostly all green hands. "Suppose they Swam ashore," noted the log keeper.[29] And in the next few days six more men disappeared. Desertions in the Azores were not a major problem for masters, however, as it was relatively simple to replace them. At midnight several days later a boat came off the island to the *Morning Star* with seven men to replenish the forecastle. And the next month seven more were shipped at Brava.[30] Such commonly was the turnover of crew at this stage of the industry.

It should not be assumed, however, that the Portuguese government was not concerned with the flight of its young men, all subject at the time to universal military service. Just on the island of Flores four security stations had been set up at various points to prevent such escape. And once offshore anyone was subject to being interdicted by a patrol boat manned by sailors from the Portuguese mainland, who would be less inclined to sympathize with the plight of the local islanders.

While most clandestine emigration was conducted by boat, it was not unusual for young men to attempt to swim out to an awaiting vessel. And so, in June of 1901, a most tragic case occurred when a youth from Fajã Grande, Flores, was killed by a fusillade of bullets fired from a patrol boat while attempting to reach a whaling vessel.[31]

Manuel Estaus Costa (1849-1914), like so many of his contemporaries, was born on the island of Faial and sailed away at the age of thirteen.[32] His long career as a master mariner began with four voyages on the schooner *Eleanor B. Conwell* of New Bedford, commencing in 1879 and ending with those of the *Bertha D. Nickerson* (Jun 1906-Jul 1908) and the *T. Towner* (Aug 1908-Aug 1911). The crew list of the *Bertha D. Nickerson* shows the increasing importance of other islanders in the industry: Azores (0), Cape Verde (2), Caribbean (9), St. Helena (1), Philippines (1), besides the typical Anglo green hands, most likely obtained from Boston.[33] Costa's vessels hunted the cachelot almost exclusively, his last two voyages bringing home from their cruises in the Atlantic, 2130 and 3280 barrels of sperm oil respectively. Costa was often accompanied on his voyages by his wife Philomena.[34]

Occasionally logs were kept in Portuguese by mates who felt more comfortable with their native language; but it was more common for them to attempt English.[35] Such was the case of the log keeper of the bark *A. R. Tucker* of New Bedford, sailing under Captain Costa from 1901 to 1903, as he made his entries in a quite phonetic, albeit comprehensible version of the language: "Comes Wette lait Wend from Soud haden on porte taque and Chorte Sail A m Varey lait Wend."[36] Similarly, in May of 1920, the keeper of the log of the bark *Wanderer* noted, in one of his phonetic observations, that "enmanin borjeda res escul espama el an loadaon tribar" (i.e. in the morning the boatheader raised a school of sparm [sperm whales] and lowered down three boats).[37]

Manuel F. Gomes was born on the island of Faial in 1864. In 1880, he signed on to the ship *Milton* of New Bedford. Gomes made several voyages as mate before accepting command of the schooner *Eleanor B. Conwell* (Provincetown, 1897; New Bedford, May 1898-Aug 1998, Oct 1898-Sep 1899, and Oct 1899-Jul 1901). The composition of the crew on Gomes's second voyage on the *Eleanor B. Conwell* is illustrative, for every one of the crew is an islander, from Cape Verde, the Azores, West Indies, or Bermuda.[38]

Gomes was next master of the bark *Bertha* out of New Bedford for its voyages of Oct 1901-Aug 1904 and Apr 1905-Jul 1907. It was on this latter that he experienced trouble with his crew. That August, after some four months at sea with almost a full hold and a portion of ambergris, Gomes made for Faial and went ashore for provisions without evidently giving the crew leave. The next morning at sea the starboard watch refused to work. When confronted by the captain they complained that they were not getting enough molasses—a not-uncommon complaint for a generally unhappy crew, this time containing many Portuguese. And so Gomes put the men in irons and placed them in the hold. Later, when the port watch also mutinied, the captain threatened all with fines and imprisonment, but finally assented to go into Faial. There virtually all deserted, thus losing any right to their share of the eventual profits. But obtaining a crew was not much more difficult than losing one, and soon Capt. Gomes had eight new men. A quick trip down to São Nicolau brought on thirteen more.[39] And so the voyage continued.

Gomes's final command was on the schooner *Cameo* (Apr 1908-Aug 1911). These last three were relatively long voyages but were quite successful, returning respectively 2670, 3100, and 3720 barrels of sperm oil. He did not finish this last command, dying in October of 1910 at the age of forty-seven from Bright's disease.[40]

Desertions at Horta became so common that repatriating the men was a serious problem for the American consul. In 1905, for example, a New York newspaper noted that the Portuguese steamer *Peninsula* had brought in as passengers eighty American sailors sent home by the consul at Faial. This does not even take into account those of foreign nationality who did not have the same right. "According to their story," stated the article, "the men have been living in secluded regions among the volcanic rocks of Fayal for periods ranging from two weeks to several months. They say that thirty more sailors are waiting to be sent home."[41]

Back in New Bedford an urban legend appeared about this time:

> Once in a great while some man shipped time after time with the whalers, despite the robberies and falsehoods of the outfitters and the hardships of the life at sea and old New Bedford whalemen tell an amusing story of one such character. This man, a typical vagrant whose only resting place is a bench on Boston's Common, has appeared annually for fifteen years to ship as a whaleman from New Bedford but no sooner does his ship reach Fayal than he deserts. For a few months he lives an indolent, easy life in the Azores, then he appeals to the American consul, is shipped home "like a bloomin' lord" as he expresses it and again resumes his life at the "Hub" until the next whaling season.[42]

Antonio C. Corvello (1879-1920) was born on the island of Flores and came to the U.S. at the age of fourteen aboard the *Frances A. Barstow*.[43] During the 1907-09 voyage of the *Pedro Varela*, Corvello replaced Henry Mandly Sr. as master. He subsequently sailed again as master of the *Pedro Varela* (Apr 1910-Aug 1912) and then made three voyages in command of the bark *Greyhound* of New Bedford (May 1913-Sep 1916, Aug 1917-Jun 1918, and Sep 1918-Aug 1919), the first of which was especially profitable, yielding 2580 barrels of sperm oil. He will be remembered as the master of the *Pedro Varela* during its well-known mutiny, which did not seem to mar his career.

One method of avoiding desertion was simply not to call at any port. That was the tactic of Capt. Corvello when he kept the schooner *Pedro Varela* away from port in the Atlantic for six months, only to contribute to the dissatisfaction that led ultimately to the so-called "mutiny" of 1911. As was typical at the time, the captain, officers, and boatsteerers were Azorean and Cape Verdean, while the greenhorns "were a motley gang of Americans, small-time criminals, a self-confessed murderer, a drug addict and peddler, and a burglar." Discontented with the cramped quarters in the forecastle and the quality of the food, these men, after complaining in vain to the captain—and being placed on short rations for it—took it upon themselves to throw overboard most of the whaling implements as well as smashing the windlass and the tryworks. Thus the captain was forced to head for Faial, from where the "mutineers" were eventually taken by a Navy ship to the U.S. where they were tried and convicted, but shortly thereafter released from prison—for this had not technically been a mutiny, since no one had refused to take orders, nor had they attacked anyone, nor even planned to take the ship by force.[44]

On a later voyage, Captain Corvello was fortunate to survive a true mutiny. It was August, 1917, and U-boats were prowling the Atlantic—not a good time to gather a crew for an extended cruise in those seas. Just a few days after Corvello sailed the bark *Greyhound* out of New Bedford, a pod of whales was sighted and three boats lowered. Only five men plus the captain remained on the vessel. As the cook and mess boy were at work in the galley they were attacked by two men, and the former suffered a serious knife wound. Corvello, who was occupied in training his glasses on the boats, heard a noise and catching sight of someone brandishing a large knife, made for the rigging—there, as Manuel José dos Reis before him—he was able to set the flags informing the boats to return immediately. The mutineers were soon subdued and put in irons. The *Greyhound* continued to whale for three more months, before coming into Barbados, whereupon the mutineers reportedly slipped away before they could be handed over to the authorities.[45]

If it was necessary to make port about this time, especially to transship oil, many vessels now preferred to sail into Las Palmas in the Canary Islands because the fees had continued to increase at Faial.[46] There, as in Faial or the West Indies, they might be met by the schooner *Richard W. Clark* of New Bedford, which would take on their oil and also deliver needed supplies. When the schooner sailed in August of 1914 it carried supplies furnished by Capt. John P. Praro Jr., now employed as a shipping agent.[47]

Predominance of the Portuguese

After the turn of the century only three ports regularly sent out whaling vessels: New Bedford, Provincetown, and San Francisco. By this time the Portuguese presence in the two Massachusetts ports was marked. One local historian, writing about Provincetown in 1922, stated that "half the town is Portuguese."[48] The major occupation of the Portuguese there at that time was fishing, as it was for those who had settled in Gloucester.

Of the 325 whaling voyages that left American ports under the command of Azorean or Cape Verdean masters (or in a few cases Portuguese Americans), virtually half (160) took place after 1900. But whaling was now a relatively minor industry. In 1900, there were only nine sailings of whaling vessels out of New Bedford (excluding an aborted seven day voyage of the *William A. Grozier*). Six of the vessels were registered in that port and three in Provincetown. Some were sailing short, certainly with the expectation of signing on more men during the voyage. Of the total of 169 crew members, including masters, 86 were Portuguese— at 45.5% almost one half of the total. Now, except on the quarterdeck, the Cape Verdeans clearly dominated: Cape Verde, 74; Azores, 11 (five of the nine voyages were under the command of Azoreans).[49]

Given their dominant presence at this time, the prohibition against speaking Portuguese on deck now gave way to its general usage aboard many a vessel. As the *Boston Globe* noted in an interview with John T. Edwards, master of the New Bedford schooner *Cameo* (1912-15), now "Portuguese was the working language on the Cameo." Finally, in 1919, the last significant year of whaling out of New Bedford, the Cape Verdeans even dominated as captains, as they commanded seven of the fourteen voyages that year, compared with three for Azoreans and one for a Portuguese American.[50]

Meanwhile, in New Bedford at the beginning of the century we can still see the Portuguese connected to businesses peripheral to the maritime industry. Francis and Silvia were operating a clothing store at 160 Acushnet Avenue. Thomas Luce and Company was still operating in the cooperage business at 278 North Water and 2 Merrill's Wharf. J. S. Figueiredo was a tailor at 83 Howland; and there were three Portuguese-run liquor stores along South Water Street.[51]

More Portuguese Masters

Benjamin Costa was born in the Cape Verde Islands, ca. 1878. In 1921, he took out the old bark *Greyhound*, which had whaled for the first time in 1851, on its final whaling voyage (Aug 1921-?). The *Greyhound* was later sold as a Brava packet; and Costa, himself, continued as a captain in the packet trade.[52] Joseph Gaspar (da Conceição) (1867-1931), also a Cape Verdean, took the schooner *Pilgrim* out of New Bedford on two cruises (Jun 1902-Jun 1903) and again in July of 1903. Unfortunately the vessel was lost at Cape Verde that September.[53]

The Edwards (Eduardo) family was especially prominent in the last years of the industry. Joseph T. Edwards (1856-1913) was the first of the family to migrate to New Bedford. He arrived in New England in 1876 aboard the bark *A. R. Tucker*, having signed on two years earlier at Flores. His first commands were aboard the bark *President II* of New Bedford (Oct 1887-Jun 1890 and Jul 1890-Aug 1893). On several voyages he was accompanied by his wife. He later served as captain of the brig *Frances A. Barstow* out of San Francisco (Jun 1894-May 1897) and the bark *Greyhound* (New Bedford, Jun 1898-Jul 1902 and May 1903-Apr 1907). On this latter voyage the *Greyhound* returned, with his nephew Antone T. Edwards as replacement master, credited with 4625 barrels of sperm oil. Joseph's final command was on the schooner *John R. Manta* (Provincetown, Jul 1907-Sep 1908). The difficulty of filling out a crew with men who were either experienced or prospective career whalemen is obvious from the crew list of this latter voyage. While the officers, boatsteerers, and senior seamen are all from the Azores, Cape Verde, or the West Indies, most of the others—judging from their provenance (Ireland, Canada, and New England)—must have been recruited in Boston and would have had little foreknowledge of what they were about to experience.[54]

Joseph T. Edwards had many interests. He was owner of the whaling schooner *Cameo* and agent of the whaling schooner *Mystic*. Whereas a few Portuguese whaling masters may have been deficient in their navigational skills, Edwards ran a popular navigation school in New Bedford. The bark *Greyhound*, under his command, was called the "schoolship" by some of his friends. At one point it occurred to him that installing a steam engine in a whaleboat could save a boat's crew the agonizing work of towing a dead whale at times for miles to the mother ship. And so he installed such an engine in one of the *Greyhound*'s boats before he sailed. Unfortunately he never had a chance to try it out, since a merchant captain spied him zipping around at one of his ports of call and purchased it from him.[55]

Edwards was innovative in other ways as well. When it was suggested, in the early twentieth century, that steel tanks be employed on board instead of wooden casks, "the skippers of the New Bedford fleet at

Capt. Antone T. Edwards and partial crew aboard the bark *Wanderer*, 1922. Courtesy of the New Bedford Whaling Museum.

first were skeptical about changing from their old methods." But Joseph T. Edwards agreed to give them a trial and had them installed on his 259-ton schooner *Mystic*, which his brother Antone took out of New Bedford in October of 1910. Soon after, they began to be adopted on many of the old whalers. An advantage of these new containers was that the oil did not have to be cooled down before they were filled, as was the case with the old wooden casks. Also, with their large diameter openings, in the event of an approaching storm the blubber could be stored inside until a more convenient time for trying out.[56]

The hazards of whaling were not all from seas and cetaceans. Various illnesses claimed many victims. There was of course the danger of venereal disease in some foreign port, or an exotic tropical sickness—but whalemen were particularly susceptible to diseases related to their living conditions on board ship, such as scurvy, beriberi, and intestinal disorders due to their poor diet, and consumption from the confined quarters in the forecastle. Joseph T. Edwards was the unfortunate victim of beriberi, which forced him into retirement at the age of fifty-two. Three years later he succumbed to the effects of the disease.

Antone T. Edwards (1882-1936), Joseph's nephew, was born on the island of Flores and arrived in the U.S. at the age of ten, after having served as cabin boy aboard the bark *President II* under his uncle.[57] At the age of sixteen he shipped as a foremast hand under his uncle Joseph on the bark *Greyhound* (Jun 1898-Jul 1902). He commanded the bark *Wanderer* during more than a decade (seven voyages from 1913-24). His last, aborted, whaling voyage symbolized to many the end of the industry. In a manner this was true, for it was the last square-rigger to leave an American port intent on a whaling voyage. On August 25, 1924, the *Wanderer* was towed out into Buzzards Bay and anchored, awaiting a filling out of its crew. By the next day, a gale was blowing, and the bark began to drag her anchor. The captain and master were ashore, and without a sufficient number of men to manage the vessel, it ended up on the rocks on Sow and Pigs Reef off Elizabeth Island. Antone continued to sail after the end of whaling, serving as master of the schooner *Burkeland,* one of the so-called Brava packets, on its voyage to Brava (Nov 1931-Jun 1932)[58]

John T. Edwards (1884-1957) and Joseph F. Edwards (1886-1933) were younger brothers of Antone. Both had immigrated from Flores with their parents around 1895. In 1908, John sailed as mate on the New Bedford schooner *Cameo* under Capt. Manuel F. Gomes (Apr 1908-Aug 1911). He then took over command on the schooner's next voyage (Mar 1912-Aug 1915). His last command was the *Arthur V. S. Woodruff* (May 1917-Oct 1919). The *Cameo* had especially valuable catches, bringing in 3720 barrels of sperm oil in 1911 and 5300 in 1915.

In 1913, Joseph F. Edwards shipped out as boatsteerer under his brother Antone for a three-year voyage aboard the bark *Wanderer* of New Bedford. Under Antone's tutelage, Joseph studied navigation and other arts of command.[59] Two years after the vessel's return, Joseph became master of the *Charles W. Morgan* (Jul 1918-Sep 1919 and Oct 1919-Jul 1920). Subsequently he commanded the schooners *Margarett* (Mar 1921-Sep 1921, Apr 1922-Aug 1922, and Mar 1923-Aug 1924) and *John R. Manta* (Jun 1927-Jul 1927).

Given the difficulty of obtaining crew members locally by this time, or even in the Azores, it became a pattern to sail with a skeleton crew in the expectation of filling out the billets at Cape Verde or perhaps in the West Indies. When Captain Joseph F. Edwards took the *Margarett* out in August of 1923 there were just six men besides him: two mates, a steward, and three seamen. Similarly, when Antone J. Mandly sailed from New Bedford in April of that year in the *John R. Manta*, he carried on board three mates, a steward, cook, and two boatsteerers. When the *Margarett* sailed in 1923 it was prepared to seek blackfish (pilot whales) and sperm whales. Refined blackfish oil was in demand at the time for the lubrication of watches and chronometers, and Edwards planned to hunt these pilot whales off the coast of Africa.[60]

Theophilus M. Freitas was the only Portuguese master of a Fall River whaleship. Born on the island of São Nicolau, Cape Verde, in 1878, he arrived in the U.S. aboard the whaling brig *Rosa Baker* of Boston in 1895. He served as replacement master for Capt. William C. Hegarty on two voyages of the brig *Sullivan*, first out of Norwich, Connecticut in 1908-11, and then from Fall River, Massachusetts, during its voyage of 1912-13. The first was a most felicitous voyage, for the *Sullivan* returned with an exceptional cargo of 4600 barrels of sperm oil. After sailing from Fall River on its subsequent voyage, the vessel was lost the following year at Faial after it dragged its anchor.[61] Freitas had one command of his own, that of the schooner *Pedro Varela* (Oct 1917-Feb 1918). He had the additional recognition of serving as harpooner during the filming of the 1922 silent film *Down to the Sea in Ships*, about a New Bedford whaling magnate and his family, starring Clara Bow.

It is interesting to note that although both Azoreans and Cape Verdeans participated in whaling out of Fall River, many more whaled out of New Bedford; and—with the exception of Theophilus Freitas—they seldom, if ever, rose above the rank of boatsteerer.[62]

John T. Gonsalves was born on the island of Brava in 1858. At the age of eleven, he shipped as cabin boy on the bark *Roman II* out of New London. For many years he followed the sea in whalers and even spent a stint coasting and barging, and participating in the Cape Verde packet trade. Captain Gonsalves began an exceptionally long career as a whaling master when he sailed out of Provincetown Harbor as captain of the schooner *Rising Sun* (Dec 1890-Aug 1891). Twenty-four voyages later he ended his career as whaling master aboard the schooner *William A. Graber* in 1922.

The Atlantic was an especially dangerous place during World War I with the threat from German U-boats. In June of 1918, Capt. Gon-

Capt. John T. Gonsalves. Courtesy of Daniel M. Rodrigues.

salves abruptly brought in the schooner *A. M. Nicholson* after a frightening experience with one. As he related his experience to a reporter:

> The submarine glided alongside the Nicholson and then shot across my bows. I gave him the American flag but he didn't answer and submerged. I thought he was an American craft, but when he submerged without answering my flag I grew suspicious, and when he went down I went to the south'ard to try to get out of his way. I had not sailed very far when up came the submarine again, and I gave him the American flag again.
>
> This time the submarine gave me back the German flag with another flag beneath it, and ordered us to heave to at once. He gave us a gun across our bows to emphasize the fact that he wanted us to stop, and we were not long in going into the wind, and dropping our boats. We put all our crew, some 25 men into two whaleboats, and according to instructions, pulled alongside the submarine, after pulling alongside our vessel for some time. We didn't take time to let the sails down, thinking that we might get a shot or a torpedo any minute.
>
> The submarine steamed alongside my boats, and her captain sang out: "What is your vessel doing?"
>
> "Catching sperm whales," I replied.
>
> "Catching any other fish?" he asked me, and I told him we were not.

> "For God's sake, captain," I said, "don't sink this vessel, I am a poor man and it will ruin me, as I am a big owner in her."[63]
>
> With that, the officer, and I think he was the mate, laughed, and said he would report to the captain. He saluted another officer who proved to be the captain, who hailed me, and said, "Don't you know that it is a poor time to buy vessel property when people are at war?"
>
> "I bought this craft before the war started," I answered.[64]

Noting Gonsalves's accent, the German captain inquired as to his nationality. When Gonsalves mentioned that he was Portuguese, the German softened his stance, praising the Portuguese as great mariners.[65] Then, as Gonsalves noted:

> The German captain caught sight of the *Ellen A. Swift*, Captain Dunham, whaling a little ways from me and he asked me what the craft was, and what she was doing. I told him it was a whaler, like us, and he waved his hand at me, and said, "You get aboard your vessel, and get home as quick as you can, and tell the other vessel to go in with you, and don't you let me catch you out this way again."
>
> Well, you can believe that I didn't stop to whale it any more. I spoke the *Swift* the next day, and we came home together. After leaving me, the submarine made over towards the *Swift*, and just as she was about to speak that whaler she sighted a big steamer (supposed to have been the Norwegian steamer sank last Wednesday), and leaving the *Swift* she went over and sank the steamer. The *Swift* and the *Nicholson* kept company and came in together.[66]

Fortunately as well for Captain Gonsalves and his crew, the German U-boat commander seemed to understand little of the whaling industry and the valuable cargoes that these small vessels might be carrying. The *A. M. Nicholson* returned from its aborted voyage carrying 625 barrels of sperm oil, worth some $30,000 at the time. Although most of his career was spent aboard schooners, John T. Gonsalves was the last to command the famous bark *Charles W. Morgan* (Provincetown, Sep 1920-May 1921), bringing in 2702 barrels of sperm oil.

Joseph Luis (1875-1918) was born on the island of Faial. Just when he arrived in the U.S. is not known, but in 1908 he sailed as master of the schooner *John R. Manta* out of Provincetown (Oct 1908-Sep 1911). His next command was on the schooner *Arthur V. S. Woodruff* out of New Bedford on a long four-year voyage in the Atlantic (Oct 1912-Sep 1916) during which he was relieved in the summers of 1913 and 1914 by Capt. James A. Tilton. The voyage returned 5090 barrels of sperm oil. While not on the *Arthur V. S. Woodruff* he managed to command the schooner *Valkyria* out of New Bedford (Jul 1914-Sep 1916). Luis's next, and tragically last, vessel was the brig *Viola*, built in 1910 for Captain John A. Cook, the well-known western Arctic whaling master. The *Viola* sailed on a one-year voyage from New Bedford in September 1916 and returned the following August. One month later, on September 5, 1917, the *Viola* again set sail, this time with Mrs. Luis and their five-year-old daughter. But the brig never returned, a victim perhaps of an Atlantic gale or even a German U-boat.[67]

Frank M. Lopes was born in Brava, Cape Verde, in 1884, and first came to the U.S. in 1901 aboard the packet schooner *Zulmira*.[68] He too was destined for a tragic end to his career as master of the star-crossed schooner *Pedro Varela*. After a successful voyage in the vessel (Jul 1918-?), Lopes set out again on an Atlantic cruise on February 4, 1919, and was never heard from again.[69]

Louis M. Lopes was born on the island of Fogo in 1876 and was master of the schooner *A. E. Whyland* (Jul 1912-Aug 1914 and Mar 1915-Sep 15). He later commanded two voyages in the schooner *Margarett*. The first was aborted after two weeks in April 1917; the second lasted only three months, April-July 1918; and he returned with 400 barrels of sperm oil. The crew members on his return were predominantly British West Indian, most having shipped at St. Kitts.[70]

Occasionally these New Bedford vessels, when not whaling, sailed as packets, carrying passengers to and from the Cape Verde Islands. Even on a whaling voyage some might work off their passage as seamen. When the bark *Greyhound*, under the command of Antonio C. Corvello, arrived at Cape Verde in September 1917, after having departed New Bedford the month previous, an entirely new complement of seamen was signed on for the remainder of the voyage.[71]

To combat this type of illegal immigration, Congress passed a series of regulations under the Immigration Act of February 15, 1917, stating in part that "upon arrival of any vessel in the United States from any foreign port or place it shall be the duty of the owner, agent, consignee, or master thereof to deliver to the principal immigration officer in charge of the port of arrival lists containing the names of all aliens employed on such vessel, stating the positions they respectively hold in the ship's company, when and where they were respectively shipped or engaged, and specifying those to be paid off and discharged in the port of arrival."[72] Furthermore it was noted in the Act that it was unlawful "to pay off or discharge any alien employed on board any vessel arriving in the United States from any foreign port or place, unless duly admitted pursuant to the laws and treaties of the United States regulating the immigration of aliens." An exception was made for anyone who intended "to reship on board any other vessel bound to any foreign port or place, he shall be allowed to land for the purpose of so reshipping."[73]

As outfitter Morris Sederholm explained, it was simply a matter of the men paying $3.00 each at the Immigration Office in New Bedford with the understanding—with which they had no intention of complying—that they would ship out on the next available vessel.[74] But it was not long before enforcement became more stringent.

The Mandlys, along with the Edwards brothers, are the best known Azorean whaling families based in New Bedford. Henry Mandly Sr. continued to command a series of whaling schooners during the first decade of the new century: the *Mary E. Simmons*, the *John R. Manta*, and the *Pedro Varela*. Henry's Mandly Sr.'s half-brother, Antone, had one of the longest careers of a whaling captain, spanning some forty years. During the new century Antone J. Mandly was master of the schooners *Pearl Nelson*, *Golden City*, *Ellen A. Swift*, *Bertha D. Nickerson*, and the *John R. Manta*. Antone Mandly was a very well-respected whaling captain; but the mate on his first voyage as master disagreed, for in November of 1901 he complained, "cap mandly has the nasty is ways of any man that ever was master."[75]

Antone Mandly's 1925 voyage in the *John R. Manta*, from May 2 to August 20 of that year, was the last successful traditional American whaling voyage, as he returned with 300 barrels of sperm oil. But it was a difficult start. Now it was hard to find not only seamen from Boston, but with the new restrictive immigration laws, Azorean and Cape Verdean aliens who were residing in New England were chary of taking a chance on being excluded upon their return.[76] For this same reason, the once-profitable packet trade between New Bedford and Cape Verde also began to suffer.

Henry Mandly Sr.'s son, Henry Jr., was born in Provincetown in 1879. At the age of four he accompanied his father on a whaling voyage in the *Mary E. Simmons*, and then later sailed under his father from 1898-1911.[77] He soon had his own command, taking a series of schooners out of New Bedford: the *Margarett*, the *William A. Graber*, and the *Valkyria*.

Although a whaling master for a relatively short span, Valentine Roza, born on the island of Maio, Cape Verde, ca. 1870, was one of the better-known whalemen of the time. Contrary to most other contemporary Azorean and Cape Verdean whaling masters, he specialized in barks instead of schooners. While sailing as mate on the bark *Josephine* of New Bedford, under Captain Horace P. Smith, in 1903-05, he was instructed in the art of navigation by the skipper's wife, Marian, who was serving as assistant navigator on the voyage.[78]

Roza's four commands were aboard the New Bedford barks *Canton II* (Apr 1907-Aug 1909 and Oct 1909-Nov 1910) and *Morning Star* (May 1910-Aug 1912 and Oct 1912-Sep 1914). His first voyage on board the *Canton II* was quite successful, bringing in 2100 barrels of sperm oil and 300 of whale-oil. The second ended in disaster when the vessel went ashore on the island of Maio in the Cape Verdes in November of 1910. On board were Mrs. Roza, two daughters and a son; but fortunately family and crew were saved.[79] At the time the *Canton II* was the oldest whaling vessel operating out of the port of New Bedford, having been built in Baltimore in 1835 for the merchant service.

Roza's two voyages on the *Morning Star* were especially profitable, yielding 3050 and 2650 barrels of sperm oil respectively. When the *Morning Star* sailed in 1910 it carried a crew of thirty plus the captain. Except for one crew member from British Guiana, all were islanders: Cape Verde—Maio, 15; São Nicolau, 6; Fogo, 1; São Vicente, 1; from St. Helena, 4; Dominica, 1; and Guam, 1. The total payout to the crew (thirty-two men, excluding the captain and the three top mates, who do not appear in the document), however, was only a little over $3,000. Other than the payments to the cooper ($480) and the fourth mate ($356), payouts averaged less than $100, with several men walking away with the J. and W. R. Wing minimum of $5.00, after the deduction of expenses.[80]

A Lost Whaleboat

One of the great fears of the men in a whaleboat was becoming lost from the mother ship, especially after being hauled away on a Nantucket sleigh ride. Such occurred during the first command of Manuel Francisco Santos, on board the schooner *Charles H. Hodgdon* in December of 1900. After boatsteerer Mathew Samuel, of St. Helena, and the only man in the whaleboat not from Cape Verde, struck a whale at about two o'clock in the afternoon off the Cape Verde Islands, second mate Antonio Silva's boat was hauled by the leviathan for some three and a half miles away from the schooner. As they were towing the whale back toward the *Charles H. Hodgdon*, it began to get dark, and so they cut from the whale. But they did not have light and could not espy any from the schooner. By morning there was nothing in sight; and the *Charles H. Hodgdon* would search in vain for the lost boat. Taking stock of what they had, which included three little packages of crackers and a small keg of water, which they somehow would make to last for seven days, they continued to sail about in the hopes of seeing the schooner—or any other vessel. On the morning of the seventh day the men were jubilant as they hailed a steamer, which came close alongside and then inexplicably sailed away. The devastated crew of the little boat had to wait another four days through the hot sun with no rain, until finally they caught the attention of a French bark. In very weakened condition the men were brought on board and carried to the ship's destination in Genoa. All survived except the second mate, who died several days after being rescued.[81]

Another disaster, not uncommon in the trade, occurred the following June. When mate Frank Almeda fired a bomb lance into a whale, the charge exploded but it evidently missed the vital organs; and the whale, turning, immediately brought down its flukes before the cry of "Stern all" could be made. As a reporter described it, "Almeda, who was standing in the bow of the whaleboat, received nearly the whole force of the blow from the monster's tail, which crushed in the bow of the boat as if it was an eggshell. His body was driven down into the wreckage, and besides having his right arm broken, his shoulder blade was either broken or dislocated and a piece of broken timber from the boat was driven up into his jaw, which was also broken. Besides this, he was injured internally."[82]

Another of the boat's occupants received a partial blow from the whale's flukes but did not have any broken bones. In his log entries, Capt. Santos seemed to have little hope that the mate, who could not even talk, would survive. But he immediately made for home, which was fortunately not far. The mate was then taken by ambulance to St. Luke's Hospital in New Bedford and survived.[83] Such were the highlights of one short whaling cruise.

A century and a half after Portuguese first began their participation in the American whaling industry little had changed. True, the hand harpoon had gone, but the vessels were still under sail, the men continued to hunt from their small boats and to diet on salt horse, hardtack, and molasses. And they still had not learned to swim. While the bark *Greyhound* was cruising around the West Indies in 1904, under the command of Joseph T. Edwards, "one sailor Jose Rufino Fortes fell overboard and was drowned we lowered the boat for him as quickly as possible but as he did not know how to swim a stroke he quickly went down and was seen only once afterwards."[84]

Shanghaied in San Francisco

As has been seen in the previous chapter, the men who sailed on whaleships from San Francisco were often ill-prepared for the experience that lay ahead.[85] The recruiting practices of the new century were now, if anything, even more egregious. While the Azoreans or Cape Verdeans tended to be officers or boatsteerers operating under an American captain, the men in the forecastle were frequently the naïve, the shanghaied, or the drifters and criminal types anxious to get away from land. And these were the first to attempt to desert, especially if their initial realization that they were on a whaler destined for the remote western Arctic was awakening in a bunk with an acute hangover. It was this that the personnel aboard the federal tugboat *Golden Gate* in March of 1908 feared when they became aware of the clandestine departure of the steam bark *Belvedere* under Captain Stephen F. Cottle. Sailing as mate on this voyage was Portuguese American whaling master William F. Joseph; and all the boatsteerers were Cape Verdean. On the other hand, almost all the ordinary seamen were northern Europeans. It was among these foremast hands that there were likely victims of shanghaiing.[86]

Due to the recent increased activity of crimps it was judged that there were on board various shanghaied individuals. Upon approaching the *Belvedere* as it sailed out of the Golden Gate, the tug raised a signal for the whaler to halt. At that moment those on the tug descried several individuals on deck, all waving furiously. "This is the government tugboat *Golden Gate*," shouted an officer from the pursuing vessel. "If by chance you have on board anyone who has been shanghaied you must let them go." At that point someone was seen moving among the gesticulating sailors and soon all was quiet. Since the seas were too rough to permit boarding the whaler the federal agents were obliged to accept the captain's shouted words, that all aboard were there of their own volition, and the vessel was permitted to continue on its voyage.[87]

Shanghaiing was a last resort for filling out a crew for a whaling voyage. More common was persuasion of one sort or another. One of the returning crew members of the steam bark *William Baylies* spoke with a reporter in San Francisco after the vessel returned to port there in November of 1904:

> One of the green hands told yesterday of the gilded dream which deluded him into seeking easy money in the Arctic. He had been clerk of a hotel at San Luis Obispo, and was on a visit to San Francisco when he met the runner who was gathering a crew for the Baylies.
>
> "You don't have to know nothing," the runner told him. "Being a clerk, the captain will most likely want you to keep his books and tally the whales as the boys herd them alongside. You'll make $100 a month clear, anyhow, and it's more likely to be $200 or $300 a month. You'll see the world and when you come back you can buy a hotel of your own."
>
> The green hand is wide awake now, and if his share of the bone and oil amounts enough to buy him a suit of clothes and a ticket to San Luis Obispo he will leave San Francisco satisfied.[88]

That same month the *Call* described in detail the exploitation of the inexperienced men who filled a whaling vessel's forecastle. The *William Baylies*, under captain H. H. Bodfish, had just returned from a successful season in the Arctic (Mar-Nov 1905), having captured sixteen bowheads that yielded 16,000 pounds of bone, worth some seven to eight dollars per pound. "Captain Bodfish, master of the *William Bay-*

lies," the article explained, "will receive more than $16,000 [almost $400,000 in today's dollars] for his eight months' work. His mate will get about $8,000, second mate $5,000, and so on down the line. The green hands will receive about $200 each, less the amount charged against them in the slop chest account. In the latter case the process used in fixing the value of the bone is reversed. For bone worth $8 the green hand is paid $1.50 a pound [for the value of the whalebone was determined arbitrarily at the start of the voyage and stated in the seaman's contract], and for shirts worth 50 cents he is charged $5."[89]

This and other notices in the San Francisco dailies were part of a campaign against the abuses at the time related to the whaling industry. The financial machinations that tended to guarantee to the master of the vessel a significant profit and an adequate one to the various officers, and which denied virtually anything to the rest are well summarized in this commentary relating to the return of the bark *Andrew Hicks*, which appeared in the *San Francisco Call* in 1905:

> There will be profit for owners and captain, but only $1 apiece for the men who did the hard work and the dirty work and who, between spells of toil, slept in a forecastle that a self-respecting chicken or a well-reared dog would forsake as uninhabitable....
>
> "I shipped for what I was told would be a three months' trading cruise in the South Seas," said one of the green hands. "I was told I could make $400 without any difficulty and probably twice as much. I signed a receipt for an advance of $35. The man who shipped me drew the money and bought for me about $4 worth of shoddy clothes."[90]

There had been some truth in what the unnamed green hand had been told before signing on. After leaving San Francisco in December 1904, the vessel did head for Hawaii and Micronesia, where it spent a few months cruising before turning north for bowheads in the Sea of Okhotsk and then returning home almost a year after sailing.[91]

Morris Sederholm was an outfitter in the port of New Bedford during the final years of the industry who worked with his father-in-law, Samuel Horvitz, owner of a firm known as the Whaling Outfitters. In an interview years later he described the hierarchical and somewhat amorphous system of obtaining crew members. As an outfitter, he considered himself "the agent's agent," signing on the crew and witnessing their signatures on the shipping articles. He would determine their lays, based on their past experience and provide them with a sea bag with the necessities for the voyage. He would then place it on the vessel for them; and only when the vessel was towed into the stream and anchored would the men be placed on board, so as to avoid their flight. But how would he find the men? Here was the role of individuals such as the Cape Verdean boarding master Joe Gold, known as "The Necessary Evil." For Joe would put up the arriving crew members in boardinghouses, at times quite removed in Rochester, Wareham or on the Cape. And he "kept them there and either got them jobs, or kept them drunk or something, or had a woman living with them, so that next spring he knew where to find them." And when Sederholm would go to him and say, for instance, "Joe, I need a third mate for this and this ship, or I need a second mate, or I need a boatsteerer," Gold would produce the man.[92]

Gold was possibly the same "Shanghai Joe" who brought in nine men at the last minute to allow the bark *Greyhound* to sail out of New Bedford on time in 1913 under Captain Antonio C. Corvello. As Joseph Gomes, who shipped as boatsteerer on the voyage, later recalled, as the vessel headed out of the harbor, "the last nine men that Shanghai Joe had brought aboard started to crawl up on deck. What a sight! Shanghai Joe had gotten them drunk and kept them that way for a couple of days. One was a young fellow who had just passed the Bar examination. They were all from Gloucester, Massachusetts. Shanghai Joe had lived up to his name."[93] Unfortunately for the captain, this sort of recruitment often did not bear great results. In this case when the *Greyhound* called at Faial, the young lawyer was carried off ill and seven of the others fled to the hills from where, after the departure of the vessel, they intended to report to the American consul who would have them shipped back to the U.S.[94]

Cape Verdeans in the Western Arctic

The voyage itself could be an extremely unhappy and violent experience, especially for foreigners and green hands. A Dutch green hand by the name of Faber recalled how Captain John A. Cook, of the steam bark

Bowhead II, had ordered the mate to punch him in the head, and that on another occasion during the two-and-one-half-year voyage to the western Arctic, Cook knocked him down to the deck and "stamped upon him with both feet."[95] To be black could make life even more difficult. As Arthur J. Allen noted in *A Whaler and Trader in the Arctic*, in describing the living arrangements of crewmen from several shipwrecked whaling vessels who were being sheltered at the Cape Smyth Whaling and Trading Station at Barrow, Alaska, in 1897, "About two-thirds of the men living in the bunkhouse were white, the others colored. The whites had the northern end of the building, and the colored men had the southern end."[96]

This latter was evidently a carry-over from the informal segregation that at times took place on board ship. As Frank T. Bullen recalled when he went into the forecastle of the bark *Cachalot* of New Bedford in the 1850s, "The whole space was undivided by partition, but I saw at once that black men and white had separated themselves, the blacks taking the port side and the whites the starboard."[97]

In its most extreme, racism tainted reportage of the whaling experience. When the *Bowhead II* returned to San Francisco in the fall of 1908 after an eight-month stint in the western Arctic, it brought back one of the crew, a southerner by the name of J. C. Gardner, "charged with the murder on the high seas of Joseph Cadoza [*sic*], a negro." As the *Call* noted, Gardner was "well educated and a boy of evident refinement," who "comes of a good family in Kentucky." He was charged with the murder of Cape Verdean boatsteerer Joseph Cardoza by hitting him on the head with a piece of scantling after "the negro grabbed the white boy" and gave him a beating. The jury in San Francisco evidently concurred in sympathizing with the young Southerner, for they acquitted him, citing "justifiable homicide."[98]

But aboard the western Arctic whaling vessels the reality of Cape Verdean officers having responsibility over white Americans and northern Europeans was generally not nearly so racist. Douglas Macmillan remembered with fondness a Cape Verdean boatsteerer on the steam bark *Fearless* out of San Francisco in 1901. Sam Brown, according to Macmillan, was almost seven feet tall and for this reason was known as the "Portuguese Man-of-War." He had spent considerable time with the young novice, teaching him how to use the bomb lance, employing a rope and barrel for practice. Macmillan noted as well a differ-

Capt. Henry J. Gonzales in the Western Arctic, ca. 1915. From Harold Noice, *With Stefansson in the Arctic.*

ence between western Arctic and Atlantic whaling at the time. For the *Fearless* had no interest in oil, only whalebone and the ever-increasing trading.[99] In 1907, for instance, the *Beluga* returned to San Francisco from a short cruise in the western Arctic with 16,000 pounds of bone, but also seventy-two fox skins and one bear skin.[100]

Henry John Gonsalves was born in Cape Verde ca. 1883 and had a long history in the western Arctic. Sometime around 1910-12 his name became Hispanicized to Gonzales. He was described thus by one of the men who sailed with him: "The mate, Henry Gonzales, had been born in the Cape Verde Islands and brought up on the whaling fleet of the Arctic. He spoke with a slight Portuguese accent, was tall, handsome, servile to the captain, and domineering with every one else, except the Eskimo damsels, with whom he was a regular *matiné* idol."[101]

Gonzales served as boatsteerer on the San Francisco barks *Andrew Hicks* (Jan-Nov 1903) and the *Gayhead II* (Apr-Aug 1909). He was also fourth mate on the steam brigantine *Karluk* (Mar 1907-Nov 1908).[102] He appeared as mate of the well-known schooner *Polar Bear* with the Canadian Arctic Expedition of 1913-18, led by explorer Vihljamur Stefansson, and soon took over as captain of the vessel. Gonzales did not have the best relations with some other members of the expedition. One suspects, from observations such as the following, that prejudice was involved. When George H. Wilkins, the expedition's official photographer, met him he noted, "He is something like I expected to see him, though even more gentlemanly. He is probably quick tempered, quick to decide, but rather unstable in purpose. A good man, no doubt, in charge of Latin people, but unsuitable for the more methodical Europeans."[103]

The story of Gonzales's participation in the Expedition is too long to narrate here, but Harold Noice, who accompanied him, told one tale worth repeating. Gonzales, on a trading trip with a few other men, journeyed by dogsled to a remote Inupiat village whose inhabitants were angry about a perceived insult sometime prior. It was clear that they were preparing to murder the men but hesitated when Gonzales presented pots, cloths, and other items they had never seen before. The danger was not over though, and that evening Gonzales retired to his snow house and sent away the two natives who had accompanied him among the residents, telling them that their visitor was a powerful shaman: "Presently Gonzales looked up from the wood-alcohol lamp over which he was cooking his supper and saw staring faces, barely distinguishable in the dim light, massed in his doorway. The natives had crept in over the snow that silenced all sound of their approach; they filled the alleyway. Then suddenly *they* saw something. Gonzales rose up in the shadows. His waving hands burned with an unearthly blue fire. He thrust his blazing fingers into his mouth and breathed out gusts of flame. The natives fled headlong." His fire-eating trick had worked—the flaming fuel from his alcohol lamp had saved their lives.[104]

While on the Expedition Gonzales married Violet Mamayauk, the teenaged daughter of a whaler and a native woman.[105] In the winter of 1917-18 while on Baillie Island, in the Canadian Northwest Territory, he was discharged by Stefansson and commenced building a cabin for himself and his wife.[106] Evidently the arrangement did not last long, for in 1920 we find him living in Berkeley, California with ex-whaling captain Ayres J. Senna, also a Cape Verdean. He is listed as single.[107] Beginning in 1919 Gonzales served as first mate on a series of Arctic vessels: diesel schooners *Herman* and *Carolyn Frances* (later rechristened the *Charles Brower*), schooner *Arctic*, and the diesel schooner *Nanuk*. Finally, he took command of two voyages on the *Charles Brower* (May-

Sep 1926) and (Jun-Jul 1927).[108] By 1930, with his whaling career at an end, Gonzales was living in Berkeley with his Portuguese wife, Candida, and working as a deckhand on the Golden Gate ferry.[109]

Another well-known Cape Verdean settler in the western Arctic was Peter Lopez (1869-1938). Lopez sailed for many years on whalers out of San Francisco, principally on the *Charles W. Morgan* during the 1890s. By 1900 he had settled in a community of shore whalers near Point Hope, Alaska, known as Jabbertown, along with other compatriots.[110] At that time he was married to a native woman named Nettua and had a daughter, Annie. As did other whalemen who settled in the region, he probably continued to serve a season aboard American whalers and took his turn at shore whaling out of Point Hope and elsewhere. In fact, a story has been told by the natives of how Lopez once jumped upon the back of a whale during a hunt. Peter Lopez is next recorded as being aboard the small whaling schooner *Rosie H.* at Baillie Island in the Canadian Northwest Territory in 1914. The following year he took over as master of the vessel, and caught at least two whales that season.[111]

Lopez left the *Rosie H.* at Baillie Island in August of 1915 to serve with the Canadian Arctic Expedition aboard the *Northstar*. In 1917, he and his wife joined Henry Gonzales on board the *Polar Bear*. Both the Lopezes and Gonzales left the *Polar Bear* later that year at Baillie Island/ Cape Bathurst to become trappers. Lopez eventually settled in Aklavik, on the Mackenzie River in the Canadian Northwest Territories, where he died in 1938.[112]

Continuing to whale at the time were several Cape Verdeans who had made their home on the North Slope. In 1901, Manuel Perry da Rosa, second mate on the steam bark *William Baylies*, and Theofilo Rosa, left the vessel at Point Hope with the permission of the captain. The following year the steam bark *Alexander* dropped off four of its crew—Joseph Ferreira and three other Cape Verdeans: Bernard, James, and Jerome Lopes.[113]

Jabbertown, Alaska

Remnants of Jabbertown still lie along the beach a few miles east of Point Hope. Charles Brower, who had been given the responsibility of taking the federal census for 1900 for all of northern Alaska, headed there after completing his enumerations in Point Hope, expecting to find his old friend Antone Bett. "Finished at the village," he noted in his diary:

> I had to go on the south side of the point 8 miles away, here was a settlement where a number of Portuguese were living mostly with native women, all married, The settlement had been named Jabbertown by the ships, from the noise made there when all hands got to talking at once, I hired some Eskimo men to take me there in a skin boat [umiak] through the lagoon which ended right where settlement commenced, When I got there all hands except the women were off aboard the William Baylies, anchored some miles south of them, Anton Bett had moved from Pt Barrow three years before this and was living here with the rest of the Portuguese, Bell his native wife was home. so to Bells house I went and had some thing to eat, waiting for the men to come back, When they were on their way in we could hear them two miles off, from the noise they made I think the place was rightly named....
>
> Before leaving I asked Antone what all the racket was about while they were on their way in, He told me they had been aboard all day, and no one had asked them to have a bite to eat, they were all hungry, so were just saying what they thought of that kind of a ship and Captain.[114]

The main business of Jabbertown was the hunting of the bowhead whale from the native umiaks. Since the Portuguese were essentially in competition with the natives at nearby Point Hope it was necessary for them to contract with natives from distant communities during the whaling season. Secondarily, the residents of Jabbertown engaged in trading for skins and other local goods with the community at Point Hope. A central figure as trader in the settlement was Heinrich Koenig, referred to by all as "Cooper" for his former shipboard occupation.

Other Portuguese continued to move in and out of Jabbertown during these years. Antone Bett, who relied on a friend to write for him in English, was forced at times to communicate with his friend Koenig in Portuguese. One letter begins: "Agoustou 25 de 1903, Brella manda a Deiz a um Sakoy de farinha cheiou de pernas de viadous...." (August 25, 1903. Bell [Antone's wife] sends to Daisy [Koenig's wife] a flour sack full of deer legs).[115] In February 1906, Bett again wrote to Koenig, probably from his new home in Barrow, informing him that he had five gangs

Joseph A. Ferreira (right) with family and business partner, Point Hope, Alaska, 1906. Courtesy of Elsie Casina Adams.

out hunting foxes, and that it was "a good year for Bears." Alluding to the language difficulty, he added as a p.s., "Please do not write to me in German as there is no German's up here and it takes me a long time to translate it in to Porte Goos and then in to English."[116]

For many years Antone Bett conducted a whaling operation at Barrow, employing native Inupiats and an occasional Azorean or Cape Verdean. According to his grandson, Noah Itta, "people liked to follow him because he was a good leader, and even in real times of storm in the ocean he was a real good seaman, and his voice was so loud that he didn't need a microphone; and people trusted him because they could hear him."[117] Evidently because of his Mediterranean complexion and foreign mannerisms the natives felt somewhat more comfortable with him. According to Itta, "people knew he was from the outside, but they didn't consider him really like a white man because he was a little different from the white man."[118] However, things didn't always go well for Bett. The year 1906 found him at Point Barrow, where he wrote to Koenig, complaining of his poor economic situation and expressing the wish that the following spring he might catch a whale, even if it were "blind or sick."[119]

Joseph A. Ferreira was an old hand in the Arctic. Born in Cape Verde, ca. 1869, he arrived in the U.S. when he was seven. Ferreira had served as a boatsteerer on several San Francisco vessels: the bark *Reindeer* (Dec 1885-Nov 1886), schooner *Alton* (Mar 1889-Nov 1889), and the steam bark *Alexander* (Apr 1901-Nov 1901 and Apr 1902-Nov 1902).[120] He settled in Jabbertown and married an Inupiat woman who had been given the name Fannie by the missionaries. In 1920, the couple had five children.[121] Following a pattern of Inupiat culture that Barbara Bodenhorn elucidated in "I'm Not the Great Hunter, My Wife Is," Mrs. Ferreira did "the dog teaming, hunting, and things like that," while Joe remained at home and participated in the seasonal whale hunt.[122] He also helped build the influential Episcopal Mission at Point Hope. Tragically, both he and his wife died while traveling across the Arctic tundra. Mrs. Ferreira evidently had a heart attack somewhere between Kivalina and Point Hope in the early 1920s. Joe died on the road as well. It was December 1924, and he was traveling by dogsled with his son Franklin from Point Hope toward Kotzebue. He had been warned that it was too early to attempt to traverse the various rivers, because the ocean was still open and the river mouths were dangerous; at the least if he decided to travel he should only cross well inland where the ice was thicker. But Ferreira evidently ignored the advice, for just before Christmas the two were found frozen to death on the banks of Kisimilok Creek, midway between Point Hope and Kivalina. The sled was missing; Joe had climbed into a sleeping bag, to no avail, while his son died sitting down some distance away, surrounded by the dogs. As a reminder of the tragedy, the creek became known for a time as "Joe Ferrerra's [*sic*] River."[123]

Finally, one should note that the Portuguese presence in the western Arctic, mainly Cape Verdean, was significant enough to leave its mark for a while in the jargon that was used for communication between the whalemen and the natives, "a mixture of Eskimo, English, Portuguese, and South Sea words."[124]

The Last of the Western Arctic Whalers

As has been noted, few Azoreans and Cape Verdeans served as masters out of San Francisco. One exception was William F. Joseph, born in Fairhaven, Massachusetts, in 1859, and the son of a mariner from the island of Flores. Joseph began his long whaling career in 1879 as a foremast hand on the bark *Ocean* out of New Bedford. Although several of his early voyages extended to two or three years, by the time he became master he apparently preferred shorter cruises, often to the North Pacific from San Francisco. His first commands were aboard the bark *California* from 1901 until 1905. As often happened with inexperienced crews at this time, there was a tendency to desert whenever possible. When the *California* anchored at Hakodate, Hokkaido, Japan, in July of 1904, twelve of the crew disappeared. Authorities soon found ten of them and returned them to the vessel.[125]

In 1906, without a command of his own, Joseph shipped as mate on the steam bark *William Baylies* for the season. Word had been filtering back for some time about the treatment the whaling crews were receiving in the western Arctic, and this year the Federal authorities decided to act. As the *San Francisco Call* expressed it, somewhat diplomatically, "For years, the captains of whalers have made their own laws from the time they left here until they returned. This year Uncle Sam interfered, as some of the whalemen were entertaining and putting into practice ideas that were not in accordance with the requirements of present day civilization."[126] That fall three veteran whaling captains were in custody or had been charged by the federal government for serious offenses. They were Edwin W. Newth of the steam brigantine *Jeanette;* John A. Cook of the steam bark *Bowhead II*; and H. H. Bodfish of the *William Baylies*. So it fell to mate Joseph to bring the latter vessel back from the Arctic.[127]

In December 1908, Joseph took command of the bark *John and Winthrop*, on which he made four voyages. His relatively poor catches on both the *California* and the *John and Winthrop* show the limited possibilities at this time of whaling in the North Pacific and western Arctic. His last voyage on the *John and Winthrop*, into the Arctic Ocean from March to October 1912, was especially disappointing, resulting in the capture of only one whale yielding thirty-two barrels of sperm oil during the seven-month voyage. However, a portion of the voyage was spent in trading with the natives at Barrow. The officers on board this voyage were virtually all Cape Verdeans (first, third, and fourth mates; and three boatsteerers—the fourth being Azorean). This was the last whaling voyage for the old vessel, built in Bath, Maine, in 1876, and one of the last without auxiliary power to sail out of San Francisco. Joseph's

Bark *Gay Head II* in the stream, San Francisco, ca. 1910
Photo K7.7402.
Courtesy San Francisco Maritime NHP.

final entry in the log reads "Put Ship in Creek [Oakland Estuary] high up to Stay."[128]

As whaling slowed there, in the ensuing years Joseph sailed with more success from New Bedford. He was replacement master for Captain A. T. Edwards on the schooner *Mystic* in the last year of its voyage of Oct 1910-Aug 1914 and also was relief master for a time on the schooner *Bertha D. Nickerson* (Feb 1913-Aug 1916). In 1916 he took command of the bark *Bertha* (May-Sep 1916) out of New Bedford, and then the schooners *Valkyria* (New Bedford, Nov 1917-Aug 1918) and *Margarett* (New Bedford, Mar-Aug 1920).

A few vessels were still actively seeking the cachalot in the Pacific, but with the attendant romance, as seen by the press. In 1909, the bark *Gay Head II*, which had been cruising the waters of the Pacific for over three decades, and in prior years had been active in the western Arctic, was on one of its last voyages. It was less ambitious than many previous voyages—little more than a summer cruise in search of sperm whales, headed by a Yankee captain and Cape Verdean mates: Joseph Baptist, 1st; John Thomas, 2nd; Henry J. Gonsalves (soon to be Gonzales), 3rd;

and boatsteerers Manuel Pena, Antone Arson, Thomas Oliveira, and Urban Pena.[129] Looking back from the vantage point of a century, we can empathize with the whale, but in August of that year the adventure of the *Gay Head II* rivaled the happenings in the fledgling aviation industry. It was also the occasion of the First International Aviation Meeting, a competition taking place in Rheims, France. As reported in the *San Francisco Call*, "At least six members of the whaling bark Gayhead, first of the blubber hunters to return, will read with calmness of the happenings at Rheims. They admit willingly that the crew of an airship does take some chances, but, as Mate Joseph Baptist said yesterday: 'What's falling a few hundred feet to being tossed 80 feet in the air and being caught when you come down on the end of a whale's tail which throws you and your shipmates back into the sky. Aviators have nurse girls' jobs compared with whaling when Mr. Whale gets up his Irish.'"[130]

Captain James A. Wing, observing the scene from the masthead, noted the extraordinary size of the behemoth, the largest he had ever seen. Mate Joseph Baptist brought his boat alongside the whale and fired. The whale, in a fury, came at the boat and with a movement of his head flung men and boat straight up in the air—the captain fell thirty feet from his vantage point, but Baptist swore later that it was more like a hundred. When they landed with the boat they found themselves sitting on the whale's tail, with which he again sent them flying, while also capsizing the boat of the third mate, Henry John Gonsalves. No one was injured in the foray and all were picked up by the boat of the second mate, John Thomas. The whale escaped, but no one seemed to care.[131]

The Last Whalers

> Old New Bedford, the New Bedford along the water front, is sniffling eagerly this spring, reveling in the smell of tar and paint, and tuning its ears delightedly to the tap of the hammer, the rasp of saw and the sounds of brisk activity on the whaling ships along the docks. For this season the port is sending out more whalers than it has done before for many a season. (*Providence Sunday Journal*, 20 Mar 1910)

By the first years of the twentieth century several factors—principally the evolution of women's fashion and the creation of a flexible substitute in the form of spring steel—led to the sudden drop in the price of whalebone.[132] This, plus the scarcity of bowheads, would finally mark the end of the whale fishery in the western Arctic. And so, during the first decade of the twentieth century the number of voyages out of San Francisco began to diminish, while those of New Bedford slowly increased. In 1910, five barks, one brig, and five schooners sailed out of New Bedford. Of the latter, three made two trips each for a total of fourteen voyages, all of them relatively successful in the hunt for sperm oil, which was in great demand. That year San Francisco sent out six vessels.

From 1915 on whaling was practiced almost exclusively out of New Bedford, with an occasional voyage out of San Francisco or Provincetown. The New Bedford whalers now restricted themselves to the Atlantic, and voyages were much shorter—often just a summer season and rarely more than a year. From providing an occasional captain for American whaling vessels during the 1850s and 60s, the Portuguese contingent had grown so in relation to the diminishing trade that after 1900 over one-third of all American voyages left port under the command of Azoreans or Cape Verdeans. And from the New England ports the percentage was much higher, almost sixty percent.

Joseph H. Senna, son of Capt. Ayres J. Senna, was born on the island of Brava in April 1881 and first arrived in the United States aboard the whaling schooner *Golden City* in 1897. He followed in his father's footsteps, becoming captain of the New Bedford schooners *Adelia Chase*, *Carleton Bell*, and *Claudia*. Like many other whalemen of the time Joseph continued his maritime career after the end of whaling. In May, 1926, he arrived in New Bedford on the Brava packet *Yukon* as mate, with his younger brother Albertino as second mate. [133]

Another Cape Verdean whaling master of the time was John Da Lomba, captain of the schooner *William A. Graber* of New Bedford (Mar-Sep 1920), and also mate under Captain John T. Gonsalves during an abbreviated voyage in 1922 (Apr-Aug). Born ca. 1876 on the island of Brava, Da Lomba must have been an imposing figure at well over six feet in height. He was best known at the time as an exceptionally competent first mate, sailing on several voyages with Captain Benjamin Cleveland in that capacity. Robert Cushman Murphy, who sailed with him on the brig *Daisy* in 1912, recalled when seeing him for the first time that he was "so full of vim that I could see his personality electrify our whaleboat's crew from the instant he greeted us at the water front."[134]

But, like Joseph Enos, due to his lack of formal education the art of astronomical navigation—even with the determined lessons of Capt. Cleveland—was incomprehensible. In spite of this Da Lomba was a successful mate and master because, as Murphy noted, "in seamanship and whalecraft he is as competent as any man could possibly be." Additionally he was a skilled scrimshaw artist.[135]

Da Lomba made one of the last attempts at whaling from the port of New Bedford on the three-masted schooner *Blossom*. After serving as a mate on the vessel during a scientific expedition to the South Atlantic, Da Lomba purchased the *Blossom* and set out in the mid-1920s in search of blackfish. All his whaling equipment was stored in one of the whaleboats, which was torn loose in a violent Atlantic storm. So it was back to New Bedford and, according to a reporter, a difficult search among antique shops for the implements of the trade. Finally, in late 1927, he set out on his whaling voyage. If what is reported is true, then the *Blossom*, not the *John R. Manta*, would have the honor of being the last whaling vessel to have sailed from New Bedford. But there is no corroborating evidence. It is more likely that the *Blossom* was being used in the packet trade between New Bedford and Cape Verde, and that any idea of seeking blackfish was ancillary.[136]

Nicholas R. (Nicolau Rodrigues) Vieira (Flores, 1856-1913) came to the U.S. in his twenties. He was mate of the bark *Canton II* of New Bedford (May 1899-Sep 1900) and then took command on its next three voyages. He was also master of the bark *Bertha* of New Bedford (Apr 1910-Aug 1911). Nicholas's son, Joseph A. Vieira, accompanied his father aboard the *Bertha* as boatsteerer and fourth mate. Later, Joseph had his own command of the New Bedford schooner *Claudia* (Apr 1918-Aug 1918) and the *Valkyria* (which was registered in Portland, Maine, but sailed out of New Bedford, Oct 1918-Feb 1919). Vieira made one more voyage as master of the *Valkyria* during the summer of 1919, returning on August 15 with a largely West Indian crew.[137]

Whereas the short voyage of the *Gay Head II* out of San Francisco in the summer of 1909 may have evoked something of the old popular romance of traditional whaling, at least in the press, the 1914 voyage of the bark *Alice Knowles* of New Bedford and its unfortunate end sent a very different signal. The first tragedy of the voyage was a harbinger of trouble to come. William C. Hegarty, a veteran whaling master, sailed short to São Nicolau, in the Cape Verde islands, with the express intent of signing on as first mate Antonio T. Pina, then fifty-nine years old, with many years experience as a mate on various New Bedford whaling vessels, and, as the *New Bedford Evening Standard* noted, "one of the best known men among the whaling fraternity" of the old whaling port.[138]

By the end of 1915, the voyage had been quite successful in obtaining sperm oil, but its luck was about to change. In February of 1916, after a whale was struck by a harpooner, it turned on the whaleboat and, striking it with its flukes, instantly killed Pina and another sailor.[139] In September of the following year, portents of a greater disaster reached New Bedford, when Captain Henry Mandly Jr. of the *William A. Graber* returned with the disturbing news that, after a heavy gale, he had come across floating debris clearly recognizable as belonging to the *Alice Knowles*. In November, news of the loss of the vessel reached New Bedford by way of Brazil. But it would take another six months before the full details of the tragedy would be learned.[140]

On the night of September 2, 1917, the gale-force winds that had been tossing the bark about increased to hurricane intensity. At about 3:00 a. m., with all the men called on deck, a powerful gust sent the *Alice Knowles* on its beam ends. About five minutes later a giant wave engulfed the capsized vessel, sweeping men, casks, boats, and all into the sea. Jules Duarte, a boatsteerer who had been signed on at São Vicente and one of only two survivors, recalled later how he found himself alone in the raging sea. Fortunately he was able to grab on to the stern portion of a whaleboat that earlier had been stove by a whale—presumably that in which Pina was killed. Soon he discovered another survivor, Quinton Degrasse, a seaman who had also been signed on at São Vicente.

When dawn broke the men looked around and saw only an empty cask that was being carried away by the waves—but no other signs of life. And so, removing their clothes to increase their buoyancy, they clung to the piece of whaleboat and prepared to face their destiny without food or water in the vast Atlantic. During three nights they held on to the wreckage. On the fourth day, just as they were about to give up hope and accept their fate, they were sighted and rescued by the merchant schooner *Fred W. Thayer*, on its way to Pernambuco, Brazil. It was from there that the news of their rescue was dispatched to the *Alice Knowles*'s home port.[141]

The two men were finally sent on to New Bedford. Unfortunately, Duarte contracted pneumonia and died soon after his arrival. With no desire to go to sea again, not even for a return home, Degrasse settled in New Bedford. A few years later he tried to sell his story to a magazine, but was told he did not have enough material. Looking back, he surmised, "I think my accent and being black had something to do with it." Later he was offered an opportunity to perform in a motion-picture version of *Moby Dick* that required him to jump overboard. Degrasse recalled that the film agent "came two or three times, and kept giving me fine cigars, and offered to pay me $125 per month. I said, 'Look mister, you could pay me a thousand dollars, but I'm not going to jump overboard from any boat.'"[142]

And yet the ethos of whaling was still deeply ingrained in the natives of New Bedford. In early 1922, a judge of the Third District Court resolved a dispute between the parents of seventeen-year-old Henry J. Souza, charged by his father with being stubborn and a roughneck. He wanted to send his son to sea on a whaler, to which Henry appeared to agree; but his mother was set against it. "Well, I'm going to overrule the mother's objection," said Judge Millikan. "The father is right. It would be the best thing possible for this young man whom you describe as a roughneck to go on a whaling voyage. It's just what he needs." The judge then placed Henry on probation "with the understanding that he sail with the first whaler leaving port." However, there is no indication that Henry followed through with the judge's injunction. Besides the few possibilities still available in the trade, being ordered by a judge rather than being opposed by a mother presented a different reality, and Henry subsequently expressed a preference for work in a shoe-shine parlor.[143]

In May of 1925 Capt. Antone J. Mandly took the schooner *John R. Manta*, with a heavily Yankee crew, on the last successful whaling voyage out of New Bedford. Two years later, on June 22, 1927, Capt. Joseph F. Edwards sailed out of New Bedford aboard the *John R. Manta* with a skeleton crew of six, intent on filling out its complement at the Cape Verde Islands. Almost from the beginning of the voyage, they encountered heavy weather, and the vessel began to leak seriously. On July 3 the captain decided to return to port, arriving on the 16th. And thus ended traditional American pelagic whaling.[144]

Whaling had ended somewhat earlier out of the port of San Francisco. According to John Bockstoce, the last bowhead was taken in 1921 by the diesel schooner *Herman*, Capt. C. T. Pederson, in 1921.[145] There were still voyages to the western Arctic, nominally for whaling, but actually for trading and supply, as well as, presumably, for the occasional seal, since the majority of the vessels' owner/agents had names like Swenson Fur Company and H. Liebes & Company (San Francisco furriers). One would assume that the lone boatsteerer, with the only lay among the crew on some of these latter voyages, was expected to earn a portion of his income from the sealskins he might obtain. William F. Joseph made his last voyages as master in this trade out of San Francisco. In 1922, he took the schooner *Fox* for a brief four-month run to Alaska (May-Sep). The following year he took the gas schooner *Arctic* on a similar voyage (May-Oct 1923), with Henry J. Gonzales as mate. While the *Fox* had not even a boatsteerer on board, the *Arctic* did carry one, forty-nine-year-old Antonio Nicolau of the Azores, who was also the only one of the crew to be contracted on a lay (1/100).[146]

Veteran Cape Verdean whaleman Henry J. Gonzales was destined to make two of these latter voyages, aboard the *Charles Brower*. The 569-ton vessel had been completed in nearby Alameda in 1918 and christened the *Carolyn Frances*. It was a three-masted wooden schooner soon to be powered by an auxiliary diesel engine. Before being sold to H. Liebes, who renamed it the *Charles Brower*, it had made two voyages to the western Arctic for the Western Whaling Company.[147]

In 1926, Gonzales took the schooner to Alaska. On this voyage there were three Cape Verdeans, and a Portuguese American born in Massachusetts. Among the crew was one "boatsteerer/seaman," a "purser and trader," an assistant purser, and a freight clerk. In 1927, Gonzales sailed again as master of the *Charles Brower*, carrying, among the crew, four Cape Verdeans. There was now no boatsteerer; instead there were three traders, including Charles D. Brower, and a stewardess. Only one man in the crew was signed on for a lay; all the rest were on monthly salary. On the way north, as the vessel approached St. Lawrence Island, it struck a rock, tearing away parts of its keel and damaging the rudder. The schooner eventually made it to Pt. Barrow, but not long after was sold to the Russians to operate off the Kamchatka Peninsula.[148]

The Fate of the Old Whaleships

Given the large number of Cape Verdeans participating in whaling out of New Bedford in the last decades of the industry, it became common practice to sign some men on who were simply seeking transportation back to their island, or perhaps seasonal employment in New England. Finally, as the traditional whaling trade neared its demise, several of these New Bedford whaling barks and schooners converted to the packet trade, carrying Cape Verdeans to work in the New Bedford mills or in the cranberry bogs of Cape Cod, and thus becoming known as Brava packets. Such was the fate of the *Sunbeam* (1909), the *Carleton Bell* (1916), the *Bertha* (1917), the *Greyhound* (1921), the *A. M. Nicholson* and *William A. Graber* (1922), and the *Margarett* (1925).

Ex-whaling masters, such as Antone T. Edwards and John Da Lomba, gravitated to this industry after the waning of traditional whaling. Others who had been active earlier on whalers as mates now moved up to command these ex-whaling vessels in their new trade. Cape Verdean Isaac Azulay, for example, took over command of the schooner *A. E. Whyland*, and in May of 1917 arrived at New Bedford with some twenty-seven "seamen"—most of whom were passengers prepared to disembark for summer employment on shore.[149]

Unfortunately, the vessel was torpedoed the following year by a German submarine off the island of Tenerife in the Canaries.[150] Azulay was born on the island of Brava in the 1860s and continued sailing in the packet trade, commanding the schooner *Julia* on its voyage from Fogo to New Bedford in 1922.[151]

In 1922, between whaling voyages, the *John R. Manta* was used to film whaling scenes for the film *Down to the Sea in Ships*. After its last aborted whaling voyage in 1927 it, too, became a Brava packet and was lost with all hands in 1934. The bark *Morning Star*, which first sailed from New Bedford in 1853, met its ignominious fate upon returning from its 1912-14 voyage under Captain Valentine Roza, as it was converted to a barge for carrying coal.[152]

Meanwhile the West Coast whaling vessels suffered a varied fate. In 1917, the bark *Andrew Hicks*, which had made its first sail out of Westport, Massachusetts, in 1867 and its last voyage in 1914-16, was pulled out of the mud of the Oakland Estuary—where the western Arctic fleet generally spent its idle months—and sent back to Massachusetts as a freighter with its first load of merchandise after a half century as a whaler. Unfortunately, its life as a merchant vessel was shortlived, as it foundered later that year off the Virginia coast with a cargo of coal destined for Santo Domingo.[153] Meanwhile, the steam bark *Bowhead II*, with its interesting history in the western Arctic under captains John A. Cook and James and George Tilton, suffered a unique demise. During World War I, some ten years after the end of its last voyage, it became the target of Navy warships and a cinema crew, who disguised it as a battleship and faked a battle in San Diego Harbor. But showing the resilience of these old wooden craft—and symbolically of this antiquated industry—neither shot, nor shells, nor torpedoes succeeded in sinking the old bark; and so it was finally towed to shore and burned.[154]

Toward the end of the second decade of the last century, as the American whaling industry was fading, whaleman Joseph Gomes made the transition to steamships and later remarked on the great differences in living aboard the latter. Water now came out of a faucet instead of an old whale oil cask. Hardtack and salt pork gave way to fresh meat and vegetables. And the omnipresent cockroaches, bedbugs, and straw mattresses were now a thing of the past.[155] No wonder it had been so difficult to find crew members in these latter years.

From being a very minor element in American whaling a century before, the Portuguese had come to predominate in early twentieth-century New England whaling. Meanwhile, Arctic whaling, especially in Alaska and western Canada, had essentially become a sideline to trading for furs. Involved in this and in native shore whaling, several Portuguese elected to remain in the region. There is perhaps no better example of the adaptability of these Atlantic islanders, and their penchant for adventure, than the stories of those—especially the Cape Verdeans—who settled in the western Arctic, in so distinctly different a climate from that of their native land, and thrived on the experience.

Notes

[1] *New York Times*, 14 Sep 1902. Unfortunately there are no extant records of crew earnings for this voyage.

[2] *Boston Sunday Herald*, 2 Oct 1902.

[3] Ibid.

[4] Bark *Canton II*, Vieira, New Bedford, Mass., 11 Apr 1901, Seamen's Register, ODHS.

[5] Schooner *Adelia Chase*, Senna, New Bedford, Mass., 1901-02, Log Wh A229 1901L v. 1, Nicholson Whaling Coll., PPL; U.S. Census, Bristol County, Mass., 1900.

[6] Schooner *Bertha D. Nickerson*, Santos, New Bedford, 1909-11, Seamen's Register, ODHS.

[7] Lopes, *Vida dos Marinheiros*.

[8] "Falam em ir á baleia, / como que, não seja nada, / agora vos vou contar / a minha vida passada! // Tinha 16 annos de idade! / Um moço bem educado. / Fugi a minha mãe, / para não ir para soldado. // Da maneira em que fugi, / foi n'um navio balieiro; / Eu tinha bastante roupa, / mas não tinha dinheiro!" Ibid., 7. Translated by the author.

[9] Bark *Greyhound*, Edwards, New Bedford, 1898-1902, Log Wh G845 1898L, Nicholson Whaling Coll., PPL.

[10] Lopes, *Vida dos Marinheiros*, 9.

[11] Ibid., 11.

[12] Record of Births at or Near Cape Fullerton, Hudson Bay; 1877-1911 (1900), George Comer Coll., MSS 102, box 1, f. 4, MSM.

[13] Lopes, *Vida dos Marinheiros*, 19. Actually this payoff was more generous than many others.

[14] Hanbury, *Sport and Travel*, 62.

[15] Ibid., 99.

[16] Schooner *Era*, Comer, New Bedford, Mass., 26 Jul 1902 ff., Log 86, ODHS.

[17] Bark *Kathleen*, Jenkins, New Bedford, Mass., 1901-02, Seamen's Register, ODHS; Jenkins, *Bark Kathleen*, 4.

[18] *New York Journal*, n. d. (ca. 25 Apr 1902), quoted in Jenkins, *Bark Kathleen*, 3.

[19] Jenkins, *Bark Kathleen*, 18.

[20] Ibid., 28.

[21] *New York Times*, 25 April 1902. There is confusion in the various accounts as to the number of crewmen in Viera's boat. Most likely there were nine.

[22] *New York American and Journal*, 25 Apr 1902.

[23] *New York Times*, 25 April 1902.

[24] Ibid.

[25] Ibid. When Viera spoke to another reporter that day, the incident became even more exaggerated: As the whale was hauling his boat toward the bark, "the line, singing through the groove, set the bow of the whaleboat on fire"; and upon the whale's impact, "the whole side of the bark was crushed in." ("Whale-Wrecked Seamen Ashore," *New York World*, 25 Apr 1902.)

[26] "Whale Hunts the Whaler," *New York Sun*, 25 Apr 1902.

[27] Bark *Sunbeam*, Higgins, New Bedford, Mass., 1904-06, Seamen's Register, ODHS; schooner *Ellen A. Swift*, Gibbons, Provincetown, Mass., 1902-1904, Log, Whaling Logs Collection, YUL.

[28] Unidentified newspaper, 29 Apr 1905, Scrapbook 2:289, ODHS; bark *Ellen A. Swift*, Provincetown, 1905-05, Seamen's Register, ODHS.

[29] Bark *Morning Star*, King, New Bedford, Mass., 25 Oct 1898, Log 3388, NBL. The temptation to swim ashore from the vessel might be great but not without risk. The keeper of the log on the bark *A. R. Tucker* of New Bedford noted in 1871, "Monday night five of our men jumped over board to swim on shore lowered a boat & picked up one & one swam back the others (Henry Spring Fred Smith & Andrew Coffin) got clear do not know if they ever reached the shore or not as we did not hear any thing from them but think the chances was against them as we Lay near a mile from the shore and there was a strong tide at the time." (Bark *A. R. Tucker*, Ricketson, New Bedford, Mass., 3 Sep 1871, Log 456A-E, ODHS.)

[30] Some forty years previous the owners of the bark *Elizabeth Swift* of New Bedford had written to its captain, Josiah E. Chase, that "it did not grieve us that your men run at Western Islands where you could get a plenty more but Falcnand [Falkland Islands] has got to be a very bad place on that account." Swift and Allen to Capt. Josiah E. Chase, bark *Elizabeth Swift*, New Bedford, Mass., 5 Dec 1859, Swift and Allen Papers, MSS 5, s-g 3, Ser. F, S-s 2, f. 1, v. 4, ODHS; bark *Morning Star*, 17 Nov and 11 Dec 1898.

[31] Gomes, *Subsídios*, 108.

[32] Amaral, *They Ploughed the Seas*, 17.

[33] Schooner *Bertha D. Nickerson*, Costa, New Bedford, Mass., 1906-08, Seamen's Register, ODHS.

[34] See pp. 231-32.

[35] Stuart Sherman lists several logs written in Portuguese: Bark *Mary Frazier*, 1843-44 (actually in Spanish, not Portuguese); bark *A. R. Tucker*, 1899-1901; brig *Daisy*, 1910-12; and bark *Morning Star*, 1912-14. Sherman, *Whaling Logbooks and Journals*, passim.

[36] (Comes with light wind from south, heading on port tack and short sail, a.m. very light wind). Bark *A. R. Tucker*, Costa, New Bedford, Mass., 29 Jan 1902, Log 1:2, NBL.

[37] Bark *Wanderer*, A. T. Edwards, New Bedford, 20 May 1920, Log 644, KWM.

[38] Amaral states that he was born on Faial in 1864 (*They Ploughed the Seas*, 73); an obituary states that he was born on Flores in 1863 (Manuel F. Gomes, Obituary, unidentified newspaper, 20 Oct 1910, Scrapbook 2:329, ODHS); schooner *Eleanor B. Conwell*, Gomes, New Bedford, Mass., 1898, Seamen's Register, ODHS.

[39] Bark *Bertha*, Gomes, New Bedford, Mass., Aug-Oct 1905, Log Wh B538.1905L, Nicholson Whaling Coll., PPL; "Mutiny on the Whaler Bertha," *New Bedford Morning Mercury*, 30 Oct 1905.

[40] Gomes, Obituary.

[41] "Mutiny on the Whaler Bertha."

[42] Verrill, *Real Story*, 93-94.

[43] Antonio C. Corvello, U.S. Passport Application, 21 Oct 1916, RG 59, NAW.

[44] Poole, "Mutinous But Not Mutiny," 135, 139-40.

[45] Poole, "Antone Fortes, Whaleman," 148-49.

[46] Ibid., 147; Gomes, *Captain Joe*, 220.

[47] Shipping Order for the schooner *Richard W. Clark*, Joseph T. Edwards Business Records Coll., MSS 83, Ser. A, S-s 2, f. 5, ODHS.

[48] Smith, *Provincetown Book*, 97.

[49] Seamen's Register, ODHS.

[50] *Boston Sunday Globe*, 29 Aug 1915; Seamen's Register, ODHS.

[51] *New Bedford and Fairhaven Directory*, (1901), 637, 644, 663, and 679.

[52] Schooner *Yukon*, Costa, New Bedford, Mass., List or Manifest of Aliens Employed on the Vessel as Members of Crew, Crew Lists of Vessels Arriving at New Bedford, Massachusetts, 1917-1943, Records of the Immigration and Naturalization Service, RG 85, NAW (mic, NBL). Referred to subsequently as "Arriving Crew List, NBL."

[53] Coli, "Cape Verdean-Americans," 16; Hegarty, *Returns of Whaling Vessels*, 37.

[54] Schooner *John R. Manta*, Edwards, Provincetown, Mass., 1907-08, Seamen's Register, ODHS.

[55] *New Bedford Evening Standard*, 4 Jan 1913.

[56] "Modern Substitutes for Whale Oil Casks." Unidentified newspaper article, ca. 1915. Scrapbook T-5:50, ODHS.

[57] Undated newspaper article, ca. 1919, Scrapbook T3:12, ODHS.

[58] Schooner *Burkeland*, Edwards, New Bedford, Mass., Arriving Crew List, 8 Jun 1932, NBL.

[59] Undated newspaper article, ca. 1919, Scrapbook T3:12, ODHS.

[60] Schooner *Margarett*, Edwards, New Bedford, Mass., 1923-24, Seamen's Register, ODHS; schooner *John R. Manta*, Edwards, New Bedford, Mass., 1923, Seamen's Register, ODHS; Amaral, *They Ploughed the Seas*, 54.

[61] The spelling of his surname, in both whaling and other documents, alternates between the original "Freitas" and the Americanized "Frates." The former is used here. (In one passport document issued to Theophilus M. Frates he has signed as Theophilus M. Freitas [Passport Application, 26 Dec 1919, General Records of the Department of State, RG 59, NAW]); Emergency Passport Application, Funchal, Madeira, Portugal, 8 Sep 1906, General Records of the Department of State, RG 59, NAW. Federal Writers' Project, *Whaling Masters*, 307; Brig *Sullivan*, Hegarty, Fall River, Mass., Crew List, NAB; Hegarty, *Returns of Whaling Vessels*, 42. (Hegarty lists the port as Somerset.)

[62] Based on available records (Crew Lists, Fall River, Mass., NAB).

[63] Gonsalves had a twenty-five percent interest in the vessel. Upon returning he sold his share. (*Ship Registers of New Bedford*, 3:2.)

[64] *New Bedford Morning Mercury*, 15 Mar 1928.

[65] Daniel M. Rodrigues, interview by Laura Orleans, New Bedford, Mass., 13 Aug 1999, Faces of Whaling Oral History Project, ODHS; Daniel M. Rodrigues, personal communication.

[66] *New Bedford Morning Mercury*, 15 Mar 1928.

[67] He was referred to at various times by some combination of his Portuguese name, "José Luis" and his American, "Joseph Lewis." "Joseph Luis" seems to have been the most common; Logbook index file 811 A-H, ODHS; schooner *Valkyria*, Portland, Maine, Whalemen's Shipping Paper, 29 Jun 1914, mic 178, NBL; Hegarty, *Returns of Whaling Vessels*, 44.

[68] Frank M. Lopes, U.S. Passport Application, 27 Jan 1917, RG 59, NAW.

[69] Hegarty, *Returns of Whaling Vessels*, 45.

[70] Schooner *Margarett*, Lopes, New Bedford, Mass., Arriving Crew List, 22 Jul 1918, NBL.

[71] Bark *Greyhound*, Corvello, New Bedford, Seamen's Register, ODHS; and Arriving Crew List, NBL.

[72] Immigration Laws (Act of February 5, 1917), Section 36, U.S. Department of Labor, Bureau of Immigration.

[73] Ibid., Sec. 33.

[74] Morris Sederholm, interview by James Healy and Reginald Hegarty, New Bedford, Mass., 17 Oct 1962, ODHS.

[75] Schooner *Pearl Nelson*, Mandly, New Bedford, Mass., 11 Nov 1900, Log 77, ODHS.

[76] Cf. *New Bedford Morning Mercury*, 29 Apr 1925.

[77] Amaral, *They Ploughed the Seas*, 97.

[78] Schooner *Clara L. Sparks*, Benton, New Bedford, 1891-95, Seamen's Register, ODHS; *Boston Herald*, 18 Jun 1908; bark *Josephine*, Smith, New Bedford, Mass., Seamen's Register, ODHS, 19 Sep 1903.

[79] Bark *Canton II*, Roza, New Bedford, Mass., 1912-14, Seamen's Register, ODHS; Hegarty, *Returns of Whaling Vessels*, 40.

[80] Bark *Morning Star*, Roza, New Bedford, Mass., 1910-12, Seamen's Register, ODHS; J. and W. R. Wing Coll., MSS 35, Ser. B, S-s 16, ODHS.

[81] *New Bedford Evening Standard*, 5 April 1901.

[82] Ibid., 10 Jun 1901.

[83] Schooner *Charles H. Hodgdon*, Santos, New Bedford, Mass., 1900-01, Log 442, ODHS.

[84] Bark *Greyhound*, Edwards, New Bedford, Mass., 9 May 1904, Log Wh G845 1903L, Nicholson Whaling Coll., PPL.

[85] See pp. 230-35.

[86] Steam bark *Belvedere*, Cottle, San Francisco, Calif., 1908-08, Crew List, NASB.

[87] *San Francisco Chronicle*, 2 Mar 1908.

[88] *San Francisco Call*, 30 Nov 1904.

[89] Ibid., 9 Nov 1905.

[90] Ibid., 16 Nov 1905.

[91] Bark *Andrew Hicks*, Scullun, San Francisco, Calif., 1904-05, Log 314, NBL.

[92] Sederholm interview.

[93] Gomes, *Captain Joe*, 95.

[94] Ibid., 139-40.

[95] *San Francisco Call*, 27 Dec 1906.

[96] Allen, *Whaler and Trader*, 62.

[97] Bullen, *Cruise of the "Cachalot,"* 9.

[98] *San Francisco Call*, 15 Nov 1908; steam bark *Bowhead II*, Tilton, San Francisco, Calif., 1908-08, Crew List, NASB; Lute Pease, "The yarn of Moses Hornsby," Lute Pease Papers, MSS HG 63, HLCA.

[99] Douglas S. MacMillan, interview by Edouard A. Stackpole, 23 Mar 1965, MSM.

[100] *San Francisco Call*, 7 Nov 1907.

[101] U.S. Census, Alameda County, Calif., 1920; Noice, *With Stefansson*, 14.

[102] Bark *Andrew Hicks*, Williams, San Francisco, Calif., 1903-03; steam brigantine *Karluk*, Wing, San Francisco, Calif., 1907; bark *Gayhead II*, Wing, San Francisco, Calif., 1909-09, Crew Lists, NASB. See also pp. xx for more on this voyage of the *Gayhead II*.

[103] Noice, *With Stefansson*, 40, 260; quoted in Jenness, *Making of an Explorer*, 316.

[104] Noice, *With Stefansson*, 68.

[105] Mendes, "Portuguese Whale Hunters," 139.

[106] Noice, *With Stefansson*, 269.

[107] U.S. Census, Alameda Co., Calif., 1920.

[108] See p. 294.

[109] U.S. Census, Alameda Co., Calif., 1930.

[110] Given the propensity for describing spoken Portuguese as "jabbering," it is easy to see how this settlement, with its considerable Portuguese population, derived its name.

[111] Whaling crew list card file, NBL; U.S. Census, 1900, Tigara, Point Hope, Alaska; "The Story of the Canadian Arctic Expedition: The People of the CAE," Online Exhibition, Canadian Museum of Civilization, http://www.civilization.ca/cmc/exhibitions/hist/cae/peo614e.shtml (accessed 29 Mar 2009); Nuligak, *I, Nuligak*, 118-19.

[112] "The Story of the Canadian Arctic Expedition."

[113] Steam bark *William Baylies*, Cottle, San Francisco, Calif., 13 Oct 1901, Log 205, ODHS; steam bark *Alexander*, Tilton, San Francisco, Calif., 1902-02, Crew List, NASB.

[114] Brower, "Autobiography," 546-47.

[115] Bett to Koenig, 25 Aug 1903, Franz Heinrich Koenig Coll., UAF. Translated by the author.

[116] Ibid., 15 Aug 1906.

[117] Noah and Molly Itta, interview by Donald Warrin, Barrow, Alaska, 16 Aug 2005.

[118] Ibid.

[119] Bockstoce, *Whales, Ice, and Men*, 252.

[120] Bark *Reindeer*, Baker, San Francisco, Calif., 1885-86; schooner *Alton*, Newth, San Francisco, Calif., 1889-89; steam bark *Alexander*, Tilton, San Francisco, Calif., 1901-01 and 1902-02, Crew Lists, NASB.

[121] U.S. Census, Noatak-Kobuk, Alaska Territory, 1920.

[122] Elsie Casina Adams, interview by Donald Warrin, 3 Aug 2005, Anchorage, Alaska.

[123] Ibid; *Orth, Dictionary of Alaska Place Names*, 525.

[124] Noice, *With Stefansson*, 71.

[125] Bark *California*, Joseph, San Francisco, Calif., 14 Jul 1904, Log 668, KWM.

[126] *San Francisco Call*, 26 Sep 1906.

[127] Ibid.

[128] Bark *John and Winthrop*, 28 Oct 1912, Log 669, ODHS; Crew List, NASB.

[129] Bark *Gay Head II*, Wing, San Francisco, Calif., Crew List, Laflin Records, SFM.

[130] *San Francisco Call*, 26 Aug 1909.

[131] Ibid.

[132] And yet, as late as 1907 whalebone was still being touted for its unique qualities. A government publication that year noted that, "The economic value of whalebone is due to its combined qualities of lightness, elasticity or springiness, and flexibility even when split into very thin strips. It has also the property of permanently retaining any shape that may be given to it when it is heated and then cooled under compression. Although many substitutes have been introduced, such as steel, cattle horn, and turkey quills, nothing has yet been found that competes with it in a combination of all the qualities above noted. It is therefore unrivaled as material for use in whips, corsets, for dress stays, and similar purposes." (Charles H. Stevenson, *Whalebone: Its Productions and Utilization* [Washington, D.C.: Department of Commerce and Labor, Bureau of Fisheries, 1907], 8.)

[133] Joseph H. Senna, U.S. Passport Application, 13 Nov 1911, RG 59, NAW; schooner *Yukon*, Costa, New Bedford, Mass., Arriving Crew List, 20 May 1926, NBL.

[134] Schooner *William. A. Graber*, Gonsalves, New Bedford, Mass., 1922, Seamen's Register, ODHS; schooner *William A. Graber*, Da Lomba, New Bedford, Mass., Arriving Crew List, 7 Aug 1922, NBL; Murphy, *Logbook for Grace*, 30. There are later references to him as John D. Lomba; but on the Whalemen's Shipping Paper for his voyage as master of the *William A. Graber*, he signed as John Da Lomba (mic 178, NBL). The original name would have been João da Lomba.

[135] Murphy, *Logbook for Grace*, 63.

[136] Carl T. Robertson, "Reminiscenses of a Cape Verde Captain," *Cleveland Plain Dealer*, n.d., reprinted in *New Bedford Daily Mercury*, 1 Dec 1927.

[137] Schooner *Valkyria*, Vieira, New Bedford, Mass., Arriving Crew List, 15 Aug 1919, NBL.

[138] *New Bedford Evening Standard*, 6 Mar 1916.

[139] Ibid.

[140] Tripp, *True Story*.

[141] Ibid; *New Bedford Evening Standard*, 13 Mar 1918; *New Bedford Morning Mercury*, 13 Mar 1918. Reginald B. Hegarty, who later would contribute substantially to the history of the American whale fishery, had sailed previously aboard the *Alice Knowles* as a child with his father, mother, and two brothers; but fortunately he, his mother, and his brother Elton were not on board. Besides losing his father on the *Alice Knowles*, Reginald also lost an older brother.

[142] Anderson, "Quinton Degrasse," 102.

[143] "Judge Overrules Mother and Orders Youth to Go Whaling" and "Shoe Shining Stand Eclipses Henry's Desire to Go Whaling" Unidentified newspaper articles, 1922, Scrapbook T3:96B-97, ODHS.

[144] Schooner *John R. Manta*, Mandly, New Bedford, 1 May 1925, Seamen's Register, ODHS; schooner *John R. Manta*, Edwards, New Bedford, Log 952, KWM.

[145] Bockstoce, *Whales, Ice, and Men*, 340.

[146] Crew list and Whalemen's Shipping Paper, NASB. Hegarty states erroneously that the *Arctic* sailed under John Bertonccini, and that Joseph was relief master (Hegarty, *Returns of Whaling Vessels*, 46).

[147] *Lloyd's Register, 1924-1925*, 1: n. p.; Hegarty, *Returns of Whaling Vessels*, 46; "Maritime Memories," 21.

[148] Diesel schooner *Charles Brower*, Gonzales, San Francisco, Calif., Crew List and Whalemen's Shipping Paper, 1926 and 1927, NASB; Brower, *Fifty Years Below Zero*, 272-74; *Lloyd's Register, 1928-1929*, 1:n.p.

[149] Schooner *A. E. Whyland*, Azulay, New Bedford, Mass., Arriving Crew List, 25 May 1917, NBL.

[150] *Ship Registers of New Bedford*, 3:4.

[151] Passenger Lists of Vessels Arriving at New Bedford, Massachusetts, 1902-1942, mic. T944, RG85, NAW.

[152] Hegarty, *Returns of Whaling Vessels*, 42-47.

[153] *San Francisco Examiner*, 23 Jul 1917; *New Bedford Morning Mercury*, 15 Dec 1917.

[154] "Fail to Sink Whaler," unidentified newspaper article, ca. Oct 1918, Scrapbook T3:2, ODHS

[155] Gomes, *Captain Joe*, 234.

CHAPTER 6

Conclusion

The American whaling industry changed remarkably little from the time it began sending sloops out into the Atlantic, until its final days. Vessels had grown in size and number of masts; but they were still constructed of sturdy oak in an age of steel, and powered, at least partially, by sail. Men still approached the leviathan in small boats, even though they were now equipped with power weapons to subdue it. Later voyages were once again of shorter duration than they had been at the height of the industry. And, as always, men did not know what would be their final remuneration—if any—until the vessel had returned to its home port. Meanwhile, a new whaling industry had been developing abroad since the last decades of the nineteenth century, one that was beginning to dwarf the results of traditional American whaling.

Portuguese participation in the American whaling industry had begun modestly on vessels out of Nantucket by at least 1765. Whether initially these were Atlantic islanders is not known. But certainly those natives of the Azores and Cape Verde soon came to dominate. Save for those documented on the pre-Revolutionary voyages out of Martha's Vineyard and Nantucket, it is difficult to identify their presence in the industry until the beginning decades of the nineteenth century. As whaling activities greatly increased during the 1830s, so did the participation of these Portuguese Atlantic islanders. Slowly they rose through the ranks of the whaling hierarchy—the majority, however, due primarily to their lack of formal education, reaching a ceiling at boatsteerer or one of the lower classes of mate. Nevertheless, after mid-century the percentage of Portuguese whaling masters, and even ship owners, began to grow—a phenomenon that would continue to increase until the end of traditional whaling.

While the percentage of blacks engaged in the maritime industry diminished in the latter half of the nineteenth century, due to heightened racism that manifested itself, for example, in segregation by maritime unions and restricted opportunities in the U.S. Navy, the whaling industry was less affected. Although the participation by American blacks had greatly decreased, those of African descent from Cape Verde and the West Indies found, on the contrary, greater chances of employment.

By the last quarter of the nineteenth century, Azoreans and Cape Verdeans formed a major portion of the industry. Their presence aboard American whaleships afforded these men the opportunity to partake of the American experience, first on board ship and later, if they so desired, on land. But the implications of this activity were much broader. To begin with, it established the bulk of the settlement patterns of the Portuguese in the United States that still hold today, i.e. New England, California, and Hawaii. The early establishment of Portuguese communities in these three regions led to the more massive immigration that followed: to the mills of New England, the agricultural pursuits of California, and the sugarcane plantations of Hawaii. And in a more global perspective, it has led to historical Portuguese settlements in places as far-reaching as New Zealand, Australia, and the North Slope of Alaska.

Scrimshaw – Whaling's Artistic Legacy

Cruising about for whales, often unseen for long periods, afforded the whaleman considerable leisure time. While much of it could be filled with talk and tobacco, and at times song and dance, there was plenty of time to unleash the creative spirit. Sewing was particularly popular; others might attempt to draw or to maintain a journal of the voyage. But the most unique art form to emerge from the whaling era was scrimshaw, the engraving of images and words upon the teeth of the sperm whale or walrus tusk. The work of some of these scrimshanders reached a high level of artisanship.

Of all the various Azoreans and Cape Verdeans who practiced the art of scrimshaw, Manuel Enos was probably the most accomplished. In fact, in the words of Stuart M. Frank, "he deserves to be listed among the select few, as one of the outstanding scrimshaw makers and arguably the greatest Azorean whaleman-artist of all time."[1] In the Whaling

Manuel Enos scrimshaw, "Rebecca at the Well." Courtesy of the New Bedford Whaling Museum.

Museum at Cold Spring Harbor are a carved pair of walrus tusks and a pair of beautifully engraved sperm whale teeth, each with the figure of a woman. There are also two sperm whale teeth in the collection of the New Bedford Whaling Museum that recently have been identified as being the work of Captain Enos in 1862, during his first voyage on the *Java*, one of which represents, in color, Rebecca at the Well, wife of the patriarch Isaac of the Old Testament. The other, also in color, depicts Columbia wrapped in a flowing American flag.[2]

Other whaling masters were also active scrimshanders. Frank C. Morris was born on the island of Faial and commanded several New England voyages between 1887 and 1907. Two others with extant works are Cape Verdean John Da Lomba and Antone T. Edwards, from the island of Flores.[3]

Other Aspects of the Portuguese Participation in Whaling

The history of shore whaling in California and the Azores has been detailed by writers such as David Bertão and Robert Clarke.[4] But there is much more. Portuguese also served on British and French whalers and even settled around the French whaling ports. One Azorean whaleman,

Jean De Souza Galorte, born on the island of São Jorge in 1804, served for many years with the French whaling fleet sailing out of Le Havre. By 1838 he had married a French woman and had been naturalized as a citizen of France. From 1840 until 1863 he served as master of several French whaling vessels.[5]

With the end of traditional whaling, the Portuguese in America continued to follow the sea. Cape Verdeans maintained the packet service carrying their countrymen between the islands and New England; and Azoreans took to fishing at ports such as Gloucester, Massachusetts; and Stonington, Connecticut. They also fished on the West Coast, in such diverse ports as San Francisco, Monterey, and San Diego, while Cape Verdeans worked on the docks in San Francisco Bay. And on both coasts Portuguese could be found in the coastal trade and in other maritime pursuits. They also served in the Hawaiian whaling fleet during the 1860s and early 70s.

The Portuguese conducted some pelagic whaling voyages. Deep-sea whaling vessels sailed occasionally out of their home port of Faial; and of course there was the attempt at South Seas whaling noted by d'Urville at New Zealand.[6] Robert Clarke documents several attempts at deep-sea whaling that took place from Lisbon beginning in 1798 and from the Azores commencing in 1875. But the costs of outfitting these voyages were not sustainable, and they were short lived.[7] It was in shore whaling, however, that the Portuguese found more success.

Whaling was apparently conducted from shore along the coast of Portugal, at least from the thirteenth through the fifteenth centuries, before the emphasis shifted to Brazil, where whales were at that time more plentiful.[8] Shore whaling by Azoreans, using traditional methods, was conducted principally from the islands of Pico and Faial, commencing on a small scale on the latter in 1832, possibly with the help of Consul Dabney, but really had its start in the 1850s and continued until 1987.[9] In the succeeding decades of the nineteenth century all the other islands, at one time or another, mounted their own stations. Since only the sperm whale was hunted, the figures are impressive for some years: in 1910, 72.3 percent of the world's catch, and during the 1920s and 30s through the mid-40s, over ten percent.[10] Traditional shore whaling was practiced as well on the island of Madeira from 1941 to 1981. One enthusiast of this activity was John Huston, director of the film *Moby Dick*, starring Gregory Peck as Ahab, who published an article in *Sports*

New Bedford whaling masters at Horvitz Outfitters Store. First row, left to right: Samuel Horvitz (kneeling), Capt. Henry Mandly Sr., Capt. Henry Mandly Jr., Capt. Joseph A. Vieira, Capt. James F. Avery, Capt. George W. Smith. Second row, left to right: Capt. Joseph H. Senna, Capt. Antonio C. Corvello, Capt James A. Tilton, Capt. Andrew A. Cory, Capt. Antone J. Mandly, Capt. Antone T. Edwards, and two unknown. Courtesy of the New Bedford Whaling Museum.

Illustrated on the topic of Madeiran shore whaling entitled "Harpoons Away!" Additionally, modern whaling was conducted from Setúbal on the mainland of Portugal for a short period in the mid-1920s by a Norwegian company, but with limited success. Shore whaling was practiced there again briefly from 1944 to 1951.[11]

The Cape Verde islands have a long history of shore whaling. By the late eighteenth century whaling was taking place from Brava. In 1874 shore whaling commenced at São Nicolau, and on Sal in 1883. There was also whaling from Maio. As in the Azores, these activities certainly benefited from the wide experience of so many Cape Verdeans on American, British, and French whaleships. But pelagic whaling soon decimated the whale population in these waters; and shore whaling all but ceased by 1914.[12]

Meanwhile, in New Bedford, whaling lived on in the form of the New Bedford Whalemen's Club, which grew out of the informal whalemen gatherings at the Horvitz outfitting store. Chartered in 1921, with James

A. Tilton as its "captain" and president and John T. Edwards as "first mate" and secretary, its goal was to keep alive the spirit of whaling and its traditions. It was hoped that the centerpiece of this program would be the venerable old *Charles W. Morgan*, to rest in the harbor and serve as a living museum. Unfortunately, these dreams never materialized; and although the *Morgan* is very much present to this day, it is as a reminder of the old whaling days at the Mystic Seaport Museum in Connecticut.[13]

Notes

[1] Frank, "Many Mysteries of Manuel Enos," 16.

[2] Cf. Frank, *More Scrimshaw Artists*, 44-49.

[3] Ibid., 86; Frank, *Dictionary of Scrimshaw Artists*, 82; Gomes Vieira, *Man and the Sea*, 400. For other Portuguese scrimshanders consult these three volumes.

[4] Bertão, *Portuguese Shore Whalers*; Clarke, "Open Boat Whaling."

[5] Du Passier, *Les Baleiniers français*, 230.

[6] For Azorean pelagic whaling see Afonso, "Baleação pelos Açores," 1289, and Gomes, *A Ilha das Flores*, 659; for the South Seas reference see p. 146.

[7] Clarke, "Open Boat Whaling," 295.

[8] Serrão, ed., *Dicionário de História de Portugal*, 1:282.

[9] Clarke mentions the possibility of much earlier shore whaling, predating the arrival of the Americans and the result of Basque influence, but the data are lacking. Myriam Ellis speaks as well of early seventeenth-century whaling in Cape Verde, but offers no documentation. (Ibid., 296; Ellis, *A Baleia no Brasil Colonial*, 25).

[10] Clarke, "Open Boat Whaling," 284.

[11] Tønnessen and Johnsen, *History of Modern Whaling*, 322-23, 476-77, and 647; Cristina Brito, "Assessment of Catch Statistics during the Land-Based Whaling in Portugal," JMBA2 Biodiversity Records on line, 5930, 1-5:2, http://www.mba.ac.uk/jmba/jmba2biodiversityrecords.php?5930 (consulted 2 Jul 2009); Huston, "Harpoons Away!"

[12] Hazevoet and Wenzel, "Whales and Dolphins," 198.

[13] William H. Tripp, "Whalemen's Club, Ten Years Old, 1931: Brief History of the Whalemen's Club," Local History Coll., MSS 77, s-g 1, Ser. F, S-s 39, ODHS; "Whalemen Who Still Sail On," *Dearborn (Michigan) Independent*, 25 Oct 1924.

CHAPTER 7

Epilogue

> "Modern" whaling… bears little resemblance to the whaling of Herman Melville's day. Moby-Dick's victorious struggle had its basis in fact: by accident, or through the malevolence of the whale, men were often overpowered by the very creature they hoped to kill. Whaleboats were swamped, or overturned, or smashed; and whalemen were crushed, or drowned, or, like Ahab, wrapped in the coils of their industry and hauled into the deep.
> (Robert Lloyd Webb, "Power Whaling: Industrial Shore Whaling on the West Coast, 1905-1972")[1]

From the foregoing chapters one would imagine that commercial whaling was about to reach its demise, as an outdated and unprofitable enterprise. But that is anything but true. Toward the end of the nineteenth century, while American schooners were prowling the Atlantic for sperm whales, and vessels out of San Francisco were predominantly seeking whalebone and trades with the Arctic natives, a new industry was developing abroad to supply the growing demand for whale oil. It was an enterprise for which the Americans were ill equipped. For the prey were the rorquals—especially the blue and fin whales—which were too swift and difficult to catch with traditional American methods.

Modern whaling involved fast, steam-powered (and later diesel) "catcher boats" of 300-700 gross tons and 100 feet or more in length, with a harpoon cannon mounted in the bow—along with a large mother ship: a floating factory capable of processing the whale carcass thoroughly at sea and delivering finished products to market. From an estimated 1654 whales caught in 1900 by this method, the number grew to 23,409 in 1914. Already by 1911 modern whaling had surpassed the

Anglo Norse of London at Elizabeth, N.J., in 1949, with 9300 tons of sperm oil. Courtesy of the New Bedford Whaling Museum.

annual record output of the American whaling fleet during its heyday in the 1840s and 50s. The result, ultimately, was the near devastation of these species—something that was only comparable in American whaling to the hunting of the gray and the bowhead with modern weapons in traditional whaling's later days.[2]

A look at the statistics of the Norwegian factory steamer *Kosmos*, 17,801 gross tons, with tanks to hold 120,000 barrels of oil, furnishes an idea of the contrast between the old wooden boat method and modern whaling. The *Kosmos* carried seven catcher boats and even an airplane to search for whales. In its first season in the Antarctic in 1929-30, it is credited with catching 1822 whales that produced 119,400 barrels of oil.[3]

Following the Civil War, New England capital had seen much greater promise in manufacturing industries such as shoes and cotton; and the nation in general focused on the opening of the American West. And so it fell to relatively tiny Norway and its pioneer in modern whaling techniques, Svend Foyn, to lead the way. What kept this activity growing in the twentieth century were the techniques that were developed to hydrogenate whale oil, thus removing its odor. In the United Kingdom, for example, by the 1930s some forty percent of margarine and thirty percent of lard were derived from whale oil—produced, of course, by the new factory-ship methods employed in these floating processing plants.[4] Industrialization had now permeated the industry; and the distance from the relatively modest catches on the old wooden vessels to those of the new factory ships was immense. As a scientist at the Food Research Institute at Stanford University observed soon after World War II, "During the two decades between World War I and II whale oil moved from an insignificant position to one of primary importance in the world's fats and oils economy. Output rose from 68,000 long tons in 1919-20 [most of which at this time was already produced by modern whaling methods] to 615,000 in 1930-31...."[5]

The use of sperm oil was widespread in the U.S. in the years after World War II. When the 16,000-ton British factory ship *Anglo Norse* arrived at Elizabeth, N.J. in February of 1949, it carried 9300 tons of sperm oil derived from some 3000 sperm whales, destined for the Werner G. Smith Company. There the oil would be processed for a variety of special uses.[6] In 1957, an industrial publication boasted that "a whale is one of the world's hugest chemical factories. Its body builds up molecules that industry can't afford to synthesize directly—even when it knows how. Instead, Man uses his laboratories to recover, refine and modify the natural products."[7] And it went on to detail the many ways in which the oil of the sperm whale was utilized in this country:

> Another whaling season has begun....
>
> To a generation of Americans, whaling died with "Moby Dick." But to thousands of scientists and businessmen the world over, it has as many lives as the proverbial cat.
>
> Today, a fleet only a third as big as the one that sailed from New England in the peak years of Yankee whaling brings back six times as much oil....
>
> Let's introduce ourselves to Betty Smith, typical young American typist of the Jones Company. She doesn't know it, but whaling plays an important part in everything she does.
>
> Her lipstick has a base of spermaceti—that strange wax from a whale's head, that flows smoothly under the gentlest pressure. The same stuff gives the opalescent sheen to her face cream and her boss's brushless shave. It's used to shape the graphite rod in her pencil, in stamping out the metal caps on the bottle of Coke she drinks, and in the ointment she rubs on her sunburn.

> Purified sperm oil lubricates Miss Smith's new electric typewriter and makes the ink spread smoothly on the ribbon. It in the plasticizer on her carbon paper, in the ink on her magazine, and it keeps her new leather jacket soft and pliable. Hydrogenated sperm oil prevents the aluminum base of the bulb in her bedside lamp from corroding fast to the brass socket and, as a textile lubricant, gives her dress its smart drape. Sulfonated, this same oil goes into the detergent Betty uses on her nylons, and into the tanning compound that made her white calfskin gloves. Sulfurized, it controls the viscosity and "EP" (extreme pressure) properties of the hydraulic fluid in her boy friend's car. Completely hydrolized, oxidized and polymerized, it guards the sheen of her fur coat.[8]

An attempt apparently was made to introduce modern Antarctic whaling in the United States when the *S.S. Frango* of the American Whaling Company (with Norwegian financing) sailed from New York to the Antarctic in 1930 with a Norwegian crew. There was now a doctor on board, and the men were to be paid a fixed overtime wage for hours worked over ten. The *Frango* carried three catcher boats armed with harpoon guns. But the timing was poor, as the nation faced the Great Depression; and over the next decades, it fell to the Norwegians and others to maintain the fishery using these new techniques.[9]

In the first decades of the twentieth century, just as traditional whaling was in its last throes, modern whaling arrived in the United States in the form of power shore whaling, initially on the Northwest coast and later in California. Catcher boats, employing harpoon-cannons, would now tow the whale carcass into modern shore stations equipped with machinery that efficiently and rapidly "not only processed oil but also manufactured fertilizers, bone meal, animal fodder, glue, and eventually edible meat, vitamins, and the stuff of lipsticks, perfumes, soaps, crayons, and even stringing for tennis racquets."[10]

Commercial interest in whaling in the U.S. slowly waned; and in the midst of international pressure, in 1972 the Nixon administration ordered the last remaining American whaling station, located at Point San Pablo in San Francisco Bay, to close. The only whaling now practiced from American shores is that of the Inupiats on Alaska's North Slope, on a very minor scale.

Notes

[1] Webb, "Power Whaling," 30.

[2] Tønnessen and Johnsen, *History of Modern Whaling*, 241; Bjørn L. Basberg, "Two Hegemonies – Two Technological Regimes: American and Norwegian Whaling in the 19th and 20th Century." In Discussion Papers, Norwegian School of Economics and Business Administration, Department of Economics (Sep 2006) http://hdl.handle.net/2330/1471.

[3] Unfortunately the airplane was lost with its two-man crew, and the experiment there was never repeated, although in later years helicopters were often employed. Tønnessen and Johnsen, *History of Modern Whaling*, 276, 372, 377-78, 381, 445, 698; Townsend, "Twentieth Century Whaling," 6.

[4] Davis, Gallman, and Gleiter, *In Pursuit of Leviathan*, 30.

[5] Brandt, *Whaling and Whale Oil*, 1.

[6] *Elizabeth (NJ) Daily Journal*, 22 Feb 1949.

[7] "Leviathan in the Lab," 34-35.

[8] Ibid.

[9] J. W. Mannix typescript, Carlton D. Morse Whaling Coll., MSS 79.22, ser. 4, box 9, f.1, JHL; *Time*, 15 Sep 1930. (In spite of the Portuguese-sounding name, there is apparently no relationship between the *Frango* and Portugal.)

[10] Webb, *On the Northwest*, 143.

APPENDIX I.

American Whaling Ports by Region

REGION	PORT	VOYAGES	FROM	THROUGH
Buzzards Bay, Mass.	Fairhaven	619	1743	1878
	New Bedford	5146	1754	1927
	Dartmouth	149	1756	1882
	Rochester	94	1771	1860
	Acushnet	1	1772	1772
	Wareham	32	1775	1861
	Westport	326	1775	1861
	Mattapoisett	143	1840	1864
	Sippican	66	1840	1866
	Marion	45	1866	1885
TOTAL		6621		
Nantucket, Mass.	Nantucket	2223	1698	c1887
TOTAL		2223		
New London, Conn.	New London	1116	1718	1908
	Saybrook	1	1753	c1753
	Groton	33	1790	1869
	East Haddam	4	1795	1838
	Norwich	39	1799	1910
	Pawcatuck	2	1820	c1821
	Stonington	264	1818	1892
	Mystic	103	1825	c1871
	Lyme	2	1844	1848
TOTAL		1564		
Cape Cod, Mass.	Barnstable	2	1733	1741
	Eastham	1	1737	1737
	Wellfleet	9	1754	1869
	Falmouth	71	1768	1864
	Unspecified	11	1768	1789
	Sandwich	17	1769	1863
	Truro	6	1774	1853
	Provincetown	1156	1789	1915
	Yarmouth	4	1846	1850
	Orleans	35	1850	1864
TOTAL		1312		

REGION	PORT	VOYAGES	FROM	THROUGH
San Francisco, Calif.	San Francisco	1171	1849	1928
TOTAL		1171		
Long Island, N. Y.	Sag Harbor	807	1667	1874
	Southampton	1	1667	c1667
	Unspecified	4	1769	c1769
	East Hampton	2	1797	c1900
	Greenport	126	1824	1860
	Cold Spring Harbor	50	1836	1862
	New Suffolk	11	1837	1852
	Amagansett	8	c1900	
	Wainscott	1	c1900	
TOTAL		1010		
Boston, Mass.	Boston	350	1733	1903
	Gloucester	19	1737	1922
	Hingham	3	1744	c1786
	Marblehead	3	1769	c1822
	Salem	99	1784	1871
	Newburyport	20	1785	1870
	Braintree	1	1786	c1786
	Lynn	38	1830	1857
	Dorchester	8	1833	1841
	Quincy	8	1840	1849
	Beverly	40	1849	1874
TOTAL		589		
Rhode Island	Unspecified	6	1723	c1821
	Newport	111	1731	1860
	Warren	199	1766	1865
	Providence	66	1768	c1902
	Bristol	94	1786	1858
	East Greenwich	6	1807	1811
	Tiverton	2	1821	1823
TOTAL		484		
Martha's Vineyard, Mass.	Edgartown	237	1738	1894
	Unspecified	15	1738	c1857
	Holmes Hole	37	1815	1865
	Chilmark	1	1848	1849

REGION	PORT	VOYAGES	FROM	THROUGH
	Tisbury	2	1867	1870
TOTAL		292		
New York	Unspecified	4	1688	c1795
	New York City	115	1769	1881
TOTAL		119		
Hudson River, N.Y.	Kingston	4	1765	c1767
	Hudson	67	1785	c1845
	Newburgh	10	1832	1839
	Poughkeepsie	21	1832	1846
TOTAL		102		
Fall River, Mass.	Fall River	67	1832	1913
	Somerset	11	1839	1849
	Freetown	4	1841	1848
TOTAL		82		
Other Massachusetts	Various	12	1761	1844
	Plymouth	59	1795	1848
TOTAL		71		
Other Connecticut	Various	8	1753	1914
	New Haven	13	1796	1863
	Bridgeport	27	1833	1909
TOTAL		48		
Maine	Portland *	23	1834	c1919
	Various	9	1834	c1884
TOTAL		32		
Others				
New Jersey	Various	17	c1685	1825
Virginia	Williamsburg	1	1751	1751
Pennsylvania	Philadelphia	6	1804	1824
Maryland	Various	2	1812	1833
New Hampshire	Portsmouth	15	1827	1848
North Carolina	Various	17	1831	c1916
Delaware	Wilmington	16	1834	1845
Washington	Seattle	1	1909	1910
TOTAL		75		
GRAND TOTAL		15795		

Regions are listed by their numerical importance.
Within each region ports are listed chronologically
Data are based on port of registry, not necessarily port of sailing.
Source: Lund, *Whaling Masters* 723-24.

* As was occasionally the case with other ports, the majority of Portland's voyages actually sailed out of New Bedford.

APPENDIX 2.

Portuguese and Portuguese American Whaling Masters in American Whaling

FIRST NAME	SURNAME	BORN	LOCATION	CAPTAINCIES	DIED
Nicholas C.	Anthony	ca. 1806-10	Flores, Azores	1852-1853	
Joseph	Avilla *	ca. 1855-56	Portugal	1899-1901	
Anthony P.	Benton	ca. 1843	Maio, Cape Verde	1876-1899	
Joseph P.	Benton ***	1861	Maio, Cape Verde	1891-1901	1901
Lebanio D.	Borges			1898-1898	
Francis	Caton			1857-1862	
Joseph H.	Caton			1867-1870	
Joseph M.	Caton	ca. 1824	Massachusetts	1852-1878	
Manuel	Caton	1834	Provincetown. Mass.	1865-1866	1866
Henry	Clay	1836	Flores, Azores	1867-1878	1901
Antonio C.	Corvello	1879	Flores, Azores	1907-1919	1920
Benjamin	Costa	ca. 1878	Cape Verde	1921	
Manuel	Costa			1854-1858	
Manuel E.	Costa	1849	Faial, Azores	1879-1911	1914
Manuel	Costa *	1849	Azores	1888-1890	1912
Joseph	Cruze *			1881-85	
Antonio	Da Costa			1835	
John	Da Lomba	ca. 1876	Brava, Cape Verde	1920	1960?
Joseph	Dias Jr.	1822	Martha's Vineyard, Mass.	1853-1857	1883
Jose M.	Domingos	1878	Cape Verde	1919-1920	
Jasper M.	Ears	ca. 1821	São Nicolau, Cape Verde	1870-1882	
Antone T.	Edwards	1882	Flores, Azores	1903-1924	1936
John T.	Edwards	1884	Flores, Azores	1912-1919	1957
Joseph F.	Edwards	1886	Flores, Azores	1918-1927	1933
Joseph T.	Edwards	1856	Flores, Azores	1887-1908	1913
Joseph T.	Enos	ca. 1846	Azores	1885-1904	
Manuel	Enos	1827	Pico, Azores	1864-1869	1915
Michael A.	Ferreira	1847	Portugal	1884-1893	
Joseph	Folger *	1792	Pico, Azores	1825	
Thomas E.	Fordham	1828	Faial, Azores	1859-1890	1892

FIRST NAME	SURNAME	BORN	LOCATION	CAPTAINCIES	DIED
Joseph B.	Foster	1838	São Jorge, Azores	1873-1895	1903
Joseph F.	Francis	1847	Pico, Azores	1874-1884	1936
Anthony J.	Frates **	1844	Azores	1881-1903	1903
Edward M.	Frazier	1852	New Bedford, Mass.	1880-1885	
John P.	Freitas	ca. 1860		1908-11	
Theophilus M.	Freitas	1878	São Nicolau, Cape Verde	1908-1917	1924
Frank J.	Garcia	1863	Faial, Azores	1906-1907	
Joseph	Gaspar	1867	Cape Verde	1902-1903	1931
August P.	Gomes	ca.1884	Cape Verde	1919-1922	
Manuel F.	Gomes	1864	Faial, Azores	1897-1910	1910
John T.	Gonsalves	1858	Brava, Cape Verde	1890-1922	1928
Henry J.	Gonzales	ca. 1883	Cape Verde	1926-1927	
Francis (Frank)	Joseph	ca. 1809	Pico, Azores	1851-1852	
William F.	Joseph	1859	Fairhaven, Mass.	1901-1922	1930
Frederick	Joseph (Jose)	1817	Faial, Azores	1843-1868	1885
Jacintho S.	Leal	ca. 1857	Faial, Azores	1887-1889	
Frank M.	Lopes	1884	Brava, Cape Verde	1917-1919	
Louis M.	Lopes	1877	Brava, Cape Verde	1912-1921	
Peter	Lopez *	1869	Cape Verde	1915	
John E.	Luce	1822	Faial, Azores	1877-1879	1909
Joseph (Jose)	Luis (Lewis) **	1875	Faial, Azores	1907-1918	1918
Jose C. S.	Maciel			1859-1861	
Antone J.	Mandly	1863	Faial, Azores	1883-1925	1929
Henry	Mandly Jr.	1879	Provincetown. Mass.	1911-1921	1953
Henry	Mandly Sr.	1848	Graciosa, Azores	1876-1909	1944
Anthony	Marks		Azores	1844-1848	
Antone E.	Mello	1850	Azores	1888-1890	
Frank C.	Morris	1857	Faial, Azores	1887-1907	1931
John	Murray	1823	Graciosa, Azores	1867-1870	1899

FIRST NAME	SURNAME	BORN	LOCATION	CAPTAINCIES	DIED
Severino D.	Pierce ***?	ca. 1817	Brava, Cape Verde	1858-1871	1871 ?
John P.	Praro Jr.	1863	Pico, Azores	1894-1906	1936
John P.	Praro Sr. **	1831	Pico, Azores	1869-1876	1876
Joseph	Roderick *	ca. 1859		1909	
John	Rogers	1837	Faial, Azores	1884-1889	1907
Antone	Rose	1841	Pico, Azores	1879-1899	1923
Valentine	Roza	ca. 1873	Brava, Cape Verde	1907-1914	
Jose Antonio	Santos	ca. 1852	Flores, Azores	1885-1890	
Manuel F.	Santos	1869	Corvo, Azores	1900-1917	1919
Antonio Jose	Senna	1859	Brava, Cape Verde	1891-1894	
Ayres J.	Senna	1857	Brava, Cape Verde	1897-1905	
Joseph H.	Senna	1881	Brava, Cape Verde	1906-1922	
John D.	Silva	1828	Faial, Azores	1878-1881	1893
Joseph D.	Silva	1823	Faial, Azores	1854-1887	1901
John E.	Simmons	1816	Faial, Azores	1855-1862	1908
William H.	Swazey*	1811	Nantucket, Mass.	1842-1844	1851
Antonio	Sylvia			1879-1884	
Francis J.	Sylvia	1810	Pico, Azores	1844-1860	
Antone	Sylvia *	1877	Brava, Cape Verde	1910-1912	
Frank Manuel	Teixeira			1894-1894	
Joseph	Thomas	ca. 1805	Terceira, Azores	1838-1847	
Joseph	Thompson	1830	Faial, Azores	1874-1904	1906
Manuel	Vera			1879-1882	
Antone	Vieira			1880-1886	
Joseph A.	Vieira	1891	New Bedford, Mass.	1918-1919	1951
Nicholas R.	Vieira	1856	Flores, Azores	1901-1911	1913

* Replacement master only

** Died at sea

*** Killed by a whale

APPENDIX 3.

American Whaling Voyages under Portuguese and Portuguese American Command

Sources:

Hegarty, *Returns of Whaling Vessels*

Lund, *Whaling Masters*

Starbuck, *History*

Judith N. Lund, et al., "American Offshore Whaling Voyages: A Database," National Maritime Digital Library, http://www.nmdl.org/projects/whindex.cfm.

VESSEL NAME	PT	RIG	TONS	MASTER	BND	DEP	MO	ARR	MO	O	SBBLS	WBBLS	WBONE
Timoleon	nb	ship	346	Joseph Folger *	BR	1824	7	1825	6		72	1862	8888
George	ly	schr		Antonio DaCosta		1835							
Liverpool	nb	ship	306	Joseph Thomas	AO	1838	6	1840	4		224	1916	
Liverpool	nb	ship	306	Joseph Thomas	IO	1840	6	1842	5		263	2265	
John Adams	na	ship	296	William H. Swazey *	PO	1841	8	1845	6		540		
Peri	nb	bark	191	Frederick Joseph (Jose)	IO	1843	8	1845	12		630		
Fame	nl	bark	258	Anthony Marks *	PO	1844	6	1847		S			
Governor Troup	nb	ship	430	Francis J. Sylvia *	NW	1844	7	1847	2		120	3400	14000
Pocahontas	hh	ship	341	Joseph Dias Jr.	PO	1850	7	1853	5		320	1720	17998
Rienzi	pt	schr	109	Francis (Frank) Joseph	AO	1851	4	1851	10		115		
H. W. Williams	pt	schr	108	Francis Joseph	AO	1852	2	1854	1		81	7	
Rienzi	pt	schr	109	Joseph(M.) Caton	AO	1852	4	1852	10		170		
Topaz	nl	brig	138	Nicholas C. Anthony	AO	1852	6	1853	4	L	95		
Rienzi	pt	schr	109	Joseph (M.) Caton	AO	1853	4	1853	10		15		
St. George	nb	ship	408	Joseph Dias Jr.	NP	1853	9	1857	5		78	2158	19350
Milwood	nb	bark	254	Joseph D. Silva	IO	1854	7	1857	7		120	969	19200
Antarctic	pt	schr	136	Manuel Costa	AO	1854		1854	7		135	10	

VESSEL NAME	PT	RIG	TONS	MASTER	BND	DEP	MO	ARR	MO	O	SBBLS	WBBLS	WBONE
Rienzi	pt	schr	108	Joseph (M.) Caton	AO	1855	4	1855	11		170		
Cleone	nb	bark	373	John E. Simmons	NP	1855	8	1858	4		367	2563	15384
St. George	NB	SHIP	405	Joseph Thomas	NW	1843	7	1845	1		35	3188	29932
Rienzi	pt	schr	108	Joseph (M.) Caton	NA	1856	5	1857	4		146	52	
Antarctic	pt	schr	136	Manuel Costa	AO	1856	9	1857	9		210		
Emporium	pt	schr	80	Francis Caton	NA	1857		1857	8			80	
George	nb	bark	280	Joseph D. Silva	NP	1857	11	1861	11		433	659	5530
Emporium	pt	schr	80	Francis Caton	NA	1858	5	1858	10			128	
Magnolia	nb	ship	396	Severino D. Pierce	NP	1858	7	1862		C	330	2400	31675
Cleone	nb	bark	373	John E. Simmons	NP	1858	10	1862	8			2286	19663
Comora	nb	ship	699	Jose C. S. Maciel	PO	1859	7	1861	3				
Atlantic	nb	ship	440	Francis J. Sylvia	AO	1859	9	1860					
Milo	nb	ship	401	Thomas E. Fordham	NP	1859	11	1863	5		263	3346	34500
John Adams	pt	schr	99	Joseph M. Caton	AO	1861	2	1861		B			
George	nb	ship	280	Joseph D. Silva	PO	1862	6	1864	12		615	5	
Europa	nb	ship	380	Severino D. Pierce *	NP	1862	7	1867	9		268	1829	25200
Florida	nb	ship	330	Thomas E. Fordham	NP	1864	7	1868	9		241	1688	48680

VESSEL NAME	PT	RIG	TONS	MASTER	BND	DEP	MO	ARR	MO	O	SBBLS	WBBLS	WBONE
Java	nb	bark	295	Manuel Enos	IP	1864	8	1869	4		112	1742	18641
Sarah E. Lewis	bo	schr	140	Manuel Caton ***	AO	1865	10	1867	6		70		
Eleanor B.Conwell	pt	schr	132	Joseph M. Caton	AO	1867	1	1868	8		150	10	
Life Boat	ne	schr	88	Joseph H. Caton	AO	1867	3	1867	8		55	31	
Abbie Bradford	na	schr	114	John Murray	AO	1867	4	1868	9		404	5	
Star Castle	fa	brig	116	Henry Clay	AO	1867	5	1869		L	1020	150	
Edward Everett	nb	bark	187	Joseph D. Silva	IO	1867	6	1869			1699	20	
Life Boat	ne	schr	88	Joseph H. Caton	AO	1867	12	1870	8		127	166	
Abbie Bradford	na	schr	114	John Murray	AO	1869	5	1869	10		500	10	
Tamerlane	nb	bark	372	Thomas E. Fordham	PO	1869	7	1873	6		754	568	3500
Crowninshield	fa	bark	257	John P. Praro Sr.	PO	1869	10	1873	8	C	1319	100	
Cicero	nb	bark	226	Henry Clay	AO	1870	5	1873	10		975	375	1300
George J. Jones	fa	schr	196	Jasper M. Ears	AO	1870	6	1871	8		139	135	
Thriver	bo	schr	69	Severino D. Pierce ****?	AO	1870		1873	2	L	83	219	
George J. Jones	fa	brig	128	Joseph D. Silva	AO	1872	5	1873	4	C	278		
N. J. Knight	pt	schr	70	Joseph B. Foster	AO	1873	2	1873	9		32	210	
Edward Everett	nb	bark	187	Joseph D. Silva	AO	1873	7	1875	8		890	23	

VESSEL NAME	PT	RIG	TONS	MASTER	BND	DEP	MO	ARR	MO	O	SBBLS	WBBLS	WBONE
N. J. Knight	pt	schr	70	Joseph B. Foster	AO	1874	2	1874	9		140	210	
Rosa Baker	bo	brig	108	Joseph Thompson	AO	1874	5	1875	5		270	15	
Lydia	nb	bark	329	John P. Praro Sr. *	PO	1874	6	1877	10		1380		
E. B. Phillips	bo	bark	144	Joseph F. Francis	AO	1874	7	1876	7		450		
Peru	nb	bark	259	Jasper M. Ears	AO	1875	4	1878	8		910		
Rosa Baker	bo	brig	108	Joseph Thompson	AO	1875	6	1876	9		450		
Golden City	nb	schr	89	Henry Clay	AO	1875	12	1876	9		440	40	
N. J. Knight	pt	schr	70	Joseph B. Foster	AO	1875		1875					
N. J. Knight	pt	schr	70	Joseph B. Foster	AO	1876	1	1876	8		160	125	
Sarah	nb	bark	128	Joseph D. Silva	AO	1876	6	1878	9		720		
E. B. Phillips	nb	bark	144	Joseph F. Francis	AO	1876	11	1880	7		1365		
Pedro Varela	nb	schr	89	Anthony P. Benton	AO	1876	11	1878	9		450		
Quickstep	pt	schr	94	Henry Mandly Sr.	AO	1876	11	1877	8		450		
Seine	nb	bark	234	Henry Clay	AO	1876	11	1877			300		
N. J. Knight	pt	schr	70	Joseph B. Foster	AO	1877	2	1877	9		55	70	
Minerva II	nb	bark	337	Joseph Thompson	AO	1877	4	1880	9		2410	45	400
Golden City	nb	schr	84	Henry Clay **	AO	1877	6	1878	9		365	40	

VESSEL NAME	PT	RIG	TONS	MASTER	BND	DEP	MO	ARR	MO	O	SBBLS	WBBLS	WBONE
Quickstep	pt	schr	94	Henry Mandly Sr.	AO	1877	11	1878	9		240	130	
Fannie Byrnes	nb	schr	66	John E. Luce	AO	1877	12	1879	8		208	25	
N. J. Knight	pt	schr	70	Joseph B. Foster	AO	1878	2	1878	9		70	220	
Charles W. Morse	nb	schr	112	John D. Silva	AO	1878	5	1879	8		290	60	750
Varnum Hill	nb	brig	125	Joseph D. Silva	AO	1878	10	1880	7		380		
Peru	nb	bark	221	Anthony P. Benton	AO	1878	11	1881	9		1410		
N. J. Knight	pt	schr	70	Joseph B. Foster	AO	1879	2	1879	8		40	200	
Sarah E. Lewis	bo	schr	87	Manuel Vera *	AO	1879	3	1879	9		50		
Ellen Rodman	nb	schr	73	Jasper M. Ears	AO	1879	5	1880	8		205		
Mary E. Simmons	nb	schr	105	Henry Mandly Sr.	AO	1879	5	1880	10		300	9	
Petrel	nb	schr	61	Antone Rose	AO	1879	5	1879	?				
Eleanor B. Conwell	nb	schr	91	Manuel E. Costa	AO	1879	10	1880	9		380	20	200
Fannie Byrnes	nb	schr	66	Antonio Sylvia	AO	1879	10	1881	11		280	25	
Mattapoisett	nb	bark	110	John D. Silva	AO	1879	10	1881	10		180	5	
Surprise	nb	schr	54	Antone Rose	AO	1879	11	1880	8		65	85	1000
N. J. Knight	pt	schr	70	Joseph B. Foster	AO	1880	2	1880	9			220	
Golden City	nb	schr	84	Antone Rose *	AO	1880	6	1881	8		300	6	

VESSEL NAME	PT	RIG	TONS	MASTER	BND	DEP	MO	ARR	MO	O	SBBLS	WBBLS	WBONE
Varnum H. Hill	nb	brig	154	Joseph D. Silva	AO	1880	9	1882	9		605	80	
Lottie E. Cook	nb	schr	82	Antone Vieira	AO	1880	10	1882	4		85	125	
E. H. Hatfield	ed	schr	89	Jasper M. Ears * **	AO	1880	11	1882	8		260	10	
Eleanor B. Conwell	nb	schr	91	Manuel E. Costa	AO	1880	11	1882	10		380	190	
Lydia	nb	bark	330	Edward M. Frazier	AO	1880	11	1884	8	C	890	335	2000
Mary E. Simmons	nb	schr	105	Henry Mandly Sr.	AO	1880	11	1881	9		370		
Minerva II	nb	bark	337	Joseph Thompson	AO	1881	2			L			
N. J. Knight	pt	schr	70	Joseph B. Foster	AO	1881	2	1881	7		120	90	
John and Winthrop	nb	bark	338	Joseph Cruze *	PO	1881	4	1885	7		2750		
E. B. Phillips	nb	bark	155	Joseph F. Francis	AO	1881	5	1884	8		620		
Charles W. Morse	nb	schr	112	Antone Rose	AO	1881	10	1883	8		380	90	
Golden City	nb	schr	84	Anthony J. Frates	AO	1881	10	1882	9		100	23	
Mary E. Simmons	nb	schr	105	Henry Mandly Sr.	AO	1881	10	1883	9		770	95	
Ohio	nb	bark	206	Anthony P. Benton	AO	1881	11	1885	4		900	8	
Union	nb	schr	66	Joseph B. Foster	AO	1882	1	1883	9		130	170	
Fannie Byrnes	nb	schr	66	Antonio Sylvia	AO	1882	4	1884		C			
Lottie E. Cook	nb	schr	82	Antone Vieira	AO	1882	5	1883	4		270	90	

VESSEL NAME	PT	RIG	TONS	MASTER	BND	DEP	MO	ARR	MO	O	SBBLS	WBBLS	WBONE
Eleanor B. Conwell	nb	schr	91	Manuel E. Costa **	AO	1883	4	1884	8		280		
Lottie E. Cook	nb	schr	82	Antone Vieira	AO	1883	5	1884	4			100	935
Varnum Hill	nb	brig	154	Joseph D. Silva	AO	1883	5	1884	10		130	10	
Charles W. Morse	nb	schr	112	Antone Rose	AO	1883	10	1886	9		410	195	
Franklin	nb	schr	77	Antone J. Mandly *	AO	1883	10	1885	8		785	210	
Union	nb	schr	66	Joseph B. Foster	AO	1883	10	1883	12	L			
Mary E. Simmons	nb	schr	105	Henry Mandly Sr.	AO	1883	11	1884	8		635		
Sea Queen	nb	bark	195	Joseph Thompson	AO	1884	4	1888		L	1370	335	
Lottie E. Cook	nb	schr	82	Thomas E. Fordham	AO	1884	5	1885	8		80		
Adelia Chase	nb	schr	85	Michael A. Ferreira	AO	1884	6	1886	6		550	10	
Stafford	nb	bark	156	Antone Vieira	AO	1884	6	1886	6		800	200	
E. B. Phillips	nb	bark	155	Edward M. Frazier	AO	1884	10	1885		C			
Eleanor B. Conwell	nb	schr	91	Manuel E. Costa	AO	1884	10	1885	4				
Mary and Susan	sf	bark	327	John Rogers	NP	1884	12	1885	10			775	17950
Eleanor B. Conwell	nb	schr	91	Joseph B. Foster	AO	1885	4	1885	9		150		
Thomas Hunt	st	schr	67	Jose Antonio Santos	AO	1885	5	1890		L	850	450	
Bloomer	pt	schr	74	Antone Rose	AO	1885	6	1885	8		50		

VESSEL NAME	PT	RIG	TONS	MASTER	BND	DEP	MO	ARR	MO	O	SBBLS	WBBLS	WBONE
Mary E. Simmons	nb	schr	105	Henry Mandly Sr.	AO	1885	6	1886	8		960	40	
Wave	nb	bark	150	Anthony P. Benton	PO	1885	6	1886	9			150	
Eunice H. Adams	ed	brig	107	Manuel E. Costa	AO	1885	7	1887	9		375		
Greyhound	nb	bark	178	Joseph T. Enos	AO	1885	7	1887	8		1000	390	
Eleanor B. Conwell	nb	schr	91	Joseph B. Foster	AO	1885	10	1887	8		590	25	
Franklin	nb	schr	77	Antone J. Mandly	AO	1885	10	1887	7		310	230	
Hattie E. Smith	ed	schr	107	Joseph D. Silva *	AO	1885	10	1887	8		250		
Bloomer	pt	schr	74	Antone Rose	AO	1886	4	1886	8		50		
Adelia Chase	nb	schr	85	Michael A. Ferreira	AO	1886	10	1889	5		680		
Mary E. Simmons	nb	schr	105	Henry Mandly Sr.	AO	1887	4	1888	8		777		
Charles W. Morse	nb	schr	112	Frank C. Morris	AO	1887	5	1890	7		835	10	
Pedro Varela	nb	schr	89	Jacintho S. Leal	AO	1887	5	1889	8		260		
Eleanor B. Conwell	nb	schr	91	Joseph B. Foster	AO	1887	10	1888	10		130	5	
Eunice H. Adams	ed	brig	107	Thomas E. Fordham *	AO	1887	10	1890	3		375		
Greyhound	nb	bark	178	Joseph T. Enos	AO	1887	10	1892	7		1860	550	
President II	nb	bark	123	Joseph T. Edwards	AO	1887	10	1890	6		1260		
Tamerlane	nb	bark	372	Anthony P. Benton	NP	1888	1	1888	10		70	65	1900

VESSEL NAME	PT	RIG	TONS	MASTER	BND	DEP	MO	ARR	MO	O	SBBLS	WBBLS	WBONE
Franklin	nb	schr	77	Antone Rose	AO	1888	4	1889	7		530		
Golden City	nb	schr	84	Antone J. Mandly	AO	1888	4	1889	7		570		
George and Mary	nb	bark	105	Manuel E. Costa	AO	1888	6	1891	8		1150		
Alcyone	pt	schr	92	Antone E. Mello	AO	1888	8	1889	6		135		
Mary E. Simmons	nb	schr	105	Henry Mandly Sr.	AO	1889	3	1892	6		1580		
Tamerlane	sf	bark	372	John Rogers	NP	1889	3	1889	11			65	1900
Francis Allyn	nb	schr	106	Joseph B. Foster	AO	1889	6	1890	8		240		
Alcyone	pt	schr	92	Antone E. Mello	AO	1889	8	1890	8		200		
Franklin	nb	schr	77	Antone Rose	AO	1889	10	1891	7		760		
Golden City	nb	schr	84	Antone J. Mandly	AO	1889	10	1891	7		700	10	
Adelia Chase	nb	schr	85	Michael A. Ferreira	AO	1890	6	1893	6		1350		
President II	nb	bark	123	Joseph T. Edwards	AO	1890	7	1893	8		415		
Rising Sun	pt	schr	69	John T. Gonsalves	AO	1890	12	1891	8		350		
Charles W. Morse	nb	schr	112	Frank C. Morris	AO	1891	4	1895	5		1110		
Clara L. Sparks	nb	schr	101	Joseph P. Benton	AO	1891	7	1895	8		900	200	
Pedro Varela	nb	schr	89	Antonio Jose Senna	AO	1891	8	1894	8		1420		
David A. Small	pt	brig	119	Antone Rose	AO	1891	10	1893	7		450		

VESSEL NAME	PT	RIG	TONS	MASTER	BND	DEP	MO	ARR	MO	O	SBBLS	WBBLS	WBONE
Golden City	nb	schr	84	Antone J. Mandly	AO	1891	11	1894	8		1260	170	
Rising Sun	pt	schr	69	John T. Gonsalves	AO	1892	4	1892	9		350		
George and Mary	nb	bark	105	Manuel E. Costa	AO	1892	5	1894	8		1040		
Star King	nb	schr	64	Joseph B. Foster	AO	1892	6	1894	7		880		
Greyhound	nb	bark	178	Joseph T. Enos	AO	1892	10	1897	11		2540	260	
Rising Sun	pt	schr	69	John T. Gonsalves	AO	1892	10	1893	8		280		
Eunice H. Adams	ed	brig	107	John T. Gonsalves *	AO	1893	4	1894	9		210		
Mary E. Simmons	nb	schr	105	Henry Mandly Sr.	AO	1893	4	1894	8		980		
Pearl Nelson	nb	schr	123	Joseph Thompson	AO	1893	5	1896	9		1630		
David A. Small	pt	brig	119	Antone Rose	AO	1893	10	1894	5		130		
President II	nb	bark	123	John P. Praro Jr.	AO	1894	4	1896	8		1470		
Rising Sun	pt	schr	69	Frank M. Teixeira	AO	1894	5	1894	6				
Frances A. Barstow	sf	brig	128	Joseph T. Edwards	NP	1894	6	1897	5		1100		
Baltic	pt	schr	84	John T. Gonsalves	AO	1894	9	1895	9		330		
Charles H. Hodgdon	nb	schr	112	Antone J. Mandly	AO	1894	10	1896	7		1015		
Joseph A. Manta	pt	schr	70	Antone Rose	AO	1894	10	1895	9		430		
Golden City	nb	schr	84	Joseph B. Foster	AO	1895	7	1895	11	L	120		

VESSEL NAME	PT	RIG	TONS	MASTER	BND	DEP	MO	ARR	MO	O	SBBLS	WBBLS	WBONE
Joseph A. Manta	pt	schr	70	Antone Rose	AO	1895	10	1896	8		330		
Rosa Baker	bo	brig	108	Frank C. Morris	AO	1895	10	1899	5		1350	15	
Mary E. Simmons	nb	schr	105	Henry Mandly Sr.	AO	1896	5	1897	9		375		
Joseph A. Manta	pt	schr	70	Antone Rose	AO	1896	10	1897	8		200		
Pearl Nelson	nb	schr	123	Antone J. Mandly	AO	1896	10	1899	9		1165	105	
President II	nb	bark	123	Anthony P. Benton	AO	1896	11	1899	5		600	15	
Baltic	pt	schr	84	John T. Gonsalves	AO	1897	5	1897	8		170		
Joseph A. Manta	pt	schr	70	Antone Rose	AO	1897	10	1898	7		260		
Adelia Chase	nb	schr	85	Joseph Thompson	AO	1897	11	1898	6		170		
Golden City	nb	schr	84	Ayres J. Senna	AO	1897	11	1899	9		885	20	
Mary E. Simmons	nb	schr	105	Henry Mandly Sr.	AO	1897	11	1899	9		975		
Eleanor B. Conwell	nb	schr	91	Manuel F. Gomes	AO	1898	5	1898	8		270		
Greyhound	nb	bark	177	Joseph T. Edwards **	AO	1898	6	1902	7		3070	30	
Pedro Varela	nb	schr	89	John P. Praro Jr.	AO	1898	6	1902	7		3325	30	
Adelia Chase	nb	schr	85	Lebanio D. Borges	AO	1898	7	1898	10		150		
Charles H. Hodgdon	nb	schr	112	Manuel E. Costa	AO	1898	10	1899	8		240		
Eleanor B. Conwell	nb	schr	91	Manuel F. Gomes	AO	1898	10	1899	9		250		

VESSEL NAME	PT	RIG	TONS	MASTER	BND	DEP	MO	ARR	MO	O	SBBLS	WBBLS	WBONE
Adelia Chase	nb	schr	85	Antone Rose	AO	1898	11	1899	7		180		
Joseph A. Manta	pt	schr	70	Anthony J. Frates	AO	1898	11	1900	8		590		
A. R. Tucker	nb	bark	145	Joseph P. Benton * ****	AO	1899	6	1901	9		1950		
A. R. Tucker	nb	bark	145	Joseph Avilla *	AO	1899	6	1901	9		1950		
Adelia Chase	nb	schr	85	Joseph Thompson	AO	1899	8	1901	9		290		
President II	nb	bark	123	Joseph T. Enos	AO	1899	8	1904	5	L			
Eleanor B. Conwell	nb	schr	91	Manuel F. Gomes	AO	1899	10	1901	7		920		
Golden City	nb	schr	84	Frank C. Morris	AO	1900	5	1901	8		470		
Mary E. Simmons	nb	schr	105	Henry Mandly Sr.	AO	1900	5	1902	8		1950		
Pearl Nelson	nb	schr	123	Antone J. Mandly	AO	1900	6	1902	8		1250		
Charles H. Hodgdon	nb	schr	112	Manuel F. Santos	AO	1900	10	1901	6		230		
Joseph A. Manta	pt	schr	70	Anthony J. Frates ***	AO	1900	10	1903	10	L			
Canton II	nb	bark	238	Nicholas R. Vieira	AO	1901	4	1902	9		2100		
Francis Allyn	nb	schr	106	Manuel F. Santos	HB	1901	7	1902	7	B			
A. R. Tucker	nb	bark	145	Manuel E. Costa	AO	1901	10	1903	8		1500		
Adelia Chase	nb	schr	85	Ayres J. Senna	AO	1901	10	1902	9		305		
Bertha	nb	bark	177	Manuel F. Gomes	AO	1901	10	1904	8		2670		

VESSEL NAME	PT	RIG	TONS	MASTER	BND	DEP	MO	ARR	MO	O	SBBLS	WBBLS	WBONE
Golden City	nb	schr	84	John T. Gonsalves	AO	1901	10	1903	8		860		
California	sf	bark	366	William F. Joseph	JO	1901	11	1902	10		1500		
Eleanor B. Conwell	nb	schr	91	Joseph Thompson	AO	1902	5	1904	6		980		
Pilgrim	nb	schr	72	Joseph Gaspar	AO	1902	6	1903	6		30	45	
Adelia Chase	nb	schr	85	Ayres J. Senna	AO	1902	11	1904	8		1240		
California	sf	bark	366	William F. Joseph	JO	1902	11	1903	11		100	120	
Canton II	nb	bark	238	Nicholas R. Vieira	AO	1903	4	1904	8		2350		
Greyhound	nb	bark	177	Joseph T. Edwards **	AO	1903	5	1907	4		4625		
Greyhound	nb	bark	177	Antone T. Edwards *	AO	1903	5	1907	4		4625		
Mary E. Simmons	nb	schr	105	Henry Mandly Sr.	AO	1903	5	1904	7		650		
Pedro Varela	nb	schr	89	John P. Praro Jr.	AO	1903	5	1906	7		2790		
Pedro Varela	nb	schr	89	Manuel F. Santos *	AO	1903	5	1906	7		2790		
Pilgrim	nb	schr	72	Joseph Gaspar	AO	1903	7	1903	9	L			
Golden City	nb	schr	84	Antone J. Mandly	AO	1903	10	1904	8		380		
California	sf	bark	366	William F. Joseph *	JO	1903	12	1904	10		800		
Eleanor B. Conwell	nb	schr	91	John T. Gonsalves	AO	1904	8	1905	8		550		
Golden City	nb	schr	84	Manuel F. Santos	AO	1904	9	1905	8	C	500		

VESSEL NAME	PT	RIG	TONS	MASTER	BND	DEP	MO	ARR	MO	O	SBBLS	WBBLS	WBONE
Adelia Chase	nb	schr	85	Ayres J. Senna	AO	1904	10	1905	9		290		
California	sf	bark	366	William F. Joseph	JO	1905	2	1905	11			100	2400
Bertha	nb	bark	177	Manuel F. Gomes	AO	1905	4	1907	7		3100		
Canton II	nb	bark	238	Nicholas Vieira	AO	1905	4	1906	8		2750		
Ellen A. Swift	pt	schr	132	Antone J. Mandly	AO	1905	4	1905	8		450		
William Baylies	sf	Sbark	380	William F. Joseph *	NP	1906	3	1906	11			200	3000
Eleanor B. Conwell	nb	schr	91	John T. Gonsalves	AO	1906	4	1906	6		150		
Ellen A. Swift	pt	schr	132	Antone J. Mandly	AO	1906	4	1906	8		450		
John R. Manta	pt	schr	147	Henry Mandly Sr.	AO	1906	4	1906	7		150		
Adelia Chase	nb	schr	85	Joseph H. Senna	AO	1906	5	1906	8		310		
Bertha D. Nickerson	nb	schr	143	Manuel E. Costa	AO	1906	6	1908	7		2130		
Eleanor B. Conwell	nb	schr	91	Frank C. Morris	AO	1906	6	1907	1		125	75	
John R. Manta	pt	schr	147	Frank J. Garcia	AO	1906	10	1907	6		160		
Carleton Bell	nb	schr	139	Joseph H. Senna	AO	1906	11	1908	8		1210		
Canton II	nb	bark	238	Valentine Roza	AO	1907	4	1909	8		2100	300	
Ellen A. Swift	pt	schr	132	Antone J. Mandly	AO	1907	4	1907	8		650		
Pedro Varela	nb	schr	89	Antonio C. Corvello *	AO	1907	5	1909	7		1850		

VESSEL NAME	PT	RIG	TONS	MASTER	BND	DEP	MO	ARR	MO	O	SBBLS	WBBLS	WBONE
Pedro Varela	nb	schr	89	Henry Mandly Sr.	AO	1907	5	1909	7		1850		
Eleanor B. Conwell	nb	schr	91	John T. Gonsalves	AO	1907	6	1907	8		230		
John R. Manta	pt	schr	147	Joseph T. Edwards	AO	1907	7	1908	9		550		
John R. Manta	pt	schr	147	Joseph Luis (Lewis) *	AO	1907	7	1908	9		550		
Bertha	nb	schr	177	John T. Gonsalves	AO	1907	9	1909	8		1970		
Cameo	nb	schr	243	Manuel F. Gomes **	AO	1908	4	1911	8		3720		
Cameo	nb	schr	243	John P. Freitas *	AO	1908	4	1911	8		3720		
Ellen A. Swift	pt	schr	132	Antone J. Mandly	AO	1908	4	1908	8		350		
Sullivan	no	brig	344	Theophilus M. Freitas *	AO	1908	6	1911	10		4600		
T. Towner	nb	schr	197	Manuel E. Costa	AO	1908	8	1911	8		3280		
Bertha D. Nickerson	nb	schr	143	Manuel F. Santos	AO	1908	9	1909	7		770		
Charles W. Morgan	nb	bark	314	Joseph Roderick *	AO	1908	9	1910	9		1200	120	
John R. Manta	pt	schr	147	Joseph Luis (Lewis)	AO	1908	10	1911	9		2130		
John and Winthrop	sf	bark	338	William F. Joseph	NP	1908	12	1909	11		700	75	
Ellen A. Swift	pt	schr	132	Antone J. Mandly	AO	1909	4	1909	8		350		
Carleton Bell	nb	schr	139	Joseph H. Senna	AO	1909	5	1911	8		1980		
Bertha D. Nickerson	nb	schr	143	Manuel F. Santos *	AO	1909	10	1911	8		1740		

VESSEL NAME	PT	RIG	TONS	MASTER	BND	DEP	MO	ARR	MO	O	SBBLS	WBBLS	WBONE
Canton II	nb	bark	238	Valentine Roza	AO	1909	10	1910	11	L			
John and Winthrop	sf	bark	338	William F. Joseph	JS	1910	3	1910	10		220	200	1800
Mystic	nb	schr	259	William F. Joseph *	AO	1910	3	1914	8		3095		
Mystic	nb	schr	259	Antone T. Edwards **	AO	1910	3	1914	8		3095		
Bertha	nb	bark	177	Nicholas R. Vieira	AO	1910	4	1911	8		2000		
Ellen A. Swift	pt	schr	132	Antone J. Mandly	AO	1910	4	1910	7		600		
Pedro Varela	nb	schr	89	Antonio C. Corvello	AO	1910	4	1912	8		1400		
Morning Star	nb	bark	238	Valentine Roza	AO	1910	5	1912	8		3050		
Greyhound	nb	bark	177	Antone Sylvia *	AO	1910	9	1912	8		1140		
John and Winthrop	sf	bark	338	William F. Joseph	JS	1911	2	1911	10		500		
Ellen A. Swift	pt	schr	132	Antone J. Mandly	AO	1911	4	1911	8		450		
Margarett	nb	schr	138	Henry Mandly Jr.	AO	1911	5	1912	9		1280		
Bertha D. Nickerson	nb	schr	142	Antone J. Mandly	AO	1911	11	1912	9		475		
T. Towner	nb	schr	197	John T. Gonsalves **	AO	1911	11	1914	8		2675		
Cameo	nb	schr	243	John T. Edwards	AO	1912	3	1915	8		5300		
John and Winthrop	sf	bark	338	William F. Joseph	NP	1912	3	1912	10		32		
Carleton Bell	nb	schr	139	Joseph H. Senna **	AO	1912	5	1914	8		1695		

VESSEL NAME	PT	RIG	TONS	MASTER	BND	DEP	MO	ARR	MO	O	SBBLS	WBBLS	WBONE
John R. Manta	pt	schr	147	Manuel F. Santos	AO	1912	5	1913	8		750		
A. E. Whyland	nb	schr	130	Louis M. Lopes	AO	1912	7	1914	8		2170		
Arthur V. S. Woodruff	nb	schr	193	Joseph Luis (Lewis) * **	AO	1912	10	1916	9		5090		
Morning Star	nb	bark	238	Valentine Roza	AO	1912	10	1914	9	S	2650		
Bertha D. Nickerson	nb	schr	143	Antone J. Mandly **	AO	1913	2	1915	8		1350		
Bertha D. Nickerson	nb	schr	143	William F. Joseph *	AO	1913	2	1915	8		1350		
Margarett	nb	schr	138	Henry Mandly Jr.	AO	1913	2	1914	8		1250		
Wanderer	nb	bark	303	Antone T. Edwards	AO	1913	4	1916	9		5300		
Greyhound	nb	bark	177	Antonio C. Corvello	AO	1913	5	1916	9		2580		
John R. Manta	nb	schr	147	Manuel F. Santos	AO	1914	4	1914	8		260		
William A. Graber	nb	schr	137	John T. Gonsalves	AO	1914	4	1914	8		550		
Valkyria	po	schr	139	Joseph Luis (Lewis)	AO	1914	7	1916	9		740		
Rosie H.	sf	schr	73	Peter Lopez *	WA	1915		1915					
William A. Graber	nb	schr	137	John T. Gonsalves	AO	1915	2	1915	7		250		
A. E. Whyland	nb	schr	86	Louis M. Lopes	AO	1915	3	1915	9		500		
Carleton Bell	nb	schr	139	Joseph H. Senna	AO	1915	3	1916	9		1650		
John R. Manta	nb	schr	147	Antone J. Mandly	AO	1915	4	1915	7		450		

VESSEL NAME	PT	RIG	TONS	MASTER	BND	DEP	MO	ARR	MO	O	SBBLS	WBBLS	WBONE
Margarett	nb	schr	138	Henry Mandly Jr.	AO	1915	4	1916	9		1860		
William A. Graber	nb	schr	137	John T. Gonsalves	AO	1915	9	1916	4				
John R. Manta	nb	schr	147	Antone J. Mandly	AO	1916	4	1916	8		600		
Bertha	nb	schr	177	William F. Joseph	AO	1916	5	1916	9		700		
William A. Graber	nb	schr	137	John T. Gonsalves	AO	1916	5	1916	7		-		
Viola	po	brig	139	Joseph Luis (Lewis)	AO	1916	9	1917	8		300		
A. M. Nicholson	nb	schr	136	John T. Gonsalves	AO	1917	2	1917	6				
John R. Manta	nb	schr	147	Antone J. Mandly	AO	1917	2	1917	7				
Margarett	nb	schr	138	Louis M. Lopes	AO	1917	3	1917	4				
Valkyria	po	schr	139	William F. Joseph	AO	1917	3	1917	9		500		
John R. Manta	nb	schr	147	Antone J. Mandly	AO	1917	4	1917	8		550		
Margarett	nb	schr	138	Henry Mandly Jr.	AO	1917	4	1917	9		650		
Pedro Varela	nb	schr	85	Manuel F. Santos	AO	1917	5	1917	8				
William A. Graber	nb	schr	137	Henry Mandly Jr.	AO	1917	5	1917	9		350		
Greyhound	nb	schr	177	Antonio C. Corvello	AO	1917	8	1918	6		750		
Wanderer	nb	bark	303	Antone T. Edwards	AO	1917	8	1918	6		1150		
Viola	po	brig	139	Joseph Luis (Lewis)	AO	1917	9	1918	7		1150		

VESSEL NAME	PT	RIG	TONS	MASTER	BND	DEP	MO	ARR	MO	O	SBBLS	WBBLS	WBONE
A. M. Nicholson	nb	schr	136	John T. Gonsalves	AO	1917	10	1918	6		625		
Pedro Varela	nb	schr	85	Theophilus M. Freitas	AO	1917	10	1918	2		320		
Arthur V. S. Woodruff	nb	schr	193	John T. Edwards	AO	1917	11	1919	9		850		
Valkyria	po	schr	139	William F. Joseph	AO	1917	11	1918	8		650		
Claudia	nb	schr	113	Joseph A. Vieira	AO	1918	4	1918	8		320		
Margarett	nb	schr	138	Louis M. Lopes	AO	1918	4	1918	7		400		
John R. Manta	nb	schr	147	Antone J. Mandly	AO	1918	5	1918	7		450		
Charles W. Morgan	nb	bark	314	Joseph F. Edwards	AO	1918	7	1919	9		1150		
Pedro Varela	nb	schr	85	Frank M. Lopes	AO	1918	7						
Wanderer	nb	bark	303	Antone T. Edwards	AO	1918	8	1919	9		1550		
Wanderer	nb	bark	303	William F. Joseph *	AO	1918	8	1919	9		1550		
Greyhound	nb	schr	179	Antonio C. Corvello	AO	1918	9	1919	8		750		
Margarett	nb	schr	138	Louis M. Lopes	AO	1918	9	1919	5		400		
Viola	po	brig	139	Joseph Luis (Lewis) ***	AO	1918	9	1918		L			
A. M. Nicholson	nb	schr	136	John T. Gonsalves *	AO	1918	10	1919	8				
Valkyria	po	schr	139	Joseph A. Vieira	AO	1918	10	1919	2		260		
Valkyria	po	schr	139	Joseph A. Vieira	AO	1919	?	1919	8				

VESSEL NAME	PT	RIG	TONS	MASTER	BND	DEP	MO	ARR	MO	O	SBBLS	WBBLS	WBONE
Cameo	nb	schr	243	Jose M. Domingos	AO	1919	2	1919	9		525		
Claudia	nb	schr	113	Joseph H. Senna	AO	1919	2	1919	9				
John R. Manta	nb	schr	147	Antone J. Mandly	AO	1919	2	1919	7		530		
Pedro Varela	nb	schr	85	Frank M. Lopes ***	AO	1919	2	1919		L			
William A. Graber	nb	schr	137	Henry Mandly Jr.	AO	1919	2	1919	9		640		
Margarett	nb	schr	138	Louis M. Lopes	AO	1919	6	1919	9		400		
A. M. Nicholson	nb	schr	136	John T. Gonsalves	AO	1919	10	1920	5		650		
Cameo	nb	schr	243	Jose M. Domingos	AO	1919	10	1920	8		670		
Charles W. Morgan	nb	bark	314	Joseph F. Edwards	AO	1919	10	1920	7		750		
Claudia	nb	schr	113	Louis M. Lopes	AO	1919	10	1920	8		601		
Wanderer	nb	bark	303	Antone T. Edwards	AO	1919	10	1920	10		1800		
Arthur V. S. Woodruff	nb	schr	193	August P. Gomes	AO	1919	11	1920	4	L			
John R. Manta	nb	schr	98	Antone J. Mandly	AO	1920	3	1920	8		410		
Margarett	nb	schr	138	John T. Edwards *	AO	1920	3	1920	8		380		
Margarett	nb	schr	138	William F. Joseph **	AO	1920	3	1920	8		380		
Valkyria	nb	schr	104	Antone T. Edwards	AO	1920	3	1920	9				
William A. Graber	nb	schr	137	John Da Lomba	AO	1920	3	1920	9		517		

VESSEL NAME	PT	RIG	TONS	MASTER	BND	DEP	MO	ARR	MO	O	SBBLS	WBBLS	WBONE
Charles W. Morgan	pt	bark	314	John T. Gonsalves	AO	1920	9	1921	5		2702		
Cameo	pt	schr	200	Louis M. Lopes	AO	1920	10	1921	9		670		
Valkyria	nb	schr	104	Henry Mandly Jr.	AO	1920	10	1921	5		557		
A. M. Nicholson	nb	schr	136	August P. Gomes	AO	1920	11	1921	9		400		
John R. Manta	nb	schr	98	Antone J. Mandly	AO	1921	3	1921	8		500		
Margarett	nb	schr	138	Joseph F. Edwards	AO	1921	3	1921	9		600		
Wanderer	nb	bark	303	Antone T. Edwards	AO	1921	3	1922	4		1560		
Greyhound	nb	bark	109	Benjamin Costa	AO	1921	8			S			
Claudia	nb	schr	118	Joseph H. Senna	AO	1922	3	1922	8		500		
John R. Manta	nb	schr	98	Antone J. Mandly	AO	1922	3	1922	8		300		
A. M. Nicholson	nb	schr	136	August P. Gomes	AO	1922	4			S			
Margarett	nb	schr	138	Joseph F. Edwards	AO	1922	4	1922	8		600		
William A. Graber	nb	schr	110	John T. Gonsalves	AO	1922	4	1922	8		650		
Fox	sf	schr	343	William F. Joseph	NP	1922	5	1922	9				
Wanderer	nb	bark	303	Antone T. Edwards	AO	1922	8	1923	9		1500		
John R. Manta	nb	schr	98	Antone J. Mandly	AO	1923	2	1923	8		450		
Margarett	nb	schr	138	Joseph F. Edwards	AO	1923	3	1924	8		300		

VESSEL NAME	PT	RIG	TONS	MASTER	BND	DEP	MO	ARR	MO	O	SBBLS	WBBLS	WBONE
Arctic	sf	schr	669	William F. Joseph *	AO	1923	5	1923	10				
Wanderer	nb	bark	303	Antone T. Edwards	AO	1924	8	1924	8	L			
John R. Manta	nb	schr	98	Antone J. Mandly	AO	1925	5	1925	8		300		
Charles Brower	sf	Dschr	569	Henry J. Gonzales	NP	1926	5	1926	9				
Charles Brower	sf	Dschr	569	Henry J. Gonzales	NP	1927	6	1927		S			
John R. Manta	nb	schr	98	Joseph F. Edwards	AO	1927	6	1927	7	S			

* Replacement master
** Left the vessel, usually because of illness
*** Died during the voyage
**** Killed by a whale

Pt (Port of Registry)
bo=Boston MA
ed=Edgartown MA
fa=Fairhaven MA
nb=New Bedford MA
ne=Newburyport MA
nl=New London CT
no=Norwich CT
po=Portland ME
pr=Providence RI
pt=Provincetown MA
sf-San Francisco CA

Rig
bark=Bark
brig=Brig
Sbark=Steam Bark
schr=Schooner
Dschr=Diesel schooner
ship=Ship

Bnd (Bound)
AO=Atlantic Ocean
HB=Hudson Bay
IO=Indian Ocean
IP=Indian & Pacific Oceans
JO=Japan & Okhotsk Seas
JS=Japan Sea
NA=North Atlantic
NP=North Pacific
PO=Pacific Ocean

Dep/Mo
Year & Month of Departure

Arr/Mo
Year & Month of Arrival

O=Outcome
B=Burned
C=Condemned
L=Lost
S=Sold

Catch:
SBbls=Barrels of sperm oil
WBbls=Barrels of whale oil
WBone=Pounds of whalebone (baleen)

Glossary of Maritime Terms Encountered in the Text

after, toward the stern.

afterhold, that portion of the hold between the mainmast and the rear of the vessel.

aloft, up a mast.

amidships, in the center of a vessel.

back (v., also 'backen'), to change the direction of a square sail so that the wind blows on its front side instead of its rear in order to stop or turn the vessel.

bark, a three-masted sailing vessel, square-rigged on the fore and main sails, and fore-and-aft rigged on the mizzen.

barkentine, a vessel resembling a bark, but square-rigged on the foremast only.

beachcomber, in its most restrictive sense, a non-native drifter in the islands of the South Pacific. But see note 150, p. 165.

beam ends, a vessel is considered to be "on its beam ends" when it is lying almost totally on its side.

belaying pin, a bar used to secure a line; but capable also of being used as a weapon on board.

block and tackle, a combination of pulleys and ropes.

blubber logged, to obtain much oil.

bond (v.), during a time of war vessels that were seized were commonly burned, but occasionally some were sent home under bond (i.e. an obligation bound by the forfeit of money) to carry crews of the burned ships.

booby hatch, a small hatchway with a sliding cover, at times giving access to a locker or storeroom.

boom, a general term for a spar.

brig, a two-masted vessel square-rigged on both the fore and main masts.

bung driver, an implement to hammer in the bung, thus sealing the container.

bung, an object used to seal a barrel, similar in principle to the cork on a wine bottle.

cat-o'-nine-tails (or simply "cat"), instrument of punishment aboard ship, usually consisting of nine pieces of line or cord, knotted at the end and attached to a thick piece of rope for a handle.

caulk (v.), to make seams on a wooden vessel tight by driving in waterproofing material.

chronometer, a highly accurate clock used by navigators to determine longitude.

clean, having no oil or whalebone, as "to return clean."

colors, flags or banners that distinguish vessels of different nations.

cooper, a crew member who made and maintained the vessel's barrels.

crimp, a person who supplied crewmen for a voyage, often through illicit means.

davits, a pair of cranes for hoisting and lowering a vessel's boats.

dead reckoning, estimating a vessel's position by its course, time, and speed, without a positive fix.

deckhouse, any enclosed structure built on the top deck of a vessel.

derelict, an abandoned floating vessel.

dogwatch, took place between 4 p.m. and 8 p.m. and was split into equal two-hour watches. Dogwatch served the dual function, on the one hand, of allowing men on each watch to have a period of rest and relaxation and to partake of their evening meal, and on the other to permit the daily alternating of watch times. The other watches were of four-hour duration.

dory, a small flat-bottomed boat.

dropsy, an ancient term for edema, a condition in which excess fluid accumulates under the skin, especially noticeable in the lower legs and feet.

floe, a large mass of floating sea ice.

flukes, the two lobes of a whale's tail.

fore-and-aft, the rig of a vessel with the plane of the sails on the centerline.

fore, relating to the forward part of the vessel.

forecastle [fo'c's'l], location of the crew's quarters in the fore part of the vessel.

founder (v.), to sink due to damage, usually from adverse weather conditions.

furl (v.), to secure a sail by folding or rolling it on its yard or boom.

gale, a strong, sustained, wind of approximately 40 to 55 miles per hour.

galley, the ship's kitchen.

gam, a meeting of whaleship captains at sea; later also on land.

garboard strake, the planking next to the keel.

gear, a general term for a ship's or sailor's equipment.

Gee, a derogatory term derived from "Portugee" (pronounced with a hard 'g').
gill, a US liquid measure equal to four fluid ounces.
gimcrack, something flimsy (doodad, knickknack).
green hand, a novice sailor.
guano, partially decomposed excrement of seabirds found on some coasts or islands.
hail (v.), to call to another vessel.
half seas over, refers to a vessel stranded on a rock or reef with the seas washing over it; also a person drunk on shore, especially if they are incapacitated.
halyard, a line for hoisting a sail, flag, etc.

haul aback, to stop the vessel's forward motion by changing the orientation of the sails.
head yards, the fore-mast yards on a full-rigged vessel.
heave (v.), to haul a rope by mechanical means.
helm, the steering mechanism of a vessel, but often referring simply to the wheel.
hoist, the part of a fore-and-aft sail that lies next to the mast.
hold, a large compartment below decks for storing cargo and supplies.
Inuit [in'ooit], native of the Canadian Arctic.
Inupiat [inyoo'peeat], native of the North Slope of Alaska.
irons, a term used on whaling vessels to refer to handcuffs (see also p. 356).
jettison (v.), to throw gear, stores, etc. overboard.
larboard, the port, or left-hand, side of the vessel.
lead [leed], a narrow channel in the pack ice.
league, a distance of three nautical miles.
leeward, a direction away from that of the wind.
line, a term applied to most uses of rope aboard ship.
log, an abbreviation of "logbook," the official record of a voyage required on all commercial and government vessels.
manifest, a document signed by the vessel's master listing the goods received on board as required by customs officials.
mast, a vertical spar that supports sails and booms.
packet, a small vessel that had a regular run, carrying passengers, freight, and mail.

Portugee, a common nineteenth-century term for Portuguese, not particularly derogatory at the time.

pull (v.), to row.

put in irons (v.), to apply handcuffs. The expression derived from the ancient use of "bilboes," which consisted of an iron bar and shackles that slid along the bar, with a lock at the end to confine unruly crew members.

ramrod, a rod for ramming home a charge in a muzzle-loading weapon.

recruits, provisions.

reef (v.), to reduce the area of a sail exposed to the wind.

revictual (v.), to supply with a fresh stock of provisions.

rigging, a general term applied to the lines that support masts and sails.

run (to put in the run) (n.), afterhold.

saudade, a longing for some experience from the past with overtones of hope that it might return in the future.

schooner, a vessel of two or more masts, fore-and-aft rigged on all.

scrimshaw, artwork etched on the teeth of sperm whales, on walrus tusks, or bone; also carved items, such a canes, from these same materials.

shallop, a small open boat propelled by oars or sails and used chiefly in shallow waters.

shanghai (v.), to recruit a crew member by means of drugs, intoxication, or force.

sheathing, a skin of copper to cover a wooden hull below the waterline.

ship, a square-rigged vessel of three or more masts.

shipping articles, a master contract between crew members and the owners.

sling, a rope used to raise or lower an object.

sloop, a vessel with one mast, fore-and-aft rigged.

slop chest, a supply of clothing, tobacco, and sundries carried on board for the crew to purchase during the voyage, usually at exaggerated prices. The profits normally went to the master of the vessel.

snow, (pron. 'snoo') a variation of the brig.

so ends this day, the most common expression used for terminating each day's entry in the log that every whaling vessel was required to maintain during its voyage.

spar, a pole in the rigging such as a mast, boom, or yard.

square rigger, a vessel in which the sails hang at right angles to the mast.

squall, a sudden strong wind, with rain or snow, and sometimes thunder and lightning.
stanchion, a fixed upright support.
starboard, the right-hand side of the vessel.
stern sheets, a seat in the aft of a small boat.
stern, the after end of a vessel.
stove (v., past tense), to be pushed in or crushed, especially the hull of a boat or ship.
stream, a vessel is considered "in the stream" after it has been towed from its dock and is anchored offshore awaiting its sailing.
supercargo, an officer on a merchant vessel responsible for the purchase and sale of the cargo.
tack (v.), to turn a sailing vessel so that it takes the wind on its other side.
toggle, a short bar spliced into the end of a line.
ton, for whaleships and cargo vessels a measure of carrying capacity, not weight.
trim(v.), to haul in, as a sail.
try out (v.), to render whale blubber.
tryworks, a brick furnace with two or three large iron trypots used to try out the whale's blubber.
turn to (v.), go to work.
U-boat, a German submarine.
umiak, an open boat covered by walrus or seal skins used by the Arctic natives for both transportation and whaling.
under weigh, an incorrect form of "underway," meaning a vessel that has weighed anchor and is moving.
waist, midship section of a vessel.
weigh anchor (v.), to hoist the anchor.
wheel, the vessel's steering wheel.
windlass, a device for hoisting an anchor.
wooding, the act of collecting wood, especially for fuel.
yard, a spar set perpendicular to the mast for setting square sails. Also used to refer to any wooden support on a sailing vessel.
yarn, a tale that is hard to believe; also "yarning": telling such a tale.

Guide to Archival Collections

A sincere note of gratitude to the following entities for making their resources available for this book.

AJHS = American Jewish Historical Society, New York
BLH = Baker Library Historical Collections, Harvard Business School
CSHM = Cold Spring Harbor (N. Y.) Whaling Museum
EHL = East Hampton (N. Y.) Public Library
FML = Floyd Memorial Library, Greenport, N. Y.
HLCA = Huntington Library, San Marino, Calif.
HSA = Hawaii State Archives
JAF = J. A. Freitas Library, San Leandro, Calif.
JHL = John Hay Library, Brown University
KWM = Kendall Collection, New Bedford Whaling Museum Research Library
MSM = Manuscripts Collection, G. W. Blunt White Library, Mystic Seaport
MVM = Gale Huntington Library, Martha's Vineyard Museum
NAB = National Archives, Boston
NAL = Nantucket Athenaeum Library
NANY = National Archives, New York
NASB = National Archives, San Bruno, Calif.
NAW = National Archives, Washington, D.C.
NBL = New Bedford Free Public Library
NBWP = New Bedford Whaling National Historical Park
NHA = Nantucket Historical Association
NLHS= New London County (Conn.) Historical Society
NYHS = New-York Historical Society
ODHS = Old Dartmouth Historical Society, New Bedford Whaling Museum Research Library
PEM = Phillips Library, Peabody Essex Museum, Salem, Mass.
PPL = Providence (R.I.) Public Library
RIHS = Rhode Island Historical Society
RML = Rogers Memorial Library, Southampton, N.Y.
SCHS = Suffolk County Historical Society, Riverhead, N.Y.
SFMP = J. Porter Shaw Library, San Francisco Maritime National Historical Park
SHWM = Sag Harbor (N.Y.) Whaling and Historical Museum
SHS = Stonington (Conn.) Historical Society
UAF = University of Alaska, Fairbanks
YUL = Yale University Library

Sellected Bibliography

The bibliography contains only books, articles, and manuscript collections cited in the text. For other materials, such as logbooks, journals, crew lists, government documents, unpublished manuscripts, online documents, and interviews, consult the chapter notes.

Books:

1901 New Bedford and Fairhaven Directory. Boston: W. A. Greenough and Company, 1901.

Albuquerque, Luís de and Maria Emília M. Santos, comps. *História Geral de Cabo Verde*. 2nd ed. 3 vols. Lisbon: Centro de Estudos de História e Cartografia Antiga, 2001-02.

Allen, Arthur J. *A Whaler and Trader in the Arctic, 1895 to 1944*. Anchorage: Alaska Northwest Publishing Co., 1978.

Amaral, Pat. *They Ploughed the Seas: Profiles of Azorean Master Mariners*. St. Petersburg, Florida: Valkyrie Press, 1978.

American Activities in the Central Pacific, 1790-1870. 4 vols. Ridgewood, N.J.: Gregg Press, 1967-68.

American Lloyd's Register of American and Foreign Shipping. New York: E. & W. Blunt, 1862.

Asbury, Herbert. *The Barbary Coast: An Informal History of the San Francisco Underworld*. Garden City, New York: Garden City Publishing Co., 1933.

Ashley, Clifford. *The Yankee Whaler*. Boston: Houghton Mifflin Co., 1938.

Bancroft, Hubert H. *The* Works, vol. 24, *History of California*. San Francisco, Calif.: A. L. Bancroft and Co., 1884.

Banks, Charles E. *History of Martha's Vineyard, Dukes County, Massachusetts*. 3 vols. Edgartown, Mass.: Dukes County Historical Society, 1966.

Barbosa, Jorge. *Obra Poética*. Lisbon: Imprensa Nacional-Casa da Moeda, 2002.

Beane, Joshua F. *From Forecastle to Cabin: The Story of a Cruise in Many Seas*. New York: The Editor Publishing Co., 1905.

Becke, Louis. *The Ebbing of the Tide: South Sea Stories*. Philadelphia: J. B. Lippincott Co., 1896.

Bertão, David E. *The Portuguese Shore Whalers of California, 1854-1904*. San Jose, Calif.: Portuguese Heritage Publications of California, 2006.

Bockstoce, John R. *Whales, Ice, and Men: The History of Whaling in the Western Arctic*. Seattle: University of Washington Press, 1986.

Bradford, G. Leroy. *Barnacles and Bilge Water: A Collection of Old Whaling Yarns*. New Bedford, Mass.: Reynolds Printing, 1941.

Braginton-Smith, John and Duncan Oliver. *Cape Cod Shore Whaling: America's First Whalemen*. Yarmouth, Mass.: The Historical Society of Old Yarmouth, 2004.

Brandt, Karl. *Whaling and Whale Oil During and After World War II*. Stanford, Calif.: Food Research Institute, Stanford University, 1948.

Brown, Barbara W. and James M. Rose. *Black Roots in Southeastern Connecticut, 1650-1900*. New London, Conn.: New London County Historical Society, 2001.

Brown, Edwin Peter. *In the Wake of Whales: The Whaling Journals of Capt. Edwin Peter Brown, 1841-1847*. Constance J. Terry, ed. Orient, L.I., N.Y: Old Orient Press, 1988.

Bullen, Frank T. *The Cruise of the Cachalot: Around the World after Sperm Whales*. Santa Barbara, Calif.: The Narrative Press, 2001.

Busch, Briton Cooper, ed. *Master of Desolation: The Reminiscences of Capt. Joseph J. Fuller*. Mystic, Conn.: Mystic Seaport Museum, 1980.

———. *The War Against the Seals: A History of the North American Seal Fishery*. Kingston, Canada: McGill-Queen's University Press, 1985.

———. *Whaling Will Never Do for Me: The American Whaleman in the Nineteenth Century*. Lexington: University of Kentucky Press, 1994.

Byers, Edward. *The Nation of Nantucket: Society and Politics in an Early American Commercial Center, 1660-1820*. Boston: Northeastern University Press, 1987.

Calabretta, Fred, ed. *Fishing out of Stonington: Voices of the Fishing Families of Stonington, Connecticut*. Mystic, Conn.: Mystic Seaport Museum, 2004.

Carreira, Antonio. *Migrações das Ilhas de Cabo Verde*. Lisbon: Universidade Nova de Lisboa, 1977.

———. *The People of the Cape Verde Islands: Exploitation and Emigration*. Trans. and ed. Christopher Fyfe. London: C. Hurst & Co., 1982.

Chamisso, Adalbert von. *A Voyage around the World with the Romanzov Exploring Expeditiion in the Years 1815-1818 in the Brig Rurik, Captain Otto von Kotzebue*, trans. Henry Kratz. Honolulu: University of Hawaii Press, n.d., ca. 1986.

Chippendale, Harry A. *Sails and Whales*. Boston: Houghton Mifflin, 1951.

Chyet, Stanley F. *Lopez of Newport: Colonial American Merchant Prince*. Detroit, Mich.: Wayne State University Press, 1970.

Clarence-Smith, Gervase. *The Third Portuguese Empire, 1825-1975: A Study in Economic Imperialism*. Manchester, U.K.: Manchester University Press, 1985.

Clarke, Robert. *Open Boat Whaling in the Azores: The History and Present Methods of a Relic Industry*. Cambridge, England: Cambridge University Press, 1954.

Cloud, Enoch C. *Enoch's Voyage: Life on a Whaleship, 1851-1854*. Elizabeth McLean, ed. Wakefield, R. I.: Moyer Bell, 1994.

Comstock, William. *A Voyage to the Pacific, Descriptive of the Customs, Usages, and Suffering on Board of Nantucket Whale-Ships*. Boston: Joseph S. Hart, 1838.

Crocker-Langley San Francisco Directory. San Francisco: H. S. Crocker Co., 1907, 1910.

Dandridge, Danske. *American Prisoners of the Revolution*. Baltimore, Md.: Genealogical Publishing Co., 1967.

Davis, Lance E., Robert L. Gallman, and Karin Gleiter. *In Pursuit of Leviathan: Technology, Institutions, Productivity, and Profits in American Whaling, 1816-1906*. Chicago: University of Chicago Press, 1997.

Davis, William M. *Nimrod of the Sea: Or, The American Whaleman*. North Quincy, Mass.: The Christopher Publishing House, [1874] 1972.

Decker, Robert Owen. *Whaling Industry of New London*. York, Penn.: Liberty Cap Books, 1973.

Dexter, Elisha, *Narrative of the Wreck and Loss of the Whaling Brig* William and Joseph. Fairfield, Wash.: Ye Galleon Press, [1848] 1988.

Dias, Eduardo Mayone. *A Presença Portuguesa no Havai*. Lisbon: Assembleia Distrital de Lisboa, 1981.

Directory of the City of New Bedford, 1871. Boston, Mass.: W. A. Greenough & Co., 1871.

Directory of the City of New Bedford, 1881-1882. Boston, Mass.: W. A. Greenough & Co., 1881.

Druett, Joan. *In the Wake of Madness: The Murderous Voyage of the Whaleship* Sharon. Chapel Hill, N.C.: Algonquin Books of Chapel Hill, 2003.

Du Passier, Thierry. *Les Baleiniers français au XIXe siècle*. Grenoble, France: Terre et Mer 4 Seigneurs, 1982.

Duncan, T. Bentley. *Atlantic Islands: Madeira, the Azores and the Cape Verdes in Seventeenth-Century Commerce and Navigation*. Chicago: University of Chicago Press, 1972.

Ellis, Leonard B. *History of New Bedford and Its Vicinity, 1602-1892.* Syracuse, N.Y.: D. Mason and Company, 1892.

Ellis, Myriam. *A Baleia no Brasil Colonial.* São Paulo, Brazil: Melhoramentos, 1969.

Ely, Ben-Ezra Stiles. *"There She Blows": A Narrative of a Whaling Voyage, in the Indian and South Atlantic Oceans*. Middletown, Conn.: Wesleyan University Press, [1849] 1971.

Evans, A. G. *Fanatic Heart: A Life of John Boyle O'Reilly, 1844-1890.* Nedlands, Western Australia: University of Western Australia Press, 1997.

Federal Writers' Project. *Whaling Masters*. New Bedford, Mass.: Old Dartmouth Historical Society, 1938.

Felix, John Henry and Peter F. Senecal. *The Portuguese in Hawaii.* Honolulu: The authors, 1978.

Felix, John Henry, Leslie Nunes, and Peter F. Senecal. *The Ukulele: A Portuguese Gift to Hawaii.* Honolulu: The authors, 1980.

Ferguson, Robert. *Harpooner: A Four-Year Voyage on the Barque Kathleen, 1880-1884*. Philadelphia: University of Pennsylvania Press, 1936.

Forster, Honore, ed. *The Cruise of the "Gypsy": The Journal of John Wilson, Surgeon on a Whaling Voyage to the Pacific Ocean, 1839-1843*. Fairfield, Wash.: Ye Galleon Press, c1991.

Frank, Stuart M. comp. and ed. *Meditations from Steerage: Two Whaling Journal Fragments*. Sharon, Mass.: The Kendall Whaling Museum, 1991.

———. *Dictionary of Scrimshaw Artists*. Mystic, Conn.: Mystic Seaport Museum, 1991.

———. *More Scrimshaw Artists: A Sequel and Supplement to the Dictionary of Scrimshaw Artists*. Mystic, Conn.: Mystic Seaport Museum, 1998.

French, Hiram T. *History of Idaho*. 3 vols. Chicago: Lewis Publishing Co., 1914.

Gomes, Francisco António Nunes Pimentel. *A Ilha das Flores: Da Redescoberta à Actualidade*, 2nd ed. Lajes das Flores, Portugal: Câmara Municipal das Lajes das Flores, 2003.

———. *Subsídios para a História da Ilha das Flores*. Lajes das Flores, Portugal: Câmara Municipal das Lajes das Flores, Portugal, 1984.

Gomes, Joseph, and Don Sevrens. *Captain Joe, Whaleman from New Bedford*. New York: Vantage Press, 1960.

Gomes Vieira, João A. *Man and the Sea: The Portuguese Participation (Azorean and Cape Verdean) in American Whaling*. Lisbon: Medialand, Lda., 2007.

Greenough's Directory of New Bedford, 1881-1882. Boston, Mass.: Greenough, 1881.

Grover, Kathryn. *The Fugitive's Gibraltar: Escaping Slaves and Abolitionists in New Bedford, Massachusetts*. Amherst: University of Massachusetts Press, 2001.

Hanbury, David T. *Sport and Travel in the Northland of Canada*. New York: The MacMillan Co., 1904.

Hare, Lloyd C. M. *Salted Tories: The Story of the Whaling Fleets of San*

Francisco. Mystic, Conn.: The Marine Historical Association, Inc., 1960.

Hawaii Reports: Cases Decided in the Supreme Court of the Territory of Hawaii. Vol. 2, Robert G. Davis, *Reports of a Portion of the Decisions Rendered by the Supreme Court of the Hawaiian Islands in Law, Equity, Admiralty and Probate, 1857-1865*. Honolulu: Government Press, 1866.

Haynes, Williams. *1649-1949: Stonington Chronology: Being a Year-by-Year Record of the American Way of Life in a Connecticut Town*. Salem, Mass.: Higginson Book Company, c1949.

Hazen, Jacob A. *Five Years before the Mast*. Chicago: Belford, Clarke and Co., 1888.

Hedges, Henry P. *Records of the Town of East-Hampton*. 5 vols. Sag Harbor, N.Y.: John H. Hunt, 1887-1905.

Hedges, Henry J., William S. Pelletreau, and Edward H. Foster. *Records of the Town of Southampton*. Vol. 1 (1639-1660). Sag Harbor, N.Y.: John H. Hunt, 1874.

Hegarty, Reginald B. *Returns of Whaling Vessels Sailing from American Ports, 1876-1928*. New Bedford, Mass.: Old Dartmouth Historical Society and Whaling Museum, 1959.

Herbert, Charles. *A Relic of the Revolution*. Boston: Charles H. Peirce, 1847.

Hezel, Francis X. *The First Taint of Civilization: A History of the Caroline and Marshall Islands in Pre-Colonial Days, 1521-1885*. Honolulu: University of Hawaii Press, 1983.

Higginson, Thomas W. *Atlantic Essays*. Boston: James R. Osgood and Co., 1871.

Hohman, Elmo Paul. *The American Whaleman: A Study of Life and Labor in the Whaling Industry*. New York: Longmans, Green and Co., 1928.

Howard, Warren S. *American Slavers and the Federal Law, 1837-1862*. Berkeley: University of California Press, 1963.

Huntington, Gale. *Songs the Whalemen Sang*, 3rd ed. Mystic, Conn.: Mystic Seaport, 2005.

Huntsman, Judith and Antony Hooper. *Tokelau: A Historical Ethnography*. Honolulu: University of Hawaii Press, 1996.

Jenkins, Thomas H. *Bark Kathleen Sunk by a Whale*. New Bedford, Mass.: H. S. Hutchinson and Co., 1902.

Jenness, Stuart E. *The Making of an Explorer: George Hubert Wilkins and the Canadian Arctic Expedition, 1913-1916*. Montreal, Canada: McGill-Queen's University Press, 2004.

Kartunnen, Frances R. *The Other Islanders: People Who Pulled Nantucket's Oars*. New Bedford, Mass.: Spinner Publications, Inc., 2005.

Klengenberg, Christian. *Klengenberg of the Arctic: An Autobiography*. London: Jonathan Cape, 1932.

Knowlton, Edgar C., Jr. *Portuguese in Hawaii before 1878*. [Honolulu]: Hawaii Council on Portuguese Heritage, 1995.

Kugler, Richard C. *The Whale-Oil Trade, 1750-1775*. ODHS Historical Sketch 79. New Bedford, Mass.: Old Dartmouth Historical Society, 1980.

Kuykendall, Ralph S. *The Hawaiian Kingdom*, 3 vols. Honolulu: The University of Hawaii Press, 1938-67.

Langley's San Francisco Directory for the Year Commencing April, 1885. San Francisco: Francis, Valentine and Co., 1885.

Lloyd's Register of Shipping, from 1st July, 1924 to the 30th June, 1925. Vol. 1, *Sailing Vessels*. London: Lloyd's, 1924.

Lloyd's Register of Shipping, from 1st July, 1928 to the 30th June, 1929. Vol. 1, *Sailing Vessels*. London: Lloyd's, 1924.

Lobban, Richard and Marlene Lopes. *Historical Dictionary of the Republic of Cape Verde*. 3rd ed. Metuchen, N.J.: The Scarecrow Press, Inc., 1995.

Lopes, Baltasar. *Chiquinho*. 6th ed. Linda-a-Velha, Portugal: África Editora, 1984.

Lopes, Manuel Tomaz. *Vida dos Marinheiros Manuel Lopes, João Ventura, Francisco Marcial (das Flores) e do Piloto Manuel Francisco Santos (do Corvo)*, 2nd ed. Lajes das Flores, Portugal: Câmara Municipal das Lajes das Flores, 1988.

Lovins, Amory B. et al. *Winning the Oil Endgame: Innovation for Profits, Jobs and Security*. Snomass, Colo.: Rocky Mountain Institute, 2004.

Lund, Judith N. *Whaling Masters and Whaling Voyages Sailing from American Ports: A Compilation of Sources*. New Bedford, Mass.: New Bedford Whaling Museum, The Kendall Whaling Museum, Ten Pound Island Book Co., 2001.

Maclay, Edward S. *A History of American Privateers*. New York: D. Appleton and Company, 1899.

Macy, Obed. *The History of Nantucket*. Boston: Hilliard, Gray, and Co., 1835.

Malloy, Mary. *"Boston Men" on the Northwest Coast: The American Maritime Fur Trade, 1788-1844*. Kingston, Ontario, Canada: The Limestone Press, 1998.

Manning, Edward. *Six Months on a Slaver: A True Narrative*. New York: Harper and Bros., 1879.

Martin, Kenneth R. *Delaware Goes Whaling: 1833-1845*. Greenville, Del.: The Hagley Museum, c1974.

Mawer, Granville A. *Ahab's Trade: The Saga of South Seas Whaling*. New York: St. Martin's Press, 1999.

McCabe, Marsha, and Joseph D. Thomas, eds. *Spinner: People and Culture in Southeastern Massachusetts*. 5 vols. New Bedford, Mass., Spinner Publications, 1980-96.

McDevitt, Joseph L. *The House of Rotch: Massachusetts Whaling Merchants, 1734-1828*. New York: Garland Publishing Co., 1986.

McNab, Robert. *The Old Whaling Days: A History of Southern New Zealand from 1830 to 1840*. Auckland, New Zealand: Golden Press, [1913] 1975.

Morgan, Theodore. *Hawaii, A Century of Economic Change*. Cambridge, Mass.: Harvard University Press, 1948.

Murphy, Robert Cushman. *Logbook for Grace: Whaling Brig Daisy, 1912-1913*. New York: The Macmillan Company, 1947.

New Bedford City Directory. New Bedford, Mass.: C. and A. Taber, 1849.

New Bedford Directory for 1867-8. Boston: Dudley and Greenough, 1867.

Noice, Harold. *With Stefansson in the Arctic*. London: George G. Harrap and Co., Ltd., 1924.

Nordhoff, Charles. *Whaling and Fishing*. New York: Dodd, Mead and Co., [1856] 1895.

Nuligak. *I, Nuligak*, trans. Maurice Metayer. Toronto, Canada: Peter Martin Associates Ltd., 1966.

Orth, Donald. *Dictionary of Alaska Place Names*. Washington, D.C.: U.S. Government Printing Office, 1967.

Perry, Bela C. *A Treatise on the Human Hair, and its Diseases*. New Bedford, Mass.: C. Taber & Co., 1859.

Petrosian-Husa, Carmen C. H. *The De Brum Photo Collection – Memento Mori.* Alele Report 2005/01. Majuro Atoll, Republic of the Marshall Islands: Historic Preservation Office, 2005.

Pickelhaupt, Bill. *Shanghaied in San Francisco*. Mystic, Conn.: Mystic Seaport Museum, [1996] 2007.

Quinn, David B. and Alison M. Quinn, eds. *The English New England Voyages, 1602-1608*. London: The Hakluyt Society, 1983.

Railton, Arthur R. *The History of Martha's Vineyard*. Beverly, Mass.: Commonwealth Editions, 2006.

Richards, Rhys. *The Foveaux Whaling Yarns of Yankee Jack*. Dunedin, N.Z.: Otago Heritage Books, 1995.

———. *Whaling and Sealing at the Chatham Islands*. Canberra: Roebuck Society, 1982.

Richardson, Robert Dale. *Abraham Lincoln's Autobiography*. Boston: The Beacon Press, 1948.

Roche, James J. *Life of John Boyle O'Reilly, Together with His Complete Poems and Speeches*. New York: Cassell Publishing Co., n.d., ca. 1891.

Ross, W. Gillies. *An Arctic Diary: The Journal of Captain George Comer in Hudson Bay, 1903-1905*. Toronto: University of Toronto Press, 1984.

Santos, Fernando dos, *Os Portugueses no Hawaii*. Newark, N.J.: Luso-Americano Newspaper, 1996.

Schmitt, Frederick P. *Mark Well the Whale: Long Island Ships to Distant Seas*. Port Washington, N.Y.: Kennikat Press, 1971.

———., Cornelis de Jong, and Frank H. Winter, *Thomas Welcome Roys: America's Pioneer of Modern Whaling*. Charlottesville:

University Press of Virginia, 1980.

Serrão, Joel, António Barreto, and Maria F. Mónica, comps. *Dicionário de História de Portugal.* 9 vols. Porto, Portugal: Livraria Figueirinhas, 1981-2000.

Sherman, Stuart C., comp. *Whaling Logbooks and Journals, 1613-1927: An Inventory of Manuscript Records in Public Collections*, rev. Judith M. Downey and Virginia M. Adams. New York: Garland Publishing, Inc., 1986.

Ship Registers of New Bedford, Massachusetts, 3 vols. (Vol. 1, *1796-1850*; vol. 2, *1851-1865*; vol. 3, *1866-1939*.) Boston: The National Archives Project, 1940.

Silveira, Pedro da. *Corografias.* Lisbon: Perspectivas and Realidades, 1985.

———. *A Ilha e o Mundo: Poemas.* Lisbon: Centro Bibliográfico, 1953.

Simpson, Marcus B., Jr. and Sallie W. Simpson. *Whaling on the North Carolina Coast.* Raleigh: Division of Archives and History, North Carolina Department of Cultural Resources, 1990.

Sleight, Harry D. *The Whale Fishery on Long Island.* Bridgehampton, N.Y.: The Hampton Press, 1931.

Slocum, Joshua. *Sailing Alone around the World.* New York: Century Company, 1900.

Smith, Nancy W. Paine. *The Provincetown Book.* Brockton, Mass.: Tolman Print, 1922.

Smyth, William H. *Chapman the Sailor's Lexicon: The Classic Source for More Than 15,000 Nautical Terms.* New York: Hearst Books, [1867] 2005.

Stackpole, Edouard A. *The Sea-Hunters: The New England Whalemen During Two Centuries, 1635-1835.* Philadelphia: J. B. Lippincott Co., 1953.

———. *Whales & Destiny: The Rivalry between America, France, and Britain for Control of the Southern Whale Fishery, 1785-1825.* [Amherst]: University of Massachusetts Press, 1972.

Starbuck, Alexander, *History of the American Whale Fishery.* Secaucas, N.J.: Castle Books, [1878] 1989.

Stevens, Marian F. *An Index to the Long Islander, 1839-1864: Marriages-Deaths*. Huntington, L. I., N.Y.: Huntington Historical Society, Huntington American Revolution Bicentennial Committee, New York State Council on the Arts, 1974.

Taylor, Nathaniel W. *Life on a Whaler: Or Antarctic Adventures in the Isle of Desolation*. New London, Conn.: New London County Historical Society, 1929.

The New Bedford and Fairhaven Directory, 1892. Boston: W. A. Greenough and Company, 1892.

The Portuguese in Hawaii: A Resource Guide. Honolulu: Ethnic Research and Resource Center, n. d.

Thomas, Hugh. *The Slave Trade: The Story of the Atlantic Slave Trade: 1440-1870*. New York: Simon and Schuster, 1997.

Thompson, Isaac G. *The American Reports*. San Francisco, Calif.: Bancroft-Whitney Co., 1877.

Thompson, John. *The Life of John Thompson, a Fugitive Slave*. Chapel Hill: Academic Affairs Library, University of North Carolina at Chapel Hill, [1856] 2000.

Tønnessen, J. N. and A. O. Johnsen. *The History of Modern Whaling*, trans. R. I. Christophersen. London: C. Hurst and Company, 1982.

Tower, Walter S. *A History of the American Whale Fishery*. Philadelphia: University of Pennsylvania Press, 1907.

Tripp, William H. *True Story of the Loss of Whaling Bark Alice Knowles*. New Bedford, Mass.: Reynolds Printing, n. d.

Valentine, Harriet G. and Andrus T. *An Island's People: "One Foot in the Sea, One on Shore."* Huntington, L.I., N.Y.: Peterson Press and Associates, 1976.

Vargas, Joseph A., III. *Stonington Ice: The Story of Sylvia & Vargas, 1874-1947*. 2nd ed. Stonington, Conn.: Stonington Historical Society, 2009.

Verrill, A. Hyatt. *The Real Story of the Whaler: Whaling, Past and Present*. New York: D. Appleton and Co., 1916.

Vital Records of Edgartown, Massachusetts, to the Year 1850. Boston: New England Historic Genealogical Society, 1906.

Vital Records of Nantucket, Massachusetts, to the Year 1850, 5 vols. Boston: New England Historic Genealogical Society, 1925-28.

Vital Records of New Bedford, Massachusetts to the Year 1850. 3 vols. Vol 3, *Deaths*. Boston: New England Historical Genealogical Society, 1941.

Warrin, Donald and Geoffrey L. Gomes, *Land, As Far As the Eye Can See: Portuguese in the Old West.* Spokane, Wash.: The Arthur H. Clark Co., 2001.

Webb, Robert L. *On the Northwest: Commercial Whaling in the Pacific Northwest, 1790-1967.* Vancouver, Canada: University of British Columbia Press, 1988.

Westwood, John. *Island Stories*. Shanghai: North China Herald, 1905.

White, Lorraine C. comp. and ed. *The Barbour Collection of Connecticut Town Vital Records.* 55 vols. Baltimore, Maryland: Genealogical Publishing Co., 1994-ca. 2002.

Whitecar, William B., Jr. *Four Years aboard the Whaleship*. Philadelphia: J. B. Lippincott and Company, 1860.

Williams, Harold, ed. *One Whaling Family*. Boston: Houghton Mifflin Co., 1964.

Winius, George G., ed. *Portugal, the Pathfinder: Journeys from the Medieval World toward the Modern World, 1300-ca. 1600.* Madison, Wisconsin: Luso-Brazilian Review, 1995.

Wood, C. F. *A Yachting Cruise in the South Seas*. London: Henry S. King and Co., 1875.

Articles:

Anonymous. "Maritime Memories." *Marine Digest* 48 (18 Oct 1969): 21.

———. "The Joachim deBrum Papers." *Journal of Pacific History* 8 (1973): 181-84.

Afonso, João. "Baleação pelos Açores na Dinâmica Atlântica desde o Século XVIII." In *Os Açores e as Dinâmicas do Atlântico do Descobrimento à II Guerra Mundial*, 1275-99. Angra do Heroísmo, Portugal: Instituto Histórico da Ilha Terceira, 1989.

Alden, Dauril. "Yankee Sperm Whalers in Brazilian Waters, and the Decline of the Portuguese Whale Fishery." *Americas* 20 (Jan 1964): 267-88.

Amorim, Norberta Bettencourt. "Mortalidade e Emigração Irreversível das Gerações Nascidas numa Paróquia Açoriana entre 1845 e 1884." In *Os Açores e as Dinâmicas do Atlântico do Descobrimento à II Guerra Mundial*, 887-914. Angra do Heroísmo, Portugal: Instituto Histórico da Ilha Terceira, 1989.

Anderson, Jill. "Quinton Degrasse" (interview). In *Spinner: People and Culture in Southeastern Massachusetts* 2: 98-102.

Barbosa, Francisco de Assis. "Hipólito José da Costa e os Estados Unidos." *Lanterna Verde: Boletim da Sociedade Felipe d'Oliveira* (Aug 1943): 41-53.

Bonifácio de Andrada e Silva, José. "Memória Sobre a Pesca das Baleias e Extracção do Seu Azeite, com Algumas Reflexões a Respeito das Nossa Pescarias." In *Memórias Económicas da Academia Real das Ciências de Lisboa, para o Andiantamento da Agricultura, das Artes, e da Indústria em Portugal, e Suas Conquistas, 1789-1815*, 2:277-93. 5 vols. Lisbon: Banco de Portugal, 1990-91.

Boylan, Betsy Wade. "The Festival of the Holy Ghost." *Historical Footnotes: Bulletin of the Stonington Historical Society* 24 (Aug 1987):1-8.

Calabretta, Fred. "The Picture of Antoine DeSant." *The Log of Mystic Seaport* 44 (Spring 1993): 93-95.

Clock, Henry A. "The Narrative of a Shanghaied Whaleman," *Pacific Monthly* 24 (Nov 1910):451-63; (Dec 1910): 623-31.

Coli, Waltraud Berger. "Cape Verdean-Americans as Maritime

Entrepreneurs: Whaling and the Packet Trade." *Cimboa* 3 (Summer 1997): 16-20.

Crapo, Henry H. "The Story of Cotton and Its Manufacture into Cloth in New Bedford." *Old Dartmouth Historical Sketches* 67 (Nov 1937).

Dias, Eduardo Mayone. "Baleeiros Portugueses na América." *Coisas da LUSAlândia*. Lisbon: Instituto Português de Ensino a Distância, n.d., ca. 1981, 181-221.

Downey, Judith. "Consequences of California Mania: Nantucket and the Whaling Industry." *Historic Nantucket* 47 (Summer 1999):25-27.

Farwell, Robert D. "Manuel Enos: Cold Spring Whaleman," *Long Island Forum* (June 1981).

Fernández-Armesto, Felipe F. R. "Medieval Atlantic Exploration: The Evidence of Maps." In Winius. *Portugal, the Pathfinder*, 41-65.

Frank, Stuart M. "The Many Mysteries of Manuel Enos." *The Bulletin from Johnny Cake Hill* (Fall 2009): 16-19.

Hazevoet, Cornelis J. and Frederick W. Wenzel. "Whales and Dolphins (Mammalia, Cetacea) of the Cape Verde Islands, with Special Reference to the Humpback Whale *Megaptera Novaeangliae* (Borowski, 1781)." *Contributions to Zoology* 69 (2000): 197-211.

Hezel, Fr. Francis X. "A Yankee Trader in Yap: Crayton Philo Holcomb," *Journal of Pacific History* 10 (1975): 3-19.

Huston, John. "Harpoons Away!" *Sports Illustrated* (6 Jun 1955): 66, 69-72.

Kenny, Robert W. "Yankee Whalers at the Bay of Islands." *The American Neptune* 12 (January 1952): 22-44.

Lessa, William A. "An Evaluation of Early Descriptions of Carolinian Culture," *Ethnohistory* 9 (Autumn, 1962): 314-15.

"Leviathan in the Lab," *The Laboratory* 26 (1957): 34-38.

Lewis, Donald. "Stonington's Portuguese Fishermen." *Historical Footnotes: Bulletin of the Stonington Historical Society* 2 (August 1965): 1-2.

Littlefield, L. A. "Fitting Out a Whaler." *Old Dartmouth Historical Sketches* 14 (Jun 1906): 4-13.

McGrath, Thomas B. "The Joachim deBrum Papers." *Journal of Pacific*

History 8 (1973): 181-85.

Melville, Herman. "The 'Gees." *Harper's New Monthly Magazine* 12 (Dec. 1855-May 1856): 507-09.

Mendes, J. Tony. "Portuguese Whale Hunters in the Arctic." *Portuguese Studies Review* 11 (Winter-Spring 2004): 133-42.

Munro, Wilfred H. "The Most Successful American Privateer: An Episode of the War of 1812." *Proceedings of the American Antiquarian Society* 23 (April 1913): 12-62.

Parker, Ian. "Birth of a Nation?: A Speck in the South Pacific Ponders Independence." *New Yorker* (1 May 2006): 66-75.

Parry, Albert. "Yankee Whalers in Siberia." *Russian Review* 5 (Spring 1946): 37-38, 48.

Poole, Dorothy Cottle. "Antone Fortes, Whaleman." *The Dukes County Intelligencer* 11 (May 1970): 129-52.

———. "Mutinous But Not Mutiny." *The Dukes County Intelligencer* 18 (May 1977): 135-40.

Reilly, Kevin S. "Slavers in Disguise: American Whaling and the African Slave Trade, 1845-1862." *American Neptune* 53 (Summer 1993): 177-89.

Sandoval Hernández, Ariel. "Talcahuano y los últimos balleneros a vela." *Revista de Marina* [Chile] 89 (March-April 1972): 233-35.

Schmitt, Frederick P. "James Fennimore Cooper and his whaleship UNION." *The Lookout* (Jul-Aug 1974): 4-8.

Seager, Helen. "Portuguese Islanders and the Old Mill." *Historic Nantucket* 51 (Spring 1992): 10-14.

Sepulveda Ortiz, Jorge. "La epopeya de la industria ballenera chilena." *Revista de Marina* 6 (Nov-Dec 1997): 544-53.

Spennemann, Dirk H. R., and Jon O'Neill. "Curating Books on Remote Islands: The Fate of the De Brum Library on Likiep Atoll, Marshall Islands." *The Australian Library Journal* 53 (Nov 2004): 397-407.

Taber, Edgar F., Jr. "Condensed Data Relating to New Bedford Cotton Mills." In *The Story of Cotton and Its Manufacture in New Bedford.* New Bedford: Old Dartmouth Historical Society, 1937: 71-96.

Townsend, Charles H. "Twentieth Century Whaling." *Bulletin of the New York Zoological Society* 33 (Jan-Feb 1930): 2-32.

Valentine, Andrus T. " 'Big Manuel,' Whaling Captain." *Long Island Forum* 17 (March 1954): 49 and 56-58.

Verlinden, Charles. "European Participation in the Portuguese Discovery Era." In Winius. *Portugal, the Pathfinder*, 71-88.

Vickers, Daniel. "Nantucket Whalemen in the Deep-Sea Fishery: The Changing Anatomy of an Early American Labor Force" *The Journal of American History* 72 (Sep 1985): 277-96.

Webb, Robert Lloyd. "Power Whaling: Industrial Shore Whaling on the West Coast, 1905-1972." *Mains'l Haul* 37 (Winter 2001): 30-47.

Manuscript Collections:

Account Books Collection. MSS 10. NHA.

Agard Collection. MSS 23. ODHS.

Business Records Collection. MSS 56. ODHS.

Comer, George, Collection. MSS 210. MSM.

Cory Family Papers Collection. MSS 80. ODHS.

De Castro Family Papers. Private document.

De Wolf Papers. MSS 382. RIHS.

Dering, Charles T. Papers. EHL.

Dietrich American Foundation Collection (New Bedford Whaling Museum).

Foote, Don C. Collection. UAF.

General Records of the Department of State, 1759-1999. RG 59. NARA.

Hewlett-Jones Collection. CSHM.

Hodge Papers. HLCA.

Howland Family Papers. MSS 7. ODHS.

Laflin Records. SFM.

Lawrence and Company Papers. Coll. 25. MSM.

Local History Collection, MSS 77, ODHS.

Lopez, Aaron. Papers. AJHS.

Lopez, Aaron. Papers. RIHS.

Mayhew, Catherine M. 110 Whaling Captains Collection. MVM.

Morgan, Charles W., Collection. Baker Library Historical Collections, BLH.

Morris, Jacob. Papers. NYHS.

Morse, Carlton D., Whaling Collection. JHL.

Nicholson Whaling Collection. PPL.

Records of the Adjutant General's Office, ca. 1775-ca. 1928. RG 94. NARA.

Records of the District Courts of the US, 1685-1991. RG 21. NARA.

Records of the Foreign Service Posts of the Department of State, 1788-1964. RG 84. NARA.

Records of the US Customs Service, 1745-1982. RG 36. NARA.
Rotch Family Papers. MSS 2. KWM.
Sanford Scrapbook. SCHS.
Ships' Logs Collection. MSS 220. NHA.
Ships' Papers Collection. MSS 79. ODHS.
Stackpole, Edouard A. Collection. MSS 335. NHA.
Strout, Robert B. Papers. Coll. 210. MSM.
Swift and Allen Papers. MSS 5. ODHS.
Tripp, William H. Collection 58. MSM.
Turner-Baker Papers. NAW.
Valentine Collection. CSHM.
Vera Family Papers Collection. MSS B 97-13. ODHS.
Vertical Collections. MSS 384. NHA.
Whaling Logs Collection. Manuscripts and Archives. YUL.
Wing, J. and W. R. Records Collection. MSS 35. ODHS.

General Index

Personal Name Index

Vessel Index